Mining Latin America
Minería Latinoamericana

AF787937

Mining Latin America

Challenges in the mining industry

Minería Latinoamericana

Desafíos para la industria minera

SPRINGER-SCIENCE+BUSINESS MEDIA, B.V.

First edition 1994

© 1994 Springer Science+Business Media Dordrecht
Originally published by Chapman & Hall in 1994

ISBN 978-94-010-4533-9 ISBN 978-94-011-1216-1 (eBook)
DOI 10.1007/978-94-011-1216-1
Apart from any fair dealing for the purposes of research or private study, or criticism or review, as permitted under the UK Copyright Designs and Patents Act, 1988, this publication may not be reproduced, stored, or transmitted, in any form or by any means, without the prior permission in writing of the publishers, or in the case of reprographic reproduction only in accordance with the terms of the licences issued by the Copyright Licensing Agency in the UK, or in accordance with the terms of licences issued by the appropriate Reproduction Rights Organization outside the UK. Enquiries concerning reproduction outside the terms stated here should be sent to the publishers at the London address printed on this page.

The publisher makes no representation, express or implied, with regard to the accuracy of the information contained in this book and cannot accept any legal responsibility or liability for any errors or omissions that may be made.

A catalogue record for this book is available from the British Library

∞ Printed on permanent acid-free text paper, manufactured in accordance with ANSI/NISO Z39.48–1992 and ANSI/NISO Z39.48–1984 (Permanence of Paper).

Contents

EXPLORATION

SOLVENT EXTRACTION

MINING – TRANSPORTATION, CONVEYOR DESIGN AND PRACTICE

MINING – TECHNOLOGY

Papers presented at the international conference 'Mining Latin America/Minería Latinoamericana' organized by the Institution of Mining and Metallurgy and held in Santiago de Chile from 10 to 14 May, 1994

Organizing committee

D.C. Bailey *(Chairman) (United Kingdom)*
A.A.C. Brewis *(United Kingdom)*
J.R. Chadwick *(United Kingdom)*
Ricardo Cortés *(Chile)*
R.D.G. Crozier *(U.S.A. and Chile)*
A. González Soto *(Chile)*
Z. Jakubiak *(United Kingdom)*
J.K. Maclean *(Bolivia)*
D.K. Presgrave *(United Kingdom)*
J.E. Rickus *(United Kingdom)*
N.H. Santander *(Chile)*
T.J.A. Smith *(United Kingdom)*
J.A. Solari *(Chile)*
K.R. Suttill *(United Kingdom)*
E.A. Tidy *(Chile)*

List of Contributors

Hugo Aduvire, P., *Universidad Nacional Jorge Basadre Grohmann de Tacna, Peru*
Osvaldo Aduvire P., *Escuela Técnica Superior de Ingenieros de Minas de Madrid, Spain*
R.A. Agar, *GeoScan Pty. Ltd., Perth, Western Australia*
A.L. Almeida, *Companhia Paraibuna de Metais, Juiz de Fora, Minas Gerais, Brazil*
M.C. Athié Lambarri, *International and Policy Services, Environmental Resources Management, London, England*
B.R. Baldock, *Ausmelt, Ltd., Dandenong, Victoria, Australia*
Julio Binvignat T., *Compañía Minera Mantos de Oro, Copiapó, Chile*
R. Brough, *Minproc Engineers, Inc., Denver, Colorado, U.S.A.*
Charles 'Scottie' Bruce, *MINTEC, La Paz, Bolivia*
W. J. Bruckard, *CSIRO Division of Mineral Products, Port Melbourne, Victoria, Australia*
Mario Castillo P., *CODELCO-Chile, División Chuquicamata, Chuquicamata, Chile*
J. Cederberg, *Empresa Minera Inti Raymi, S.A., Bolivia*
J. Cochrane, *International and Policy Services, Environmental Resources Management, London, England*
K.D. Culbertson, *Design Workshop, Inc., Denver, Colorado, U.S.A.*
L.J. De Beer, *Compañía Minera Doña Inés de Collahuasi S.A., Santiago, Chile*
J.H. Decker, *Norwest Mine Services, Salt Lake City, Utah, U.S.A.*
L.A. Dick, *Compañía Minera Doña Inés de Collahuasi S.A., Santiago, Chile*
Victor Flores, *United Services Advisors, Inc., San Antonio, Texas, U.S.A.*
D.M. François-Bongarçon, *Mineral Resources Development, Inc., San Mateo, California, U.S.A.*
N.R. Fraser, *GeoScan Pty., Ltd., Perth, Western Australia*
Igor Gonzales, *Southern Peru Copper Corporation, Toquepala, Tacna, Peru*
G.A. Guorgi, *Ausmelt, Ltd., Dandenong, Victoria, Australia*
W.R. Hopkins, *Davy International, Nonferrous Sector, San Francisco, California, U.S.A.*
C.W. Hunt, *Boart SECO, Roodepoort, South Africa*
H.A. Janzon, *The Robbins Company, Kent, Washington, U.S.A.*
O.H. Jiménez, *Sección de Refinación y Moldeo, Fundición de Concentrado, División Chuquicamata, CODELCO-Chile, Chuquicamata, Chile*
Carlos López Jimeno, *Escuela Técnica Superior de Ingenieros de Minas de Madrid, Spain*
Pertti Koivunen, *Tamrock Oy, Tampere, Finland*
Carlos Leal, *ABB Industrial Systems, AB, Västerås, Sweden*
H. Lieberwirth, *Department of Mining and Materials Handling Systems, Krupp Fördertechnik GmbH, Duisburg, Germany*

N.H. Lockett, *Ashton Mining, Ltd., Perth, Western Australia*
G. May, *Minproc Engineers, Inc., Denver, Colorado, U.S.A.*
S.C. Mulshaw, *Department of Geology, Imperial College of Science, Technology and Medicine, London, England*
Jorge Omori V., *Compañía Minera Mantos de Oro, Copiapó, Chile*
Robert B. Parsons, *Price Waterhouse, Toronto, Ontario, Canada*
J.A. Porter, *Workshop, Inc., Denver, Colorado, U.S.A.*
C.A. Puig, *Armenonic del Ecuador, Quito, Ecuador*
M.M. Randall, *RTZ Technical Services, Ltd., Bristol, England*
G.A. Riveros, *Dirección Convenio I&D, Centro de Investigación Minera y Metalúrgica, Santiago, Chile*
K.R. Robilliard, *Ausmelt, Ltd., Dandenong, Victoria, Australia*
D.P. Ryan, *Signet Engineering Pty., Ltd., Perth, Western Australia*
R.I. Salas, *Dirección Convenio I&D, Centro de Investigación Minera y Metalúrgica, Santiago, Chile*
L.S. Santana, *Design Workshop, Inc., Denver, Colorado, U.S.A.*
K.W. Severing, *SpinTek Systems, Huntington Beach, California, U.S.A.*
W.E. Short, *Ausmelt, Ltd., Dandenong, Victoria, Australia*
G.J. Sparrow, *CSIRO Division of Mineral Products, Port Melbourne, Victoria, Australia*
William Torres, *Southern Peru Copper Corporation, Smelter Division, Ilo, Peru*
P.A. White, *Sandwell, Inc., Swan Wooster Division, Vancouver, Canada*
J.T. Woodcock, *CSIRO Division of Mineral Products, Port Melbourne, Victoria, Australia*
Adrzej Zablocki, *Atlas Copco Chilena S.A.C., Santiago, Chile*
S.P. Zamorano, *Bateman Ingeniería Chile, S.A., Santiago, Chile*
J.A. Zúñiga, *Dirección Convenio I&D, Centro de Investigacción Minera y Metalúrgica, Santiago, Chile*

Foreword

In November, 1986, the Institution of Mining and Metallurgy held the 'Mining Latin America/Minería Latinoamericana' conference in Chile – a conference covering a broad range of topics relevant to South America and, in that sense, complementary to its regional 'Asian Mining' and 'African Mining' series of events. This first conference proved to be a resounding success and confirmed that Chile, in particular, and South America, in general, were indeed ideal for the Institution in the pursuit of its objective of disseminating information related to the international minerals industry.

In relation to South America, the Chilean conference was followed, in 1988, by that entitled 'Silver – exploration, mining and treatment',* which was held in Mexico City; in February, 1995, Caracas will host the 'Mineral resources of Venezuela/Recursos minerales de Venezuela' conference, serious planning for which is under way as I write.

The Institution of Mining and Metallurgy, in association with the Instituto de Ingenieros de Minas de Chile, *Minería Chilena* and *Latinominería* agreed that the second 'Mining Latin America/Minería Latinoamerica' conference should be held, again in Santiago, in May, 1994, on the occasion of the 1994 Expomin show, which had been been particularly successful in attracting visitors from mining countries worldwide in 1992 and which will continue, at two-yearly intervals, to the year 2000 and, no doubt, beyond.

To ensure the long-term future of our industry and of the Institution it is essential that we continue and extend our learned-society activities, of which international conferences, as well as their associated very valuable technical tours, are an important part.

I was therefore pleased to accept the chairmanship of the Organizing Committee for 'Mining Latin America/Minería Latinoamericana' and should like to express my thanks to all the members of the Organizing Committee and to IMM staff Michael Jones and Liz Munro for their very significant contribution to the planning and implementation of the conference. I should also like to thank all those who have assisted in the compilation and refereeing of the papers that constitute the *raison d'être* of the event. I have no doubt that the 1994 conference will prove to be as successful as its predecessor and we look forward to it and to its successor in 1996.

D. C. Bailey
Chairman, Organizing Committee
February, 1994

**Silver – exploration, mining and treatment* (London: IMM, 1988), 344 p.

Investment, Potential and Evaluation

Mining investment in Argentina

J. H. Decker
Norwest Mine Services, Salt Lake City, Utah, U.S.A.

Abstract

Argentina's significant mining potential has long been recognized; however, it has traditionally been overlooked in view of the country's other tremendous natural resources. Recent world interest in mineral projects, particularly in neighbouring Chile, has caused the Argentine government to review its policy with respect to mining. This paper provides an overview of the mining environment and the changes required to encourage investment in it.

1 Introduction

The Secretaria de Minería commissioned Norwest Mine Services, Inc. (Norwest) of Salt Lake City, Utah, U.S.A. to conduct a feasibility study on the environment required to encourage mining investment in the Republic of Argentina. Work commenced in August 1992 and was completed in August 1993. This paper summarizes Norwest's appraisal of Argentina's mining environment and changes required to stimulate mining investment.

Argentina is widely regarded as a country with significant mineral resources, but with a very limited mining tradition. Mineral production currently contributes only 0.3% to the Argentine gross domestic product (excluding oil and gas). Argentina's mining law dates from 1886, and though it has been modified from time to time in an attempt to meet various regulatory needs, the modifications have not been adequate to bring about a nationwide improvement in the mining sector. Current federal legislation places ownership of mineral resources in the hands of the individual provinces. Provincial laws governing royalties, taxes, and incentives are inconsistent from province to province, and are in need of standardization.

Argentina's recently accomplished economic stabilization has provided an atmosphere that is much more conducive to high risk, long term investments such as mineral development projects. The overall objective of the feasibility study was to analyze the conditions necessary for the development of new investment in mining activities in Argentina within this new economic climate. Specific tasks required to meet the objectives of the study are summarized as follows:

Evaluation of available geological and mining information produced by previous exploration programs, and the implementation of new ideas and technology to determine specific geographic areas of interest for mineral development. This evaluation is focused on potential polymetallic ore deposits associated with Tertiary Andean volcanism, with special emphasis on gold occurrences.

Recommendation of streamlined institutional structures for fiscal, tax, financial, royalty, and economic policies as alternatives to the current national, provincial, and municipal system. This includes preparation of a draft proposal for simplified and uniform rules in all provinces.

Evaluation of laws currently in force in Argentina, and a review of laws in other countries which have encouraged and enabled mining investments. This evaluation includes not only laws which directly involve mining, but also includes foreign investment laws, tax laws, environmental laws, and legal provisions on imports and exports and on exchange transactions. An analysis of provincial laws includes recommendations on adaptations to conform with current international practices.

Recommendation of comprehensive, environmental guidelines (air, water, and soil). This specifically addresses the environmental impact of mining and all related activities. The recommendations can serve as a foundation for national and provincial legislation.

While the focus of this paper is exclusively on the mining investment aspects of Argentina, a very brief review of the geological prospects was conducted. The overall conclusion was that the potential for the discovery of new deposits of gold may be higher in Argentina than in Chile. However, the potential for new discoveries of porphyry copper deposits appears to be greater in Chile than in Argentina. Also, because the requisite conditions for formation are the same, the structural/tectonic favorability for gold in Argentina translates to highly prospective terrain for large deposits of borates, including both the tincal and hydroboracite species. Regarding the likely styles of gold mineralization to be encountered in northwest Argentina, it is concluded that high sulfidation systems of the El Indio and porphyry gold-types (Maricunga) have the greatest probability of discovery along the west margin of the north-south trending rift systems, whereas Cripple Creek style Au-Te systems related to alkalic intrusives may offer moderate probability of success along the east margin of the rift. This project also confirmed that thematic mapping is an effective exploration tool for interpreting structural geology and alteration but it must be emphasized that its greatest utility in mineral exploration can only be achieved through an interactive use of direct field investigations with satellite images. The systematic application of this approach will almost certainly lead to new mineral discoveries in Argentina because the semi-arid climate and sparse vegetative cover favor these methods and the uniquely diverse geology of this country provides a high probability of success.

2 Mining Investment

In recent years Argentina has made big improvements in fundamentally important areas: it now has a democratically elected government, the days of high inflation are over and the economy is stable. The economic improvements have been achieved by the following:

Promoting efficiency through exposing the market to greater international competition and removing regulations, subsidies and other government-imposed distortions.
Balancing the federal government budget through improved tax collection, restructured debt service requirements and the sale of the extensive system of state-owned enterprises through the privatization process.

The National government's greatest achievement is that so much has been accomplished within a truly democratic environment and many of these politically unpopular and painful measures have received the explicit approval of the Congress. However, in spite of these major improvements, Argentina continues to have an unfavorable image relating to mining investments for the following reasons:

Limited and difficult access to land for purposes of conducting exploration.
Administrative procedures that are time-consuming, expensive and non-transparent with the distinct appearance of being bureaucratically and politically manipulated.
The strong and direct role played by the Provinces through their Provincial Mining Companies in matters

of exploration and mining. In many cases, mining companies wishing to explore must negotiate with these governmental entities which were specifically created to represent the interests of the Provinces. These negotiations can take as long as 5 years to conclude and, in several well-documented cases, the mining company wishing to invest finally gives up in despair. Excessive royalty payments to the Provinces.

In support of the above picture is the fact that not a single mine has opened in Argentina since the mid-1960's. It is worth noting that within the same period of time in the US and other regions of the world, dozens of mines have opened up and shut down. The contribution of Argentina's mining sector to the GNP is very small; it is less than 0.3% compared to more than 11% in neighboring Chile. While Argentina shares many of the same regional geological trends with Chile, the lower geological potential of much of Argentina will prevent the Argentine mining sector's contribution to GNP from being as high. Nevertheless, Argentina's mining sector is performing significantly below its true potential and is certainly capable of increasing its present contribution to the GNP.

Following is a summary of our most important findings.

3 Increased Competition For Mining Investments

The competition to attract mining investments has increased significantly in recent years. Entire regions which were previously hostile to private enterprise are now actively soliciting mining investments. This includes large regions of Africa, Latin America and the former Eastern Bloc countries. Argentina is now but one competitor among many. At present there is a world-wide oversupply of mineral commodities and Argentina's potential contribution is not desperately needed. Therefore, for Argentina to successfully attract mining investments it must be clearly regarded by the international mining community as being as good as those countries against which it competes for mining investments.

Following in order of decreasing importance, are the key considerations for mining investors:

Right to mine.
Right to repatriate profits.
Management control.
Fixed tax terms and conditions.
Modern minerals legislation.
Rules fixed for the life of the mine.
Guaranteed arbitration.

The challenge for Argentina is to both satisfy the above criteria and convince the international mining community that it is at least as attractive as those countries against which it competes. Only at that point will Argentina succeed.

4 Taxation

The Mining Investment Law recently passed by the Congress will make Argentina at least as attractive from the point of view of taxation as those countries against which it competes. This conclusion is based on a comparative taxation study of Brazil, Indonesia, Mexico, the US and Canada. To be implemented in practice, the Investment Law must still be formally written into law by the provinces and to date only the provinces of Catamarca, Chubut and Jujuy have done so. However, this law should not be considered to be the only item required to stimulate new mining investment. Other equally important issues must be addressed (such as the great difficulty in gaining access to open land for exploration) before mining companies can effectively take advantage of the Law. Taken in isolation and with due consideration of other governmental restraints on the development of the mining sector, the Law must be regarded as a piecemeal action. Nevertheless, we consider the Law to be an important first step to attract mining investment to Argentina but have recommended that the following aspects of the Law be improved:

Allow mining investors to withdraw from the system if taxes are reduced in the future.

Do not limit loss carry forwards to 5 years, which may be too short under the following scenarios:
- marginal projects with low rates of return, and;
- projects with low initial profitability. Examples: Long lead times resulting from high stripping ratios to reach profitable mining horizons, or from the lack of suitable access and other infrastructure.

Under these scenarios the rapid depreciation provisions of the new law actually result in a <u>tax penalty</u>.

5 Royalties

Within the mining industry royalties are a very controversial topic. Mineral resources are by definition non-renewable and once mined can never be replaced. For that reason the owner of the mineral resource wishes (and is entitled) to obtain compensation for the removal of the resource. In theory, the royalty payments should be used to develop other investments to compensate for the irre-trievable loss of the resource. Royalty payments are most frequently structured to act as a tax on production and do not consider the effect of the imposition of the royalty on the profitability of the mining operation. In addition, royalties have the undesirable effect of increasing the mine cut-off grade with the resulting irretrievable loss of mineable reserves and reduction of mine life. These issues are at the root of the controversy. The mineral owner believes that it is his right to be paid a royalty for the mineral that is mined and profitability is not a concern to him. In fact, he might even state that if the mining operation cannot sustain the royalty payment, then mining operations should not be conducted. On the other hand, the mining company has a different opinion: it regards royalties as regressive and would prefer not to pay them. Instead it would prefer to compensate the owner of the mineral resource through a payment on the profitability of the operation. Not surprisingly, there is strong disagreement within the Argentine mining industry regarding the issue of royalties and its constitutionality has even been questioned by the legal community. Furthermore, the mining industry believes that the Provinces should look beyond the narrow issue of royalties to the overall benefit (economic growth, taxes, jobs etc) to be gained from a vibrant and profitable mining industry. Nevertheless, the mining industry will reluctantly accept royalties on condition that they are sufficiently low so that they will not measurably harm the overall profitability of the operation.

The Mining Investment Law caps the royalties at a maximum of 3%. This does not appear to be unreasonable. Norwest has recommended that:

> The Provinces agree on a standard method of royalty calculation which allows deductions for certain production costs.
> The provinces do not "stack" (i.e. add other royalties on top of the 3% statutory maximum).

6 The Problem Of High Internal Costs

One of the key factors affecting the mining industry is the high domestic cost structure, also referred to as the problem of "high internal costs". This phenomenon exists in all sectors of the economy and can be attributed to the long history of protectionism. The effects of this have not yet left the system and will take a long time to overcome. The high value of the peso (relative to the US$) has made the cost of labor and materials high in comparison to other South American countries. For example: in a certain underground mine the workforce has been reduced by over 40% in the last 2 years yet the cost of labor per tonne of ore has increased by 66%. Also, a large portion of mining costs are from materials and supplies purchased from industries where the competition is limited and costs are above those of international competitors; for example cement, fuel, oil and electricity. The producers of these commodities are now moving to a free market system but they are not yet efficient producers and, as a result, they pass on the high costs to their customers. Some mines have found that their overall costs of production have doubled over the last two years.

7 The Mining Code

In spite of being written in 1886, and being modified several times since then, the Mining Code is generally regarded as being an effective tool by those persons with expertise in its interpretation. However, for those persons without this expertise, the Mining Code is perceived as being very complex, outdated and providing few guarantees. While there are inherent structural problems with the Code, the biggest problems lie at the Provincial level where the Code is inconsistently administered. An unintended result of the Code is widespread speculation by both the Provinces and private sector which makes it very difficult for legitimate mining companies to obtain access to land for purposes of exploration. The inherent problems include:

Unrealistically small allowable size (by modern technological standards) for both the "cateos" (or exploration claims) and mining claims.
Unrealistically low cost (Canon) for both mining claims and cateos, thereby fuelling the speculation.
Prohibition on holding adjacent exploration and mining claims.

To address some of the problems posed by the Mining Code, the national government has recently taken the initiative and passed legislation (which must be accepted by the provinces) to increase the Canon. Nevertheless, Norwest has recommended that the Mining Code should be completely rewritten and updated within the next several years.

The issue of the Canon is very complex and has many implications. Under ideal conditions it should be set at a cost level that discourages speculators who have no intention of developing their properties. This would turn the land over and make it available for active exploration. At the same time the Canon should be equally affordable by both small and large investors and should not be so high that investors spend disproportionate amounts of money on the Canon instead of on active exploration. Recognizing that there is indeed a major problem with speculators who tie up prospective land, an increase in the Canon is desirable as it will act to discourage speculators. There may be other methods that could be used to discourage speculation. However, it is also recognized that an increase in the Canon could unintentionally penalize legitimate mining operations which are either small in scale or temporarily closed due to low commodity prices.

In advance of a complete rewrite of the Mining Code, the Code should be administered to allow that any claims forfeited should revert to the public domain and all associated geological information be made public.

Any future modification to the Mining Code must not dilute in any respect that single most important issue in the perception of mining investors: the right to mine. Whether setting policy or developing administrative procedures, governments must recognize this very basic issue. This concept is strongly established in the Chilean Mining Code and is regarded as one of the principal reasons for the revitalization of the Chilean mining industry.

8 The Role Played By The Provinces

It is our firm conclusion that the provinces are the root cause of the problems faced by the mining industry in Argentina.

The provinces directly influence mining in two ways:

Firstly, through their policies and procedures they administer the requirements of the federal Mining Code regarding the awarding and registration of claims and titles. The manner in which these functions is performed has historically been bureaucratic, inefficient and sometimes inaccurate and many of the administrative procedures are not transparent. The awarding of claims on non-reserved land has frequently been delayed for no stated reason although the claim applicants have clear rights under the Mining Code. The administrations have not felt compelled to act within fixed frameworks of time i.e. there are no clear deadlines for governmental action. Situations like this create severe negative impressions for mining companies, both Argentine and foreign who have dealt with the Provinces.

Secondly, the Provinces are frequently directly involved in mining activities through mining companies which are either 100% owned by the Province or a mixture of Provincial and private ownership with the Province having majority ownership. In accordance with the provisions of the Mining Code, the Provinces have acquired vast areas of land (up to 200,000 ha) with the highest geological potential and companies wishing to conduct exploration on this Provincial land are required to negotiate with the Provinces. The Provinces are frequently unreasonable in their financial expectations and demands and the frequent result is that the negotiations continue indefinitely or are never concluded at all. The demands typically include high upfront payments, purchase of vehicles, annual expenditure commitments etc. These demands frequently apply to land for which there is no valid exploration data available.

It is worth noting that the Mining Code also allows Federal entities to reserve up to 200,000 ha in each Province. While this right was occasionally exercised in the past, this is not a present problem. Nonetheless, the possibility exists under the law.

This dual Provincial government role as mine owner and Mining Authority cannot be properly performed in a completely impartial manner. Also this dual role must be considered a gross conflict of interest that is unacceptable to major mining companies.

The net result is that mineral development in Argentina is very low in comparison to its true potential. By making it unnecessarily difficult to access land for exploration and making unrealistic demands through their mining companies, the provinces have stifled the development of a strong mining sector which could make a big contribution to Provincial taxes, employment and development of infrastructure.

Privatization has achieved great success at the national level at generating greater economic efficiency in industries which were previously operated by governmental entities. We believe that this process should be continued and extended to the provincial level where similarly good results will be obtained.

Therefore, we recommended in the strongest terms that the Provincial Mining Companies be privatized as soon as possible (in a period of 2 to 3 years) and that their land holdings be made available to the private sector for the purpose of active exploration.

After the privatization process has been completed, the Provinces would focus exclusively on the impartial administration of the Mining Code and the implementation of other measures to actively encourage mining investments.

In addition, two other recommendations are presented.

9 Development And Implementation Of A Long term Mining Investment Strategy

To develop enough confidence to invest in Argentina, the mining investment community must understand the long term mineral investment strategy of the country, and believe that it will be carried out by all levels of government.

Therefore both the Provincial and Federal governments should agree on an overall strategy to develop a viable mineral industry. The development of a comprehensive strategy will be preceded by a frank and open discussion between the most senior Federal and Provincial officials and must clearly recognize the fact that for Argentina's mining industry to become viable in today's highly competitive marketplace, certain fundamental changes must be made. In other words, more of the same is simply not good enough and will continue to fail. At the very least, this strategy must take into account the following items:

The short, medium and long term goals of the two branches of government regarding the development of a viable mining industry.
The short, medium and long term fiscal expectations of the governments regarding the mining sector.
The negative effects created by the Provincial administrations and Provincial Mining Companies.
The philosophy of the governments regarding the implementation of changes to the Mining Code which will make it more applicable to modern mining requirements, and most importantly
Past problems experienced by mining companies wishing to make investments.

The cohesive and comprehensive strategy developed as a result of the above discussions will allow policies to be developed which will achieve the goals of the government and at the same time satisfy the needs of the mining companies. The creation of a situation where both sides benefit (i.e. a "win-win" situation) is the only way to ensure success. The greatest builder of confidence for the investment community will be the development and implementation of a long term plan and the belief that government will not measurably deviate from it.

If the mining investor believes a country has stable policies he can plan his own long term economic decisions with confidence.

10 Temporary Suspension Of The Current Promotional Campaigns

The piecemeal implementation of mining investment and other mining promotion initiatives will not result in the rapid and logical development of new significant mining investment for the reason that they do not address other problems which are of equal concern to mining investors. For example: mining investors now have a low confidence level in the Provincial administration of mineral titles. Unless the administrative rules relating to mineral titles are modernized, made consistent between Provinces and applied fairly, the mining investor will continue to look elsewhere for investment opportunities. Therefore, it is essential that promotional campaigns at international meetings, conferences etc. <u>follow</u> the creation of favorable investment conditions. The worst scenario would be an effective promotional campaign which results in considerable interest by companies who subsequently realize that the actual investment conditions are quite unfavorable. Such uncoordinated promotion will backfire considerably and make it more difficult to promote in the future, even if the conditions by then have improved. Please note that Argentina may already be in this position.

11 Recent Developments In Argentina

Within the last year there have been several new developments related to mining in Argentina. These developments are in both the government and the private sector.

These developments are noted below, and where appropriate, brief comments are made.

11.1 Governmental

The provinces, as owners and custodians of the mineral resource, bear a major share of the responsibility for the under-performance of Argentina's mining sector. Therefore, to comprehensively address this complicated issue, coordinated and meaningful action is required from both the national government and the Provinces.

Dr. Angel Maza, Secretary of Mines has taken the initiative to focus on the under-performance of the mining sector and address problems that reside at the federal level by introducing appropriate legislation. The most significant piece of legislation is the Mining Investment Law, previously discussed. Other legislative initiatives include the Mining Restructuring Law (Ley de Reordinamiento Minero) and the Federal Mining Agreement (Pacto Federal).

The Mining Restructuring Law does the following:

Re-establishes a national Geological Survey to create geological charts of Argentina.
Recreates a high level governmental council (Consejo Federal de Mineria) for the discussion of mining affairs.
Increases the Canon to more realistic levels.
Creates special duty-free zones for mining.
Increases the surface area allowed for exploration.

The Federal Mining Agreement is comprised of 21 separate items, all pertaining to the Mining Code. The items relate to a variety of topics including public bidding on reserve areas, taxation, publication of lapsed claims, information exchange among the Provinces and the federal government and time deadlines for the performance of work. The most important objectives of the Pacto Federal are to:

Promote closer cooperation between the federal government and the provinces.
Improve the workability of the Mining Code.
Make it easier for companies to gain access to land for purposes of exploration.

For mining companies to actually benefit from the Mining Investment Law, they must first make economic discoveries which can only be achieved by allowing them reasonable access to land for purposes of exploration. This is where the biggest problem confronting the development of Argentina's mining industry resides: the lack of available land at reasonable costs for exploration. Most of the land is legally held in accordance with the Mining Code either by the Provinces or private speculators. Therefore, at this time the Mining Investment Law can only be considered to be a partial solution to the overall problem; other partial solutions must also be made before the problem is fully addressed. In other words, for the Mining Investment Law to be fully utilized by the mining community, the Federal and Provincial governments must actively work to resolve the other problems.

The Federal Mining Council will advise the government on all mining-related matters. While it is a good idea to both elevate the stature of the mining industry in government circles, the existing mining industry in Argentina is already represented through two professional organizations, and these organizations are highly qualified to advise the government on all mining-related matters. The Federal Council will be comprised entirely of government officials most of whom will have limited prior experience in the mining industry. The

Council will tend to duplicate the functions of the existing professional organizations but will do it less effectively due to their limited knowledge. In addition, two negative messages are being conveyed to the private sector:

The advice and contribution of the private sector is not necessary.
The problems of Argentina's mining sector will be solved by increasing the role of government in mining affairs.

Mining-related environmental control legislation is presently being formulated. However, without the necessary solid understanding of the technical aspects of mining and processing, such legislation will almost certainly be cumbersome and difficult to administer. (Please note: Norwest prepared environmental guidelines and other recommendations as part of the complete Feasibility Study. These are available for use in such environmental legislation).

All of the above legislative initiatives will, to varying degrees, assist in creating the environment in which a more active mining industry will develop in Argentina.

However, in conclusion, it must be stated that, regardless of its success in passing mining-related legislation, the Federal government cannot be expected to solve all of the problems facing the Argentine mining industry. Equivalent and forceful action is required of the Provinces regarding privatization, transparency of procedures etc.

11.2 New Mining Projects

Although no new mine (as defined by conventional standards) has opened up in Argentina since the mid-1960s, international mining companies are currently active in Argentina and several projects are in varying stages of development. The level of activity on the part of international companies has increased within the last 2 years which coincides with the period of economic and currency stability. The existence of these projects indicates that the international mining community is beginning to look more favorably on Argentina as a place for mining investments.

Some of the companies are engaged in exploration and several have advanced to the feasibility study phase.

It is not known when any of these projects will be placed into production.

In recent months, as a result of all of the publicity about Argentina's mining potential, there has been an upsurge of interest and many companies (both big and small) have visited Argentina to assess the situation at first hand. However, it is reported that when these companies make detailed inquiries they learn that there are virtually no viable projects ready to go; only opportunities for long term grass-roots exploration on undeveloped land. This is, of course, the natural result of past governmental policies.

Bolivia—its potential and recent activity

Charles 'Scottie' Bruce
MINTEC, La Paz, Bolivia

During the past few years, thanks to its newfound stability and the lack of exploration during its turbulent years, Bolivia has become increasingly more attractive to the mining investor.

Bolivia may be divided up into six distinct geological regions within the Highlands, the Plains and the Precambrian Shield. The latter is located in the eastern sector of the country. Each region is discussed in sufficient detail to give an overview both of the geology and of the mineral potential.

Long thought of as a narrow-vein, silver-tin, bonanza - type producer, recent discoveries have shown that economic disseminated gold deposits occur and that the potential for the discovery of low grade precious metal deposits, ignored in the past, exists. Of considerable interest are the non-metallic prospects and the platinum potential of the Rincon del Tigre ultramafic complex.

The exploration programs which have been undertaken since Bolivia acquired stability are described with emphasis on the results obtained to date. Specific prospects where discoveries have been made and/or mining operations have started are briefly discussed. In conclusion, selected prospective areas with high potential for mineral discovery are outlined.
Keywords: Bolivia, mining developments, exploration potential.

Introduction

Of the Latin American countries, few have had as turbulent a past as that of Bolivia. In fact, only 8 years ago, just to mention the word Bolivia would elicit smiles, due to its history of political instability. Despite this, Bolivia has traditionally honored its word and agreements with the result that it also has a history of vast mining fortunes having been made in the past, largely in tin, silver and antimony. However, by 1985, the military coups and instability had taken their toll and Bolivia suddenly found itself in real economic chaos with the inflation rate at 26,000 percent.

Drastic situations call for drastic solutions! In 1985, Dr. Paz Estenssoro was elected president of the country and with his team he launched an austerity program of unequalled harshness based on a free market economy. It was highly successful firmly established the democratic process and brought to the country the long -

sought - after stability it now enjoys. In fact, today it is considered one of the safest countries among the developing nations.

There are several economic reasons for considering Bolivia for investment. It has a free exchange rate, free currency exchange, the lowest inflation rate in Latin America, the lowest income tax rate, attractive labor rates, no restrictions nor taxes on capital entering or leaving the country, profit remittance unrestricted with tax on dividends at only 13 percent, no restriction on imports and no restrictions on foreign ownership with the exception of the border zone.

The above economic factors were enacted into law by the government which followed Dr. Paz Estenssoro. They also redrafted the mining code and approved a new investment law.

With such an attractive scenario, then why is Bolivia at the cross roads? To have rapid growth, it requires foreign investment from private sources. This requires stability with a strong, long-term outlook. The recent elections have proven that these conditions will be met and, more important, the President - elect is considered, both nationally and internationally, to be a great choice. President - elect, Gonzalo Sanchez de Lozada, (Goni), was himself the president of the largest and most successful mining company in Bolivia which he built from scratch.

Bolivia is at the cross roads because this year will show if "easy - to - find" mineral deposits do indeed exist. Recent successes have been notable, e.g., the Battle Mountain - Inti Raymi project is now on stream with a 65 million ton reserve of 2.33 gm/tonne Au and 14 gm/tonne Ag. The discovery of the San Ramon gold deposit in the Precambrian by Comsur which, though relatively smaller, is mineable. The latest discovery, the Don Mario Cu - Au prospect, is also in the Precambrian and located some 360 kilometers east of San Ramon. The owners of Don Mario have entered into an agreement whereby Billiton will explore and, if results warrant, develop the property.

Bolivia is at the cross roads because the active drilling programs projected for 1993 by both major and minor mining companies, will, if sucessful, project Bolivia further into the limelight with rapid growth or, if negative, will slow-up the investment process, at least vis-a-vis grass-roots exploration.

Geological overview

Bolivia has a variety of geology which is virtually unparalleled in any other country. Within its borders there are six distinct geographic, geologic and metallogenic regions.

In the Cordillera Real, the Eastern Andes, where the glaciers reach almost 23,000 feet above sea level, the exposure of Tertiary intrusives contrast sharply with the Devonian-Silurian phyllites. It is within this sector that the richest silver and tin mines were found in the past.

To the west, the Altiplano, High Plateau, at an altitude of 13,000 feet, stretches to the Western Andes where Recent volcanics are found amidst dacites and andesites of Tertiary age. The Altiplano itself consists of red-bed sediments and volcanics.

To the east, after passing through the Sub-Andean zone, there is a wide section of alluvial cover until the Precambrian shield is reached which covers an area of some 100,000 square miles.

Within these widely varying geological environments, there is an equally wide variety of mineral deposits which have only been partially explored. In fact, it can be estimated that only 10 percent of the mineral wealth has been tapped and that the potential for discovery is unusually high.

Cordillera Real

This has been the most productive geological region thus far in terms of rich lode deposits. It includes the famed Cerro Rico de Potosí which has operated continuously for over 400 years and has yielded more than 3,000 million ounces of silver. Estimates are that a similar amount may exist as disseminated ore within the mountain, which may be amenable to modern mining techniques. While Cerro Rico is the most spectacular of the silver mines, those of Oruro, Colquechaca and Pulacayo are also impressive.

Catavi, Huanuni and Viloco were the most important of the tin deposits. The latter was famed for its large, gem quality, cassiterite crystals.

The ores were found to a large degree as veins in the sediments. Only now is attention being paid to the possibility of disseminated ores in the Tertiary rhyolite and dacite intrusives. This region is referred to as the "traditional" mining area as past operations were confined to narrow-vein mining.

Altiplano

The Altiplano is an intermontane basin some 80 miles wide by 435 miles long. It consists largely of Tertiary red-bed sediments and volcanics. There are numerous volcanic domes and some stratovolcanoes. One such volcanic complex has been developed into a successful open-pit, heap-leaching, gold operation by a Bolivian company, Inti Raymi, in joint venture with Battle Mountain.

Some 125 miles south lies Toldos, another bulk mining operation for silver. Four other precious metal prospects in various degrees of exploration and development have already been indentified.

Near Uyuni is an extensive salt lake with a high potassium content which contains the largest lithium deposit in the world. There are several smaller salt and borate deposits.

Western Cordillera

The Western Cordillera primarily consists of Tertiary to Recent volcanic rocks and sediments which have been intruded by dacite stocks. These have a high potential for the discovery of epithermal, bulk-minable type, precious metal deposits. Several such prospects have already been discovered within this environment which hold a high degree of interest for the investor.

Of secondary interest, but yet important, are the numerous native sulfur deposits which are relatively large and free of impurities.

Sub-Andean zone

The Sub-Andean Zone forms the eastern flank of the Cordillera Real. It is not commonly thought of as a separate mineral province, yet the black Devonian shales are excellent host rocks for silver and base metal deposition. Several hundred prospects have been found within this sector but only one of these, the Quioma-Asientos zinc-silver deposit, has been developed into an important mine.

Alluvials

Gold has been mined for many years from the Tipuani-Mapiri-Guanay district in the foothills of the Cordillera Real north of La Paz. One dredge, owned by COMSUR, operated in the area but, apart from this, most of the mines are labor intensive and are controlled by cooperatives.

This area is famed for its "nuggety" ore and little attention is given to the recovery of fines. To the extreme north-east, along the Bolivia-Brazil border, there are numerous small, suction dredges working in the Madera river.

The area with great potential for discovery is the Beni river and its numerous tributaries which originate in the Tipuani area. The Madre de Dios river is similar and it has several active dredging operations across the border in Peru.

An estimate of the yardage of gravels within these drainages, considered to have potential for alluvial gold, approximates 4,000 million cubic yards. While virtually no exploration was done in the past due to the lack of infrastructure, there are numerous reported near-surface gold showings. This could constitute an untapped alluvial gold field of major dimensions. Auspac, an Australian company has recently entered into a J.V. agreement with Comibol on one of the areas of interest.

Bolivian Precambrian

The Precambrian section in Bolivia is confined to the tropical plains of Eastern Bolivia. It was prospected by the Jesuits during the colonial period and there are numerous small workings on narrow veins and in stream beds.

It was not until recently that serious consideration was given to this area when the British Geological Survey, BGS, completed a ten year mapping program of the 100,000 square mile exposure of this unit in Santa Cruz, Beni and Pando. The project was undertaken jointly with the Bolivian Geological Service, GEOBOL, and was finished in 1986.

The geology comprises a Proterozoic basement of granitic gneisses and granulites, with several schist belts and mid-Proterozoic mobile belts.

The most important types of mineralization encountered so far comprise the following:

- Gold bearing greenstone belts and alluvials which have been traced over 240 miles and which were worked by the Jesuits in the 18th century.
- A major layered mafic-ultramafic intrusion with copper, nickel, chromium and platinum group elements.
- Pegmatite fields with niobium, tantalum, beryl, muscovite and amethyst.
- Specialized tin-granites similar to those of Rondonia, Brazil.
- An alkaline carbonatite province.
- Banded iron and manganese formations.

Until recently, virtually no exploration had been undertaken to follow up on the BGS study, or to delineate the economic potential, apart from gold exploration in the greenstones by a major national company, COMSUR. Utilizing modern exploration techniques, a 12.5 mile length of iron formation with shear-zone related gold mineralization in the greenstones near San Ramon was delineated.

The Precambrian is the least known of Bolivia's geological regions, but it is considered to have excellent potential.

Recent activity

There are presently 11 advanced projects underway which are briefly described as follows:

Cordillera Real

Bolivar, a semi-developed polymetallic underground mine (grade 12% Zn, 300 g Ag/tonne) was recently awarded to Comsur after a second round of bidding. Comsur is contractually obligated to have it on stream by mid - 1994.

Ubina lies within the western flank of the southern sector of the Cordillera Real. While long of interest for tin and tungsten recent studies by Newcrest, Australia has delineated an area, 1200 m x 300 m, which carries high gold and silver values. It is the first such prospect to have been identified, where the Au/Ag values occur within Devonian - Silurian calcareous sandstones and quarzites to form a bulk - mineable resource in a manto-type horizon. This prospect will be further evaluated by Phelps Dodge and will be drilled later in the year.

Altiplano

Kori Kollo, the Inti Raymi project, in early 1993, was brought on stream. Battle Mountain, which is the 86 percent joint-venture partner, selected Minproc Engineers to construct their new 14,500 tpd C.I.L. plant which was completed over two months ahead of schedule and well under budget. The oxide heap- leach section will be phased out by year-end. An increase in production capacity is already under consideration.

Iroco, a joint venture between a local co-operative and Cameco, Canada will be further evaluated by drilling in 1993. It is an attractive small gold bulk-mining project with Au values in the order of 1.7 gm/tonne.

Laurani is a vein-type deposit with known high-grade gold-silver veins and massive sulfides with enargite. The in-depth potential for bulk-mining targets is presently being evaluated by RTZ.

Packekala, a joint venture between Newmex, Australia and Minnova, Canada is a very interesting target. Initial exploration on the northern side of the complex has shown widespread gold anomalies of between 2 - 5 gms Au/tonne in brecciated dacites cemented with specularite. These outcrops proved to be shallow and more extensive exploration of the southern area showed indications of a possible Au - Cu porphyry. The data are being complemented with geophysics prior to drilling later this year.

Tiawanaku, a Joint Venture between Newmex Exploration, Minnova and a Bolivian national will be the first property to be drilled within the border zone under the new regulations which now permits foreign interests to work there through a joint venture. The target is an Au - Ag deposit with minor base metals hosted in a quartz - feldspar prophyry amenable to bulk mining. Surface sampling has shown widespread mineralization with appreciative gold - silver values recorded in areas of dense fracturing.

Western Cordillera

La Riviera, lies within the frontier zone and has been partially explored. It is held under J.V. by Central Mining, Sydney. A limited drilling program to the extreme north of the property indicated a horizon with ore-grade mineralization. The entire concession is now being mapped in detail with emphasis on the southern volcanic pile where it is believed that sufficient ore-grade material can be drilled out to justify development.

Skorpio is a small prospect on the western extremity of a large color anomaly. A silicified shear zone has been delineated which carries 5.6 gms Au/tonne on the surface over a width of 2 meters. A winze sunk on the structure by hand methods indicates that the values continue in depth. While Skorpio itself is of minor interest, its relationship to the whole (a bleeder?) justifies the on-going program.

Precambrian

San Ramon was drilled first by Comsur and secondly by RTZ. A resource of 300,000 ounces was defined in the area of Puquio Norte which is open-pittable. More important, the major structure, which is in greenstones, has proven that mineralization continues to a depth of at least 630 meters. The greenstone host structure continues on strike for at least 12 kilometers.

Don Mario was discovered by loggers in a remote area north of San Jose de Chiquitos. It was then pegged by a local mining company which, with Battle Mountain, outlined an area with a potential resource of several million tons containing 3.5 percent Cu and 2.7 gm Au. Later Billiton formed a joint venture with the local company to explore and, if justified, develop the property. As the ore is hosted in greenstone schists which are accompanied by a band of magnetite, airborne geophysics should easily delineate any major targets in an otherwise inhospitable site. Billiton hopes to have a definitive decision as to it relative merit within 12 months.

Rincon del Tigre is a major layered ultramafic intrusion with copper, nickel, chromium and Pt group elements. This property was first tested by Central Mining, Sydney who found values, both gold and PTG, in geochemical samples and in rock chip samples. Subsequent work by RTZ has shown wide-spread PTG anomalies sufficient to justify a drilling program, now in progress.

There are several companies, such as Amax, USA; Arimetco, USA; Asarco, USA; BHP, Chile; Cominco Resources, Canada; Cyprus Minerals, USA; Minproc Resources, USA; Hexagon, USA; Paranapanema, Brazil, which are actively involved with grass-roots exploration and/or evaluation. Their continued interest in numerous projects augers well for the country.

Public sector

While this paper is directed toward the private sector, mention must be made of the national mining company. Corporación Minera de Bolivia (Comibol). Starting with the government of Dr. Paz Estenssoro, it was planned to convert Comibol into a holding company. Towards this end, several properties have been leased to private companies or joint venture agreements have been signed. These are as follows:

ITOS :	Ag - Sn, tailings J.V. with BAREX Ltd.
BOLIVAR:	Ag - Zn - Sn, mine, J.V. with COMSUR
ARARAS:	Au, alluvial, J.V. with AUSPAC
MARIA LUISA:	Polymetallic, J.V. with AUSPAC
COLQUIRI:	Zn - Sn, tailings lease with MINPROC
TASNA:	Bi, Sn, Au, mine, lease with COMINESA
CATAVI:	Sn, tailings, J.V. with PARANAPANEMA
ASIENTO:	Au, prospect, J.V. with BRGM
POOPO:	Sn - Zn, mine, J.V. with TIWANACU

The Bolivian president, Mr. Sanchez de Lozada and his government have developed a new and innovative plan to capitalize the state companies. Several of these companies are quite valuable, e.g. hydrocarbons (Y.P.F.B.), communications, and electricity. Under their platform of the "Plan de Todos" or plan for everyone, they seek J.V. partners for all of the state enterprises whereby the investor would invest the required capital for 49% of the shares in the companies of his choice plus full management. The remaining 51% of the shares would be sold to the public.

This plan has been well received both by the worker and by the international investment community as an attractive alternative to direct privatization. It is indeed a novel approach to foreign investment and opens the door to new opportunities. Comibol has vast property holdings which are virtually unexplored as well as active mines which require both new technology and exploration.

Investment in mining: the Canadian experience

Robert B. Parsons
Price Waterhouse, Toronto, Ontario, Canada

Latin America has a long history of mining. Aboriginal peoples were using iron, copper, gold, silver, and coal long before the arrival of Europeans in the 15th century. Despite this tradition, Latin America has a relatively short history of modern mining. Only in recent years has the Latin American mining industry proclaimed its intention of being a major world player.

The successes of the Latin American mining industry over the past 20 years are unprecedented. In many respects, some Latin American countries have accomplished in one or two decades what took more than a century in North America, South Africa, and Australia. Fast growth cannot normally take place without some discomfort. Nevertheless, Latin American countries might be able to minimize growing pains by learning from the experience of other so-called "mature" mining countries.

On the bold assumption that Latin America can learn from the Canadian experience, this paper provides an overview of Canadian mineral policies over the past 30 years, and seeks to identify some fundamental principles that could serve as references for the formulation of mineral policies in Latin America. In addition, this paper reports on the status of current mineral policy development in Canada. These latter policies will likely be intended, at least in part, to help the Canadian mining industry to deal with the increased competition that results from the significant successes of the Latin American industry.

1. Importance of the Canadian Industry

Canada's mining industry is important to its people. In 1991, the minerals industry accounted for approximately 17% of the country's exports. Roughly 60% of Canada's rail and sea freight transportation is used for moving minerals. Since 1982, the industry has accounted for, on average, 4.6% of gross domestic product, which is ten times the comparable percentage in the United States. Over 300,000 Canadians work in the mining industry.

In 1992, the total value of metallic mineral production exceeded $16 billion. Canada ranks first in the world in production of zinc and uranium. Canada is the second largest producer of nickel after Russia, the fourth largest producer of copper after Chile, the United States and Russia, the fifth largest gold producer, and the seventh largest producer of iron ore.

2. Framework of the Canadian Industry

Canada is the second largest country in area in the world, after Russia. Canada is 17% larger than Brazil, and is larger than all other South American countries combined.

Canada consists of ten provinces and two territories. Under the country's constitution, regulation and ownership of mining is a matter of provincial jurisdiction over lands in the ten provinces, and is a matter of federal jurisdiction in the two territories. The provinces account for some 39% of the total land area and more than 95% of the value of mineral production. The provinces have authority over land use, environmental, and health and safety matters. This allocation of mining responsibilities between federal and provincial authorities is subject to the outcome of on-going negotiations with respect to Aboriginal land claims.

Each of the federal and provincial governments imposes an income tax on corporations and individuals. In addition, each of the provinces and territories collects a mining tax, royalty, or similar severance tax from mine operators. In most cases, these levies are based on mining profits, so that a mine typically bears three levels of profit taxes: a federal income tax, a provincial income tax, and a provincial mining tax. Generally speaking, the resulting tax burden is competitive with other mining nations.

3. Pre-1972 Policies

The 1950s and the 1960s were good years for the Canadian mining industry. During this period, the tax regime was both attractive and stable. From 1948 to 1971, there were no major changes to the tax rules at either the federal or provincial levels, and there were no significant differences in the tax regimes from province to province. The combined federal/provincial income and mining tax rate was in the 50% range, but the effective rate was actually much lower than this as a result of the deduction of a percentage depletion allowance. Further, new mines were exempt from income tax for the first three years of production.

4. 1972 Tax Reform

Reform of the pre-1972 income tax system involved a process that stretched from 1962 through 1972. There were four key dates in this process:

- 1962 - appoint Royal Commission;
- 1967 - release Commission's report;
- 1969 - table government White Paper; and
- 1972 - amend legislation.

Prior to 1972, investors generally viewed the tax treatment of the Canadian mining industry in a relatively favourable light. However, during the tax reform process, it became increasingly evident that the advantageous treatment of the industry could disappear. The basis for this sense of pending doom is outlined below.

In 1962, the federal government appointed the Royal Commission on Taxation (the Carter Commission) to inquire into, and report upon, the incidence and effects of federal taxation and to make recommendations for improvement in the tax laws and their administration. The Commission's lengthy and complex report (six volumes, 2,700 pages) was released in February 1967. The Commission recommended a complete transformation of the Canadian tax system.

The proposed changes that would have affected the mining industry included:

- Capital gains would be taxed at the same rates as ordinary income. (At that time, capital gains were tax-free).
- The three-year tax exemption for new mines would be eliminated.
- The automatic 33% depletion allowance would be abolished.

The federal government considered the Commission's report, and tabled a White Paper in November 1969 to set out its major proposals for tax reform. The White Paper proposed that capital gains be taxed in full, except that only one-half of the gain realized on a sale of a widely-held share would be taxed. The three-year exemption for new mines would be phased out by January 1, 1974, and automatic depletion would continue until the end of 1975 when it would be replaced by an "earned depletion" system.

In June 1971, the federal government tabled a Notice of Ways and Means Motion to implement its final tax reform proposals. Those most important to the mining industry included:

- establish a general rule whereby one-half of all capital gains would be included in income, as compared to the earlier proposal whereby the entire capital gain would be taxed;
- phase out the three-year tax exemption by the end of 1973; and
- replace the automatic depletion allowance with an earned depletion allowance at the beginning of 1976.

In 1972, the June 1971 proposals were enacted.

The 1972 tax reform process left a cloud of gloom hanging over Canada's mining industry. Prior to the tax reform process, there was a general perception domestically and abroad that Canada's fiscal system made the country an attractive place to invest in mineral exploration and development. 1972 tax reform tarnished that reputation. Moreover, in reality, tax reform resulted in a higher tax burden on the mining industry. The perception and the reality of the tax reform process both discouraged decisions to explore and develop in Canada.

5. Federal/Provincial Quarrel

During 1974 and 1975, Canada's mining industry faced unprecedented tax increases. For example, 1974 changes to tax legislation in British Columbia could have resulted in more than 100% taxation of revenue derived from a metal price increase. The mining industry reacted to these changes with alarm. "Capricious" and "bizarre" were among the more polite terms the industry used to describe the tax changes.

The reasons for the extent and pace of the changes were complex. In 1973, escalating world prices for minerals swelled mining company profits. Both the federal and provincial governments reacted by demanding an increased share of the industry's "windfall" gains. Provincial governments took the position that they were entitled to a substantial minimum return for the exploitation of provincial mineral resources, regardless of the level of industry profits. In addition, public and government reaction against the tax concessions that had been available to the mining industry in the past was strong.

The worldwide oil crisis heightened concern that depletable natural resources needed to be preserved. Provincial governments were increasingly of the view that it is not always desirable to encourage the most rapid development of their province's natural resources.

Each level of government imposed new taxes apparently without regard for the position of the other and without regard for the overall effect on the mining industry. The governments in British Columbia, Manitoba, and Ontario started the "grab" for a bigger piece of the mining pie. Their actions triggered a reaction from the federal government, which involved detrimental tax changes. The other provinces soon jumped into the fray with higher taxes. Starting in 1977, the provinces began to recognize the harmful effects of their actions of 1974 and 1975, and moved to ease the tax burden on the industry.

The first in this series of events occurred in February 1974, when the province of British Columbia introduced additional taxes on mining activities in the province. Mining companies would be required to pay not only a provincial mining tax but also a unique two-tier royalty combining a basic 5% royalty with a 50% "super royalty" to be levied when the metal price collected by the mineral producer exceeded a predetermined value. The new royalties took effect on January 1, 1974. Both the basic royalty and the super royalty were determined with reference to the production value of all minerals mined in British Columbia regardless of whether a mine operated at a profit or a loss.

On March 21, 1974, the Manitoba government stated its objective to obtain for the people of Manitoba "a fair share of any added value or economic rent accruing to their mineral resources because of scarcity and/or the increased value of the minerals which are unrelated to the immediate costs of production". As an interim measure (pending the introduction of a completely new variable volumetric royalty system), Manitoba increased the basic rate of provincial mining tax from 15% to 23% effective April 1, 1974.

On April 9, 1974, Ontario announced a substantial increase in mining taxes. The previous Ontario mining tax rate of 15% was replaced by a graduated rate schedule ranging from no tax at all on the first $100,000 of mining profits to a maximum tax of 40% on mining profits in excess of $40 million.

Under the federal legislation that was in effect in 1974, provincial mining levies were deductible in computing taxable income for federal income tax. Therefore, significantly higher provincial mining taxes and royalties, which were introduced in the early months of 1974, threatened to seriously erode the federal income tax base. The May 6, 1974 federal budget dealt with this concern by immediately disallowing the deduction of provincial mining taxes and royalties. Since none of the federal and provincial income taxes and mining levies were deductible in computing the tax base to which the other taxes applied, the income and mining tax burden faced by a mining company was simply the sum of all of these taxes. In some instances, this burden could amount to more than 100% of pre-tax profits.

Canada's mining industry reacted to the heavier tax burdens by shelving exploration and development projects until some semblance of reasonableness to the tax system was restored. Offshore investors reacted by avoiding mineral exploration and development in Canada. The perception and the reality were the same: high tax rates and the uncertainty of rules made Canada an unfavourable place to invest in exploration and development.

Confidence gradually started to return beginning in 1977, when a series of provincial tax changes resulted in more moderate tax rates, which, generally speaking, continue to be in place today.

6. Flow-through shares

An important factor in the recent history of investment in the Canadian mining industry has been the use of flow-through shares. In simple terms, "flow-through" shares are common shares that are issued by a Canadian mining company in order to finance an exploration program. They get their name because the consequent tax deductions "flow through" to the investor. The investor can deduct them in computing his income tax.

The significantly higher levels of exploration expenditures from 1985 to 1988 (Figure 1) correspond to high levels of flow-through share financing. These increased exploration expenditures were largely funded by flow-through share financing, as improvements to the tax rules made flow-through shares popular.

The popularity of flow-through shares peaked in 1987 and 1988, and started to wane in 1989. The rise and fall of flow-through share funding corresponds to a series of tax changes that affected the attractiveness of investment in flow-through shares.

This series of tax changes can be summarized as follows.

- **1981** -- The Income Tax Act (the Act) was amended to repeal the rule that denied capital gains treatment to the sale of a flow-through share. Henceforth, the sale of a flow-through share would be eligible for the favourable capital gains tax rates, instead of being taxed in full as ordinary income.
- **1983** -- The depletion regulations were amended (a "mining exploration depletion allowance" - MEDA - was introduced) so that a flow-through share investor could, in effect, deduct up to 133% of the cost of acquiring a flow-through share, instead of 100% as previously.
- **1985** -- The Act was amended to provide each Canadian individual with a lifetime $500,000 capital gains exemption. Flow-through shares were eligible for this exemption.
- **1985** -- The so-called "60-day rule" was introduced to the Act, so that an investor's deductions in a particular calendar year could include exploration expenses incurred by the corporation in the 60-day period following the end of the calendar year.
- **1986** -- A technical amendment to the Act removed the concern of the flow-through share investor about his third-party liability with respect to the company's exploration programs.
- **1987** -- The federal government's White Paper on tax reform proposed to phase out MEDA (the extra 33% deduction associated with flow-through shares) by July 1, 1989. In addition, under tax reform flow-through shares would no longer be eligible for the capital gains exemption.
- **1988** -- The federal government announced that the Canadian Exploration Incentive Program (CEIP) would be introduced effective in 1989 to provide for 30% cash grants for eligible exploration expenditures. These cash grants were linked directly to flow-through shares, and were intended to compensate for the removal of MEDA.
- **1990** -- CEIP was cancelled.

As alluded to above, the tax rules relating to flow-through shares were most favourable from 1985 to 1987. All of the tax changes implemented from 1981 through

1986 were favourable. Starting with 1987 tax reform, the changes to the tax rules hurt the industry. Even the announcement of CEIP in 1988 was unable to "save" flow-through shares, mainly because the October 1987 stock market crash had such a devastating effect on share financing of all kinds, including flow-through share financing.

Many speaking on behalf of the industry observed that the flow-through share tax rules, at their best, were probably too generous, and that too much money was probably chasing too few worthwhile exploration projects. Nevertheless, the flow-through share rules were indeed responsible for unprecedented levels of exploration spending in Canada from 1985 to 1988.

7. Nationalism

Investors are generally inclined to be relatively tentative about investing in a country that demonstrates a high degree of nationalistic sentiment. The conventional wisdom is that such a sentiment raises the investment risk factor by some undefinable margin.

During the decade of the 1960s and 1970s, and into the first half of the 1980s, nationalistic sentiment in Canada grew, at least at the political level. This sentiment manifested itself in a number of ways in the country's legislative framework, including:

- the establishment of Canada Development Corporation (1966);
- the enactment of the Foreign Investment Review Act (1973); and
- the announcement of the National Energy Program (1980).

The culmination of these programs in 1980 led to less than friendly charges, threats, and accusations from foreign investors, particularly those based in the United States. Despite these brief heated exchanges of words, there is no evidence that exploration and development suffered. At the time of this nationalistic trend in Canada, other factors (excellent geology, political stability, etc.) made Canada a relatively attractive place for foreign investment.

8. 1990s

Today, the Canadian mining industry faces three major issues:

- environment;
- land access; and
- taxation.

For the past few years, the federal and provincial governments have been changing existing environmental regulations, and have been drafting new rules and regulations. Within each government, different ministries have been doing the same thing. In brief, the Canadian mining industry faces a maze of new and pending legislation, some overlapping and some contradictory. This uncoordinated scramble by thirteen federal and provincial governments, and by separate ministries within these governments, has been largely driven by governments' wish to be seen by the populace to be environmentally angelic, even if this means overkill. The current state of the legislation, together with increasing compliance standards imposed by all levels of government, has created:

- an atmosphere of instability and unpredictability;
- excessive permitting delays and costs;
- new regulations that may not be supported by scientific evidence; and
- high costs of upfront reclamation funding requirements.

Another issue confronting the mining industry in the 1990s is the question of land access. The federal and provincial governments have announced that they are committed to setting aside a minimum of 12% of Canada for parks and protected areas. As of 1992, mineral activity was prohibited in 6.4% of Canada. These lands would be closed to exploration and development. There is confusion about the process for designating these protected areas, and there is uncertainty about which lands will be affected. As evidenced by the province of British Columbia's decision to close the Windy Craggy area to mining, protected areas could include established mining camps, thereby introducing an element of confiscation to the process.

In addition to the withdrawal of lands for prescribed areas, the issue of Aboriginal land claims and Aboriginal self-government also adds to the uncertainty of land access, although these issues are probably less serious.

The main taxation issue of the 1990s relates to the steady year-by-year increase in the burden of non-profit based taxes being borne by the industry.

9. Non-profit Based Taxes

Seventy-seven years ago, Canadians were unceremoniously introduced to something called income tax. Some years later, the provinces jumped on the bandwagon with their own income taxes, and later still with mining taxes. These profit-based taxes appealed to elected officials and to a segment of the populace. The onus of the taxes varied directly with the ability to pay (in theory, at least). Previously, the principal sources of government revenues had been property taxes and customs duties, which were viewed as somewhat regressive in nature.

The mining industry has always had concerns about different aspects of income and mining taxes - the tax rate being too high or the depreciation rate being too low, and so on. Recently, though, the industry has turned its sights on the expanding array of taxes that are not based on profit, including payroll taxes, capital taxes, water taxes, sales taxes, fuel taxes, property taxes, environmental levies and user fees. The industry is seriously concerned that the rising level of non-profit based taxes is not only hampering the industry's ability to reinvest in Canada, but is also placing Canada at a disadvantage in trying to attract new investment capital.

Non-profit taxes are becoming increasingly popular with federal and provincial governments. The reasons for the inclination to widen the tax net by imposing non-profit taxes include:

- Income taxes have a high profile, as compared to non-profit taxes. When it comes to taxes, politicians like to lie low;
- Non-profit taxes can be used to snipe at specific industries or activities;
- Canada has little room left to raise its income tax rates while remaining competitive with other industrialized countries;
- Most non-profit taxes are based on the widely acceptable "user pay" principle; and,
- Non-profit taxes generally are deductible for income tax purposes; thus, if a province introduces a non-profit tax, the enterprise gets 29% of the tax back from the federal government through income tax relief.

This last point has not gone unnoticed by the federal Department of Finance. The federal government itself is sufficiently concerned about the rising level of non-profit taxes that a recent federal budget proposed to put a ceiling on the amount of such taxes that could be deducted by the enterprise for income tax purposes.

Criticisms currently being levelled at the wide range of non-profit taxes include reference to the regressive nature of the taxes, as well as the unpredictability and lack of coordination associated with their implementation.

10. International competition

Besides the domestic issues of environment, land access, and taxation, the Canadian mining industry is experiencing unprecedented competition from abroad. There is a serious concern in Canada that the restoration of economic and political stability to key Latin American countries, coupled with their attractive geology and mineral investment policies, are diverting significant investments away from Canada towards Latin America. The political and economic restructuring of Eastern Europe and the CIS also raises fears in the Canadian mining industry about the stability and level of metal prices, and the future flows of investment capital.

11. Lessons for Latin America

Despite all of the problems that have confronted Canada's mining industry over the past 30 years, the industry has nonetheless continued to grow and prosper, at least until the last few years. The Canadian experience suggests that a number of factors are necessary for the maintenance and growth of a healthy mining industry, including:

- good geological conditions and a useful geological database;
- political stability;
- responsible and stable tax and fiscal policies;
- trained and dedicated work force;
- a spirit of cooperation between industry and government; and
- basic infrastructure (rail, roads, ports).

When one of these factors is weakened or disappears, damage to the industry will be minimized as long as the other factors remain in place. Also, when one of these factors changes, a well-conceived and methodical change can be absorbed more easily by the industry than an abrupt change. A number of instances of change can be cited to illustrate these points, including the following examples.

- Although 1972 Tax Reform resulted in a higher tax burden on the industry, news of the changes was spread over more than ten years, and the new rules reflected industry input as well as transitional provisions. As a consequence, industry's initial negative reaction to this tax reform was short-lived.
- The fight between the federal and provincial governments in 1973 and 1974 over the sharing of mining profits was much more damaging to the industry, because the resultant changes were implemented without warning, there was no industry input, there were no transitional provisions, and some of the new rules were totally unreasonable. The industry took several years to recover its confidence in the federal and provincial governments. Even in these years, however, a considerable level of investment in the industry continued, because

other necessary factors were still largely in place.

- Over the past five years, investment in the Canadian mining industry has been in a steady decline due, in part, to the uncertainty and confusion created by sudden, uncoordinated, and ill-conceived environmental and land access regulations and actions. In many cases, these changes have been capricious, and have been proposed and implemented without industry consultation or input. It will likely be a few years before stability of law is reestablished in the area of environment and land access. In the meantime, to finance new projects in Canada will be more difficult.

Fiscal incentives affect investment behaviour in a significant way only where the incentive is substantial. For example, the government of the province of Ontario had to deliver a highly attractive incentive through the province's mining tax legislation in order to induce Texasgulf to construct a copper smelter at the company's Kidd Creek mine in the mid-1970s.

In the mid-1980s, the income tax rules relating to flow-through share financing were improved considerably, and became so attractive that exploration expenditures hit almost unbelievably high levels in 1987 and 1988.

These two examples of tax incentives also have another lesson: great care must be exercised in devising an incentive, if unwanted distortions are to be avoided. In the case of the Kidd Creek smelter, many industry and government officials questioned whether it was in the nation's best interest to add unnecessary smelting capacity to the Canadian industry: these observers felt that the Province of Ontario placed provincial self-interest ahead of the national interest.

In the case of the flow-through share rules, the richness of these rules was considered by some to be excessive, and was blamed for allegedly inefficient or unnecessary spending on exploration. The flow-through share rules have also been blamed for excessive focus on gold exploration (in order to attract unsophisticated investors), to the detriment of base metal exploration, which did not have the same appeal to investors.

12. Current Canadian Concerns and Policies

Over the past few years, the Canadian mining industry has become increasingly concerned about its future. As indicated above, Latin America, Eastern Europe, the CIS, and, to a lesser extent, Asia and Africa are seen as competitive threats to the Canadian industry. The industry offers a range of evidence to make their case that the industry is in jeopardy. For example:

- Exploration spending has been in a steady decline since 1987 (Figure 1);

Figure 1
Canadian Exploration Expenditures by Junior and Senior Companies

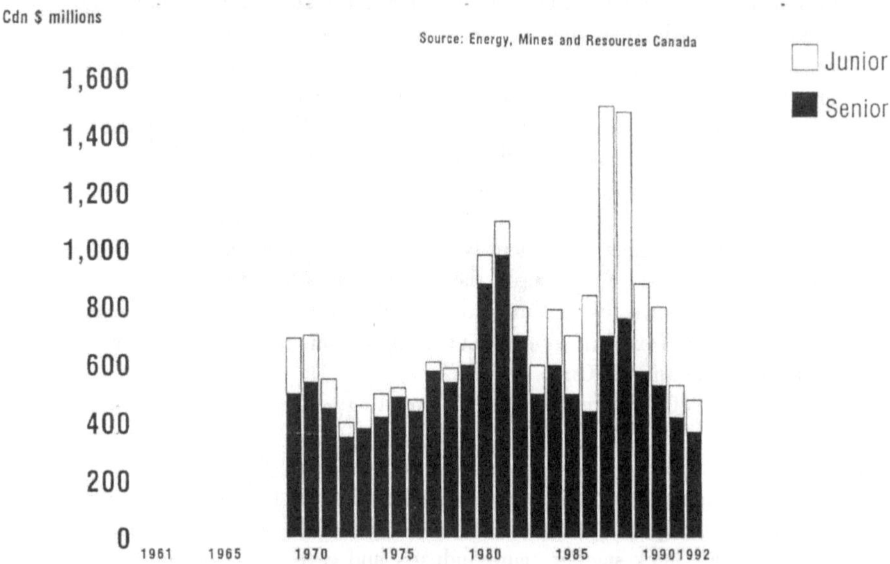

- From 1988 to 1992, the number of mine closures has exceeded the number of mine openings (Figure 2);

Figure 2
Mine Openings and Closings - Canada

	1988	1989	1990	1991	1992	Total
Number of openings & re-openings	26	23	24	18	8	99
Number of closures & suspensions	16	22	26	33	28	125
Net openings (closures)	10	1	-2	-15	-20	-26
Net effect on ore production capacity (tonnes/day)	18700	25600	-30800	-44800	-88000	-119300
Net effect on direct employment (no. of workers)	900	1850	-1760	-2060	-5800	-6870

Source: Energy, Mines and Resources Canada, Mineral Industry Quarterly Report, March 1993

- From 1986 to 1992, total capital expenditures in the industry have declined (Figure 3);

Figure 3
Total Capital Expenditures, Mining

Cdn $ millions

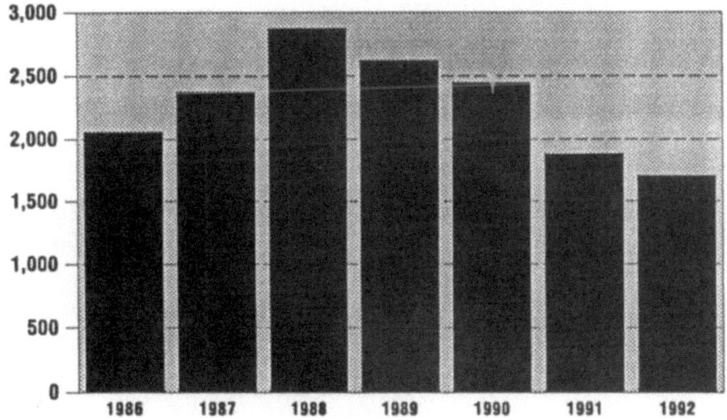

- In 1992, the per capita value of mineral production was less than in 1980 (Figure 4); and

Figure 4
Per Capita Value of Mineral Production

$ Cdn

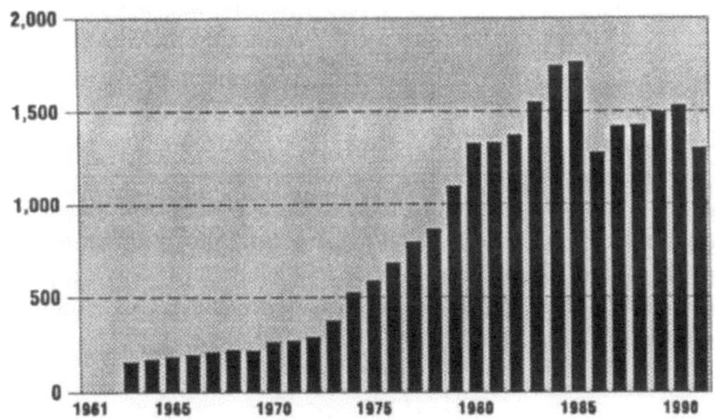

Source: Energy, Mines and Resources Canada, Mineral Industry Quarterly Report, March, 1993

- The number of people employed directly in the industry has declined steadily since 1974 (Figure 5).

Figure 5
Employment in the Mining Industry

1974=1

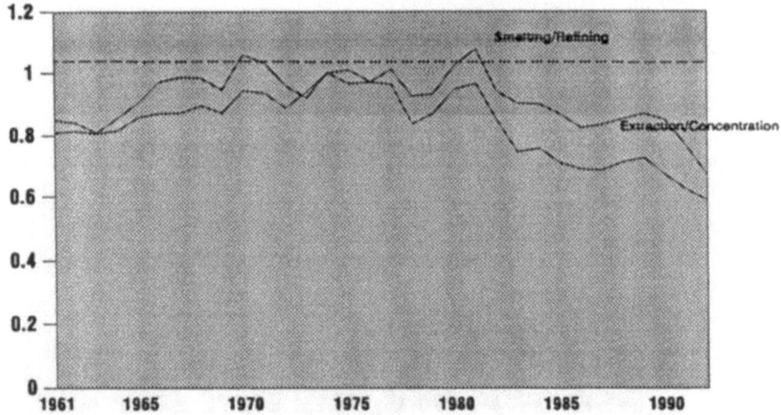

Source: Energy, Mines and Resources Canada, Mineral Industry Quarterly Report, March 1993

Source: Energy, Mines and Resources Canada, Mineral Industry Quarterly Report, March 1993

Government statistics show that Canada's known reserves of base metals have been declining steadily over the past ten years.

At the same time, Latin America has experienced a tremendous growth in actual investment in mining. From 1974 to 1991, there were no new gold and base metal projects in Canada larger than U.S.$500 million, whereas thirteen such projects were announced in Latin America.

Canada intends to reverse these trends. The country's action plan to ensure the future of Canadian mining may be summarized as follows.

At the 1991 Mines Ministers Conference, the federal and provincial ministers of mines agreed that a joint governmental/industry task force be formed to review and assess Canada's international competitiveness for mineral investment capital. The task force that was formed consisted of both government and mining industry representatives.

The task force tabled an interim report on its findings at the 1992 Mines Ministers Conference. At that conference in Whitehorse, the federal and provincial mines ministers agreed to a proposal from the Mining Association of Canada to participate in a process for the development of a new vision for the Canadian mining industry. The "Whitehorse Mining Initiative" (WMI) is a multi-stakeholder approach to dealing with the current concerns of the industry. Key segments of the Canadian society, government, industry, Aboriginal peoples, environment groups, and labour, are participating in WMI.

WMI is expected to generate:

- A better understanding among all stakeholders of issues of mutual concern;
- A general vision, set of guiding principles, or accord for reference in developing policies and legislation;
- Specific bilateral or multilateral agreements that are intended to resolve major issues and obstacles; and/or
- New processes for dealing with other issues or problems as they arise.

Under the auspices of WMI, four issue groups were formed to address the environment, land access, finance and taxation, and the workplace. At the 1993 Mines Ministers meeting, the issue groups reported on the status of their deliberations. A final report, with recommendations where appropriate, is scheduled to be tabled at the 1994 Mines Ministers Conference.

WMI is a unique initiative. It is thought to be the first exercise that brings together interested parties who frequently have strong, opposite views on different facets of the mining industry. Such a unified force could be exactly what Canada needs to deal constructively with important matters that hurt the country's competitive position. Latin America beware!

Restructuring of Mexico's mining sector: an international model

Victor Flores
United Services Advisors, Inc., San Antonio, Texas, U.S.A.

Abstract

Mexico's mining industry has benefited as a result of fiscal, administrative, and monetary policy changes implemented as part of the country's economic restructuring. Government macroeconomic policies have reduced the fiscal deficit, inflation, and interest rates. The privatisation of mining interests held by government-owned companies has dramatically reduced the government's share of the country's mineral production and altered radically the industry's structure. The mining industry, essentially closed to foreign investment since 1961, has opened its doors to foreign investment. New rules guarantee the foreign investor equal rights and responsibilities as the domestic investor. The question of the environment has become one of the key issues facing the country's mining industry, as is the lack of infrastructure. Mexico's insular mining sector will have to respond aggressively to the new competition.
Keywords: Mexico's macroeconomic restructuring, Mexico's mining history, Mexico's mining law, privatisation of Mexico's mining industry, Mexico's mining production, Mexico's environmental policies.

1 Introduction

Mexico's mining industry has responded favourably to the macroeconomic policy changes implemented by the government during the last two administrations. (Miguel de la Madrid Hurtado, 1983-1988, and Carlos Salinas de Gortari, 1989-1994.) A dramatic reversal in the role played by the state in allocating the use of productive resources and establishing the prices for goods and services has led to a sizeable increase in the amount of new investment in the mining sector. In addition, the privatisation of mining interests held by government-owned or controlled companies has reduced the government's share of the country's mineral production from 50% in 1986 to virtually zero by the end of 1993. The government has also abandoned its policy of differentiating between strategic and non-strategic minerals, and has reversed its goal of retaining a dominant market position in these mineral commodities. Finally, the government has relinquished control over millions of hectares of prospective exploration ground held under the National Mining Reserve program.

The mining industry in Mexico has also benefited as a result of fiscal, administrative, and monetary policy changes implemented as part of the country's economic restructuring. At the macroeconomic level, a reduction in corporate income taxes has had the effect predicted by the Laffer Curve: enhanced government tax revenues and reduced tax evasion. The government's commitment to bring down the rate of inflation by reducing its fiscal deficit has provided the necessary price stability to encourage new business development and has arrested the flight of capital. Falling interest rates have made Mexican businesses more internationally competitive and have helped sustain the growth of the capital markets, which will likely become an increasingly important source of funding for the mining sector. At the microeconomic level, the elimination of a prohibitive mining royalty levied on the gross value of production has restored the economic incentive to develop new mining projects. New mining regulations have substantially reduced the time required to obtain mining concessions and other permits.

The government's economic policies have created an ideal climate for foreign investment, and billions of dollars have flowed into Mexico in the past six years. The mining industry, essentially closed to foreign investment since 1961, has also opened its doors to foreign investment. New rules guarantee the foreign investor equal rights and responsibilities as the domestic investor. The results has been a flood of new investment by foreign mining companies. Mexico has provided an attractive haven for small companies (primarily from the United States and Canada) which probably could not survive increasingly restrictive business and regulatory conditions in their home countries. The creation of a North American Free Trade Zone further enhances the ability of capital to flow into regions and projects with attractive rates of return.

Mexico's insular mining sector will have to respond aggressively to the new competition. Long dominated by the government and a handful of large mining groups, the industry now faces competition from well-funded companies, many of which have a wealth of international experience. Those Mexican mining companies which are best able to establish strategic relationships with their foreign counterparts will be the survivors of this new era. Mexican mining companies are also likely to respond by becoming more international in their own right, primarily by expanding into other Latin American countries. Although the government has vowed to continue supporting small and medium-sized mining, many of these operators are being forced out of business because they do not have the resources to weather low commodity prices and low margins caused by the strength of the *peso*. Only the small producers of speciality mineral commodities are likely to survive this transition period.

2 Macroeconomic background

The changes in Mexico's economic policies are rooted in the problems experienced by the country in the first half of the 1980's. The Mexican debt crisis, which took the international financial community by surprise in August of 1982, precipitated the need for a radical rethinking of the country's macroeconomic policies. The nationalisation of the banking industry and the devaluation of the peso decimated the confidence of private industry and created the conditions for the ensuing flight of capital, which by some estimates may have reached 80 billion dollars. By the end of 1982, when outgoing president José Lopez Portillo was about to hand over the reins of power to Miguel de la Madrid Hurtado, inflation had reached 100%, the currency had lost half its value, and the government's foreign debt had climbed to almost 75 billion dollars.[1]

The six years of the de la Madrid administration were spent stabilising the economy and implementing the policy initiatives recommended by the IMF, viz., reducing government spending, increasing consumption taxes, controlling wage inflation, ending domestic price controls, and allowing the exchange rate to float freely in order to reduce imports and increase exports. The country's new economic path was outlined in the National Development Plan of 1983, whose principal architect was the country's Secretary for Budget and Planning (and now president), Carlos Salinas de Gortari. By the end of the de la Madrid administration, the government had put the economy back on track and had set in motion two key policy initiatives: the National Solidarity Plan and Mexico's entry into GATT. The National Solidarity Plan was a carefully orchestrated political program which was designed to maintain the more socially disruptive components of the government's macroeconomic policies in place without tearing the fabric of Mexican society apart. For example, in order to convince the labour unions to forego wage increases which could have led to a wage indexation spiral (such as that which plagues Brazil), the government had to convince private business to limit price increases, as well as convince labour that price gouging would be penalised. The private sector, thus assured that its wage costs would remain under control, was better able to resist the temptation to increase prices. Both sides, of course, had to be convinced that the government was serious about its anti-inflation campaign. Mexico's entry into GATT was a watershed event, as it reversed the country's long-standing fear of open, international competition and its traditional diplomatic position of avoiding agreements which would violate the principle of mutual non-intervention. It also signalled that Mexico was turning its back on the idea that the country's oil wealth could support a closed economy based on import substitution, and was instead gearing up to take advantage of its privileged geographic location as a bridge between the United States, Latin America, and the Pacific Basin.

By the beginning of the Salinas administration, the government was ready to begin its privatisation plan in earnest. One of the principal reasons why the government had such a large fiscal deficit was that it owned or controlled over 1,200 businesses, ranging from the state oil company (Pemex) to banks, night-clubs, and soft-drink bottlers. Many of these enterprises, including most of those in the mining sector, were overstaffed, inefficient, and unprofitable. The sale of these was the key to reducing the government's role in the economy and bringing the deficit under control. Lowering the deficit would, in turn, alleviate the need to add to the money supply, which would bring inflation under control. The social cost of privatisation was unquestionably one of the government's primary concerns, for it was clear that overstaffing was an endemic problem throughout the system and the primary reason why so many of these businesses were unprofitable. Once these businesses were back in private hands the government was faced not only with a labour force whose real wages had declined throughout the decade, but with the masses of the newly unemployed who put an additional strain on the system. However, despite these obstacles, once the process of privatisation began it proceeded fairly smooth and fairly quickly, as can be seen in Figure 1.

Figure 1. Mexico's state-owned companies

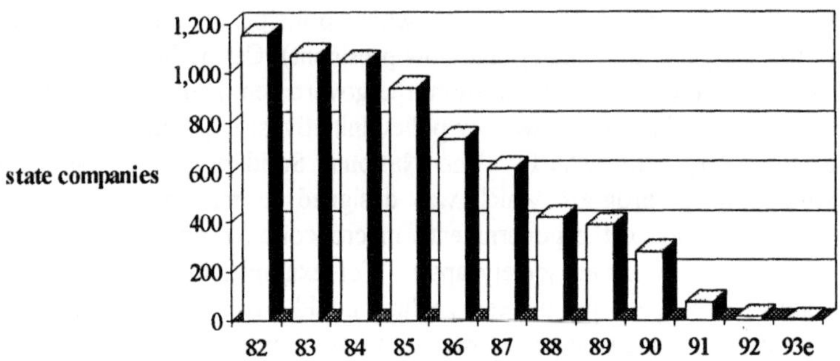

The process of bringing down the fiscal deficit progressed hand in hand with the sale of government-owned companies. In 1987 the fiscal deficit had grown to 16.0% of GDP, an alarming statistic. Government borrowing to fund the deficit was crowding out private sector borrowing and putting additional upward pressure on already-high interest rates. Finance Secretary Pedro Aspe has pursued a two-pronged approach to the question of government finances: increasing revenues and reducing expenditures. Revenues have been increased by lowering corporate and personal income tax rates and by increasing the government's percentage of revenues collected through consumption taxes. Lowering the corporate tax rate has encouraged new business investment and has reduced the level of tax evasion. At the same time, more aggressive and efficient tax collection has actually increased government tax revenues, while the widespread use of a Value Added Tax provided a stable source of revenue which was easy to collect and difficult to evade (see Figure 2). As mentioned, the privatisation program and the streamlining of government reduced government expenditures, to the point that by 1993 the government had posted a fiscal surplus (see Figure 3).

Figure 2. Mexico's tax revenues as a % of GDP

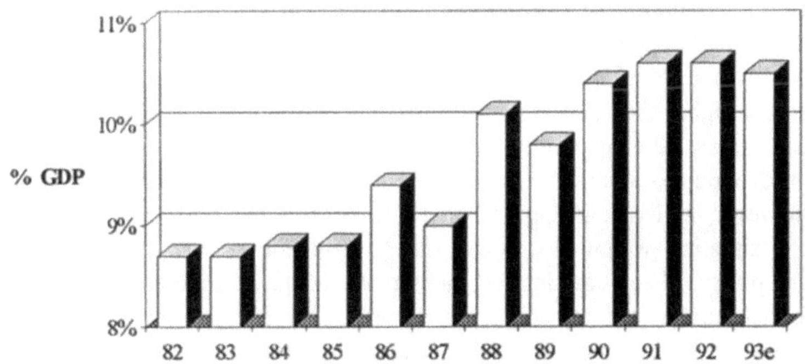

Figure 3. Mexico's fiscal deficit as a % of GDP

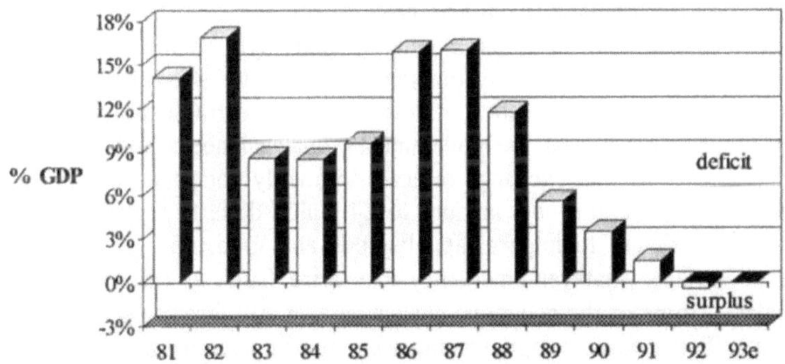

By attacking inflation on several fronts, that onerous statistic began to decline as well. Selling off state-owned enterprises brought in surplus cash and reduced the government's need to borrow, taking the burden off the Bank of Mexico to continue printing money. While Mexico never experienced hyperinflation, as was the case in several other Latin American nations, consumer prices did rise by over 150% in 1986 and have seriously and continuously eroded the purchasing power of the Mexican people throughout the decade. Furthermore, some economists argue that Mexico's inflation problem was actually worse than that of its Latin American cousins, as it was high enough to disrupt the economy yet not high enough to consume itself (i.e., Bolivia). The National Solidarity Plan helped ease the social strains caused by declining real wages and helped keep indexation from becoming endemic as it has in Brazil. By 1993 (see Figure 4) inflation in Mexico was below 10%. Inflation in the service sector, which is largely free from foreign (and in some cases, even domestic) competition, continues to put pressure on the system, as does the government's elimination of price controls remaining on basic staples such as cornmeal, bread, and fuel. And although inflation has dropped dramatically, Mexico's inflation rate is still considerably higher than that of its major trading partners.

Interest rates have responded well to improving government finances and lower inflation, dropping dramatically in the period 1986-1993 (see Figure 5). Real interest rates, however, still remain high. Paradoxically, this has both positive and negative effects, a situation which has forced the government to walk a tightrope with its monetary policy. High real interest rates make Mexican businesses less competitive because their cost of capital is higher than that of their international competitors. High rates, however, have helped keep Mexico's inflation under control, but they also have kept economic growth in check. At the same, high real interest rates are one reason that Mexico has so successfully been able to attract foreign capital. As rates have dropped world-wide--particularly on the short end of the yield curve--capital has flowed into Mexican short-term debt instruments. Thus, the government has had to carefully strike a balance between its desire to provide the private sector with optimum business conditions and its need to attract foreign capital.

By all measures Mexico's economic restructuring program has been a resounding success. The root causes of high inflation and high interest rates--uncontrolled increases in money supply and a voracious public sector--have been brought under control. New initiatives to create a truly independent central bank will finally destroy the incestuous link between the government's financial needs and monetary policy. The government's foreign exchange reserves--primarily composed of the proceeds of the privatisation program--now amount to 24 billion dollars, providing Mexico with the financial flexibility to import capital goods and foodstuffs (should this become necessary) or face whatever emergency may arise (such as the 1985 Mexico City earthquake). Some of the economic challenges that still face the government include attracting sufficient foreign capital to finance a current account deficit approaching 25 billion dollars, generating sufficient economic growth to provide a better standard of living for a growing population, preparing inefficient sectors of the economy for the challenges of international trade, improving the country's infrastructure, and paying down the government's still massive foreign debt.

Figure 4. Mexico's inflation 1990-1993

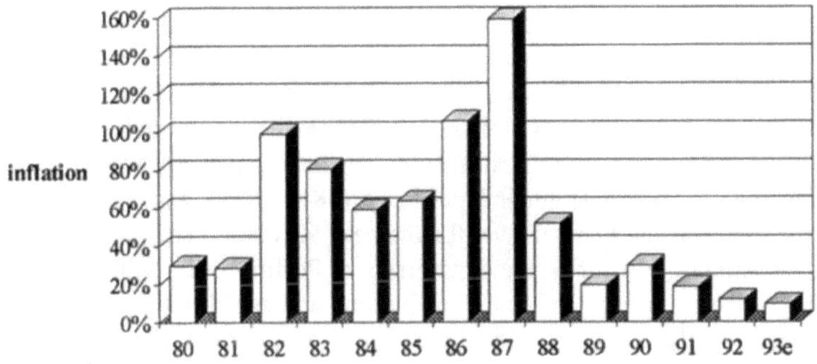

Figure 5. Mexico's interest rates (28 day gov't debt)

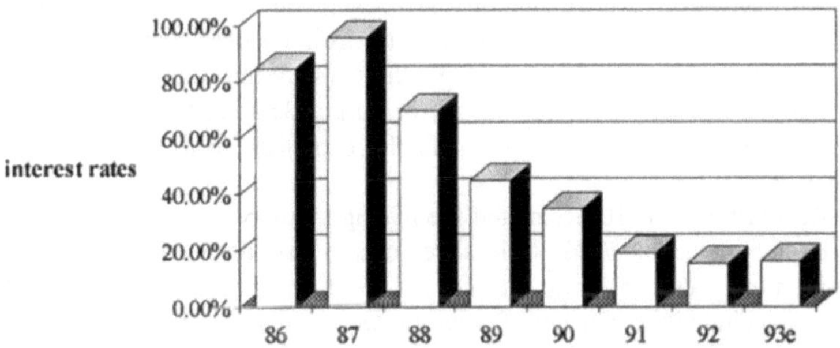

3 Mexico's mining sector-historical background

The recent changes in Mexico's economic policy have affected all areas of the economy, including the mining sector. But Mexico's mining industry has also been shaped by its history and, in order to understand the changes taking place, it is necessary to understand the historical context in which the industry developed. Mining has always played an important part in the country's economic development. Beginning in 1525, only four years after Spain vanquished the Aztecs and established the colony known as New Spain, silver was discovered in the present day state of

Jalisco. The city of Taxco, established to mine silver and other metals in 1534, is still an important centre of mining activity today. The cities of Zacatecas and Guanajuato, both located in still-active mining districts, were founded in 1546 and 1548, respectively. The foundation of Mexico's statist attitudes towards mining can be found during the colonial period, which lasted nearly 300 years until Independence from Spain in 1821. Mexico's mining wealth supported the Spanish Crown, which utilised Mexico's silver and gold to finance its wars in Europe. The Crown's *Ordenanzas* regulated and guided the colony's mining activity, and firmly established a precedent for governmental influence in the industry.[2] The mining industry was virtually wiped out during the country's war for independence (1810-1821), and little attention was paid to the sector at the Federal level for the next 50 years. The industry once again regained its former glory during the dictatorship of Porfirio Diaz. Although reviled for derailing Mexico's nascent democratic institutions and establishing a brutal autocracy which lasted nearly 35 years, Diaz did much to promote the mining industry. Favourable concessions, low taxes, and heavy investment in infrastructure helped attract investment capital for mining ventures. The *Código Nacional de Minería* of 1884 was the first Federal legislation since 1811 to regulate the mining industry. Basically an updated version of the *Ordenanzas* of 1783, the *Código* recognised both domestic and foreign investors as equals, and indeed much of the investment during this period was financed by foreigners.[3] In *The Mines of Mexico* (1905) J.R. Southworth wrote: "To understand how liberal their merits are we need but contrast them with the rules and regulations governing the United States...[in Mexico] foreigners have the same civil rights as the citizens in business matters, and are treated in the same manner in obtaining mining titles..."[4] While there is no question that mining was the domain of the private sector during this era, the State's ownership of the country's mineral resources was implicitly recognised in the *Ley Minera* (Mining Law) of 1892.

The Revolution of 1910 decimated the mining industry; by 1915 the production of gold had plummeted by 75% while silver output had declined nearly 70%.[5] The Constitution of 1917 radically altered the existing mining legislation. The new constitution included the concept of *dominio directo* (direct domain) over the country's mineral wealth. In addition, it recognised the right of the State to actively participate in the mining sector and restricted the role of foreigners. The dominance of the State was reinforced during the presidency of Plutarco Elías Calles. The *Ley de Industrias Mineras* (Mining Industry Law) of 1926 prohibited foreign companies from obtaining new mining concessions. It also barred foreigners from operating in coastal or border zones.[3]

President Lázaro Cárdenas, best known for nationalising Mexico's oil industry in 1938, also increased the government's role in the mining sector. The *Ley Minera* of 1934 created the National Mining Reserve, which reserved large tracts of land for the government and reserved certain mineral commodities for exploitation by the State.[5]

In 1961 the mining law was changed once more in what is known as the "Mexicanisation" of the industry. Foreign participation was restricted to 49% and the participation of the State was increased. The State's participation in the industry increased to 21% (by value) by 1971 and reached 49% in 1985. In 1985 the government (including companies with government participation) produced 100% of the country's iron ore, manganese, sulphur, and selenium, 82% of the country's copper, 77% of its molybdenum, 75% of its salt and 70% of its phosphates.[2] The statist policies of the Echeverría government (1970-1976) and the discovery of oil in the late 1970's helped to entrench the role of the State in all sectors of the economy. The government was not only convinced that it should play a primary role in certain key industries, it also had the capital at its disposal which to carry out its ambitious plans.

Thus was created an environment in which mining became one of several industries heavily controlled by the government. The rationale was simple: only the government could insure that the country's mineral wealth would be exploited for the benefit of all. The government's statist economic policies, including a belief in the import-substitution model, also drove it to attempt self-sufficiency in the production of mineral commodities.

The collapse of oil prices and Mexico's debt crisis of 1982 put an end to the government's ambitious development plans for mining and forced it to rethink its attitude vis a vis the sector. After all, the importance given the mining industry by the government went far beyond the sector's contribution to the economy. In 1982 mining contributed about 4% of the country's GDP; currently the figure is about 1.1%.[6] The government's primary concern were those minerals it considered strategic, including iron ore and steel, steam and coking coal, copper and molybdenum, salt, sulphur, phosphates, and radioactive minerals.

The public sector participated in the mining industry in several different ways. In certain cases the government created wholly owned subsidiaries to mine and develop the country's resources. The steel company *Sidermex*, for example, was a holding company under the control of the *Secretaría de Energía, Minas e Industria Paraestatal* (Ministry of Energy, Mines, and State-owned Industry). By law the government was also allowed to create joint venture companies in which it held a majority stake. These companies, known as *Empresas Estatales de Participación Mayoritaria* (Majority-owned State Enterprises), are allowed to sell as much as 49% of their equity to outside investors. Other companies are controlled by the *Comisión de Fomento Minero* (Commission for the Development of Mining), whose primary role is to provide credit and technical assistance for small mining enterprises. In many cases the mining enterprises are themselves subsidiaries of other government-owned companies, or owned jointly by several governmental entities.

4 Mexico's new mining law

Mexico's new mining law was published in the Official Daily newspaper on the 26th of June, 1992, and became effective on September 27 of 1992. The country's new Mining Regulation was published in March of 1993. The new law essentially codifies the changes which had already been made by presidential decree in 1990. The new law covers four principal areas: legal, deregulation, bureaucratic simplification, and the promotion of small and medium mining. The new law emphasises the importance of providing greater legal security to entities involved in the mining sector. This includes fair indemnisation for expropriation or other forms of occupation, new procedures for government inspections, and protection from abuse by the authorities. Another important part of the new law gives leaseholders the right to mine other minerals discovered on a property which were not specifically mentioned when the mining concession was granted.

The process of deregulation focuses on the right to explore for, and exploit, certain minerals and areas heretofore reserved for the state. The new law deregulates the extraction of sulphur, phosphates, potash, iron ore, and coal. In addition, the private sector is now allowed to carry out exploration along the nation's coastline, continental shelf, islands, and reefs. Furthermore, private mining companies may now also explore and develop projects on land belonging to *ejidos* (agrarian co-operatives). The new regulations also facilitate the process by which leases held by the *Consejo de Recursos Minerales*-CRM (Mineral Resource Council) are transferred to the private sector.

The measures aimed to simplify bureaucratic requirements reduce the number of bureaucratic procedures by one-third. Exploration concessions are now granted for a period of six years (previously three years), but these cannot be extended. Exploitation concessions have been doubled in length from 25 to 50 years, and can be extended for a like period. Mining and exploration leases, which could take up to several years to obtain, are now granted in about five months. In addition, leases are now freely tradable, whereas in the past government approval was required to transfer leases. Companies are no longer required to obtain a separate concession for the beneficiation of minerals extracted.

In order to promote small and medium size mining, the new law gives the CRM the ability to provide technical assistance (such as verifying reserves) to this sector.

The government has also given the mining industry fiscal incentives. The most significant of these was the elimination of a burdensome 7% production royalty. The mining industry has also benefited from a reduction in the maximum corporate tax rate, which was lowered from 50% to 35%. However, in order to reduce the number of idle leases, the government concurrently raised its lease assessment fees.

Changes in Mexico's foreign investment laws are also undoing years of isolationism in the mining industry. The long-standing rules that restricted foreign ownership of mining interests to 49% have been modified. Foreign companies may own 100% of an exploration lease and retain this interest for twelve years after mining begins. This is achieved through a trust that holds the 51% of the shares (known as "A" shares) which are supposed to be owned by Mexican nationals. The trustee, usually a bank, does not actually interfere in the mining company's affairs, thus effectively affording it total control. After twelve years the foreign company must reduce its interest to 49%, but is allowed to own up to 49% of the Mexican partner. The draft for the country's new Foreign Investment Law has provisions for increasing a foreign company's direct interest from 49% to 100% with prior approval of Mexico's Foreign Investment Commission if the project meets certain criteria. Furthermore, the North American Free Trade Agreement contains provisions which would allow 100% foreign ownership without approval of the FIC (for NAFTA signatories only) after a five year phase-in period.[7]

5 Privatisation

Despite the nationalistic sentiment associated with the mining industry in Mexico, assets in this sector were not spared during the privatisation process, and a large part of Mexico's mineral production was returned to the private sector. In the steel sector, the government privatised Sidermex, responsible for virtually all of Mexico's steel production and 60% of its iron ore production, and Altos Hornos de México, the country's largest producer of pig iron. Manganese producer Minera Autlán--only partially owned by the government--was sold in 1993. Approximately 98% of Mexico's copper production was in government hands until the two operating companies, Mexicana de Cobre and Compañía Minera de Cananea were privatised. Mexico's largest silver, zinc, and lead producer, Minera Real de Angeles, was 51% owned by the government until its stake was sold to Empresas Frisco in 1989. Silver-lead producer Minera Real del Monte y Pachuca was sold as well. In the non-metallic minerals arena, the government privatised a number of important operations, including phosphate producer Roca Fosfórica Mexicana, sulphur producers Azufrera Panamericana and Compañía Exportadora del Itsmo, and coal producer Minera Carbonífera Río Escondido. One notable exception is salt producer Exportadora de Sal, which the government says will remain in the public sector.

The structure of Mexico's mining industry has been transformed dramatically by the privatisation process. The state's overwhelming control over the country's copper, iron ore, coal, and sulphur production has now passed into the hands of private enterprise. It is worth noting the distribution of the assets privatised and its long term implications for the development of the industry. Most of the government's base metal and ferrous metal production was sold to large mining companies. This is primarily a factor of the size of the assets which were privatised. On the other hand, the country's non-metallic mineral production is now primarily in the hands of small and medium size producers (see Table 1). The country's gold production remains almost evenly split between large and small producers. The future development of the country's mineral resources is expected to develop along similar lines: large base metal projects in the hands of large domestic and international firms and speciality non-metallic mineral production in the hands of smaller producers.

The privatisation of the mining industry in Mexico has gone beyond the mere sale of assets. One of the most important steps taken in the modernisation of Mexico's mining industry has been the disincorporation of the country's *Reservas Mineras Nacionales*-RMN (National Mining Reserves). The reserves were established in the 1930's and encompassed millions of hectares of prospective exploration ground. The government's intention was to explore and develop the minerals found in RMN lands. Unfortunately, the government never had the resources with which to explore and exploit the reserves, such that much of this land has remained idle and unexplored.

Table 1. Changes in Mexico's mining structure

	Percentage Production by Sector					
	1985			1992		
Commodity	Gov't	Private sector		Gov't	Private sector	
		Large	Small		Large	Small
Gold	13.1	49.1	37.2	0	44.1	55.9
Silver	23.1	57.5	19.1	0	73.3	26.7
Copper	82.1	13.3	4.6	0	97.1	2.9
Lead	19.1	65.6	15.3	0	86.9	13.1
Zinc	8.4	88.8	2.8	0	93	7
Cadmium	0	98.5	1.5	0	91.4	18.6
Molybdenum	76.7	23.2	0	0	100	0
Tungsten	0	51.3	48.7	0	100	0
Coal	93	5.9	1.1	0	97.3	2.7
Coke	100	0	0	0	67.8	32.2
Iron	100	0	0	0	41.5	58.5
Manganese	100	0	0	0	0	100
Barite	47.8	31.9	20.3	0	66.3	33.7
Celestite	0	0	100	0	0	100
Dolomite	58.1	0	41.9	0	64.5	35.5
Feldspar	na	na	na	0	0	100
Fluorspar	0	98.9	1.1	0	0	100
Graphite	5.4	0	94.6	0	0	100
Gypsum	na	na	na	0	0	100
Kaolin	40	0	60	0	0	100
Phosphate	69.1	30.7	0.2	0	0	100
Salt	75	0	25	93.3	0	6.7
Sulphur	100	0	0	0	0	100
Wollastonite	0	0	100	0	0	100

Source: SEMIP, Chamber of Mines

The RMN system divided lands into three categories. Land included in the first category was off limits to mining. The second included minerals that are the exclusive domain of the State, primarily uranium and other radioactive minerals. The last category includes minerals that can only be mined by means of a special concession. These key mineral commodities were considered strategic and included iron ore, coal, phosphates, sulphur, and potash. The companies that mined these five commodities had to be at least 66% domestically owned.

As part of the government's effort to reduce its role in the mining sector while at the same time promoting new development, the country's National Mining Reserves are being disincorporated. Under the disincorporation scheme, the special concessions granted under the RMN system automatically become ordinary concessions. The stated goal is to disincorporate at least half of the land held in the RMN. A total of 2.5 Mha were disincorporated in 1991 and an additional 1.0 Mha were released from the RMN system in 1992.

Mexico's most prospective exploration ground, which has essentially remained idle, can now be explored and developed. In some cases the government's exploration arm, the *Consejo de Recursos Minerales*, did carry out exploration programs on RMN lands, outlining potentially economic deposits. In addition to disincorporating RMN lands, the government is making the work carried out by the CRM available to interested parties. The implications of this strategy are obvious: Mexico's mineral wealth can now be utilised more efficiently, and it is very likely that a number of deposits will be discovered as capital and modern exploration techniques are applied to these areas.

6 A review of Mexico's mining production

Mexico is an important producer of a number of mineral commodities, including copper, lead, zinc, silver, bismuth, cadmium, arsenic, mercury, manganese, molybdenum, barite, graphite, fluorspar, sulphur, salt, and strontium. Low commodity prices and the country's economic problems led to a drop in the production of most minerals during the mid 1980's, but the production of most commodities has now recovered. Table 2 details Mexico's mineral commodity production in the period 1980-1993.

The level of participation by foreign firms in the Mexican mining industry has grown as the laws and investment climate have improved and as conditions in their home countries deteriorated. In mid 1991 there were less than sixty foreign mining companies operating in Mexico; by early 1993 this number had doubled.[8] Of this number, over 75% of these are small exploration companies with no current production, although the number of mid and large size mining companies has grown as well. Figure 6 details the distribution and growth of foreign mining firms in Mexico. The primary focus of the foreign mining company has been large tonnage, low grade gold properties amenable to heap leaching, followed by porphyry copper targets with the potential for solvent extraction-electrowinning potential (see Figure 7). Given the type of targets being sought it is not surprising that most of the prospects are located in the northern states of Sonora and Chihuahua. Foreign mining companies have also benefited from the lower cost environment that Mexico provides, as mineral property acquisition costs are on average half of what they are in the United States or Canada.[8]

The entry of foreign mining companies into Mexico has also dramatically increased the level of exploration capital available for the industry. The Chamber of Mines estimates that in 1992 its membership invested approximately 316 million dollars in exploration, new projects, expansions, and other projects.[9] Given the recent influx of

Table 2. Mexican mineral production 1980-1993

	1980	1981	1982	1983	1984	1985	1986	1987	1988	1989	1990	1991	1992	1993e
Precious Metals (kg)														
Gold	6,096	6,319	6,104	6,930	7,058	7,524	7,795	7,988	9,098	8,613	8,548	8,937	10,412	10,527
Silver	1,472,557	1,654,829	1,550,221	1,910,839	1,986,690	2,152,959	2,303,142	2,414,954	2,358,907	2,306,091	2,351,561	2,223,647	2,317,382	2,335,921
Non-ferrous Metals (mt)														
Copper	175,399	230,466	239,091	206,062	189,111	167,977	174,558	230,573	268,359	249,328	298,695	284,174	279,042	295,785
Lead	145,549	157,384	145,844	167,405	183,314	206,732	182,672	177,161	171,337	163,017	177,279	160,406	172,563	195,686
Zinc	238,231	211,629	231,910	257,443	290,236	275,412	271,351	271,480	262,228	284,058	322,487	300,706	289,119	423,559
Antimony	2,176	1,800	1,565	2,519	3,064	4,266	3,337	2,839	2,185	1,906	2,614	2,752	1,064	1,468
Arsenic	5,250	4,936	3,590	3,452	4,164	4,782	5,315	5,304	5,164	5,551	4,809	4,922	4,293	4,169
Bismuth	770	656	606	545	433	925	749	1,012	958	883	733	651	807	997
Cadmium	1,791	1,433	1,444	1,341	1,135	1,140	1,183	1,249	1,726	1,439	1,973	1,797	1,879	2,439
Molybdenum	74	451	5,190	5,866	4,054	3,761	3,350	4,400	4,456	4,189	2,001	1,716	1,458	1,235
Selenium	46	12	29	24	44	42	23	29	13	20	12	3	0	0
Tin	60	28	27	50	416	380	585	369	274	159	7	15	5	3
Tungsten	266	199	99	90	274	282	294	213	206	170	183	194	162	0
Ferrous Metals and Coal (mt)														
Coal	408,464	1,237,201	785,948	1,818,489	2,215,056	2,440,350	3,677,618	4,251,715	4,210,842	4,243,838	4,219,841	4,864,733	5,059,947	5,515,342
Coke	2,409,228	2,425,530	2,450,145	2,424,826	2,375,480	2,389,971	2,050,012	2,340,265	2,332,245	2,260,480	2,337,159	2,107,589	2,033,003	1,904,923
Iron ore	5,586,693	5,748,700	5,382,239	5,306,343	5,489,343	5,161,144	4,817,410	4,965,133	5,564,492	5,373,051	5,327,890	4,976,087	5,154,046	5,813,764
Manganese	160,966	208,193	183,120	133,004	180,940	150,647	174,416	146,407	168,573	149,875	138,850	78,451	137,746	111,574
Non-metallic Minerals (mt)														
Barite	269,322	317,738	323,753	357,043	426,095	467,693	321,186	401,336	534,954	324,739	305,716	203,975	443,782	137,129
Celestite	40,761	41,344	31,676	37,506	31,991	30,482	24,289	32,407	38,974	37,841	51,306	50,000	59,088	67,242
Dolomite	378,316	372,027	353,753	357,043	426,095	467,693	321,186	401,336	534,954	324,739	305,716	203,975	443,782	592,005
Feldspar	117,214	130,826	115,559	117,518	395,046	432,031	85,019	90,908	86,258	113,438	124,188	168,143	159,718	113,080
Fluorspar	916,455	924,854	631,386	556,977	627,433	697,410	756,768	723,594	756,096	779,357	633,814	370,297	286,640	272,308
Graphite	44,506	41,142	34,370	44,327	41,529	35,378	37,780	37,946	43,831	40,246	24,916	30,579	31,470	49,156
Gypsum	1,708,924	1,882,913	1,530,564	2,127,453	2,300,413	2,366,019	2,625,179	2,457,810	2,649,290	2,898,603	2,814,439	2,338,954	2,960,126	3,344,942
Kaolin	29,542	19,195	18,630	20,149	14,745	37,975	10,067	12,826	11,633	15,693	5,403	21,172	9,403	12,073
Phosphate	283,246	330,611	512,096	498,112	518,293	645,299	660,425	633,204	666,753	625,386	557,125	445,805	338,744	90,106
Salt	6,575,302	7,953,263	5,561,187	5,703,030	5,456,438	5,450,868	5,926,762	6,393,221	6,965,173	6,942,165	7,135,121	7,532,421	7,395,152	7,239,854
Sulphur	2,102,301	2,077,177	1,815,447	1,602,029	1,825,729	2,019,753	2,050,735	2,303,775	2,138,240	2,086,333	2,122,482	1,814,555	1,484,497	884,760

Source: Mexican Chamber of Mines

Figure 6. Foreign companies operating in Mexico

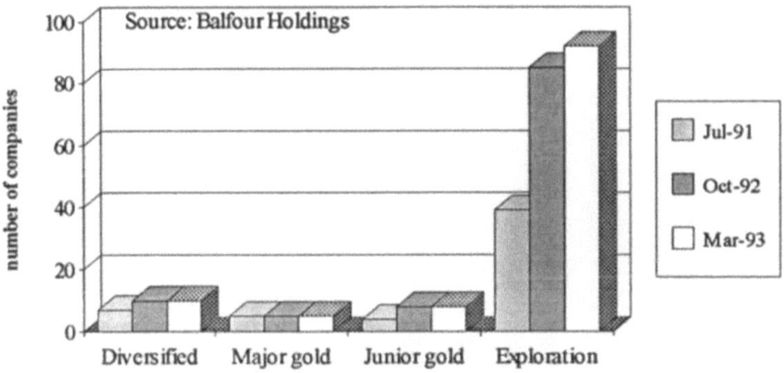

Figure 7. Foreign companies in Mexico by commodity type

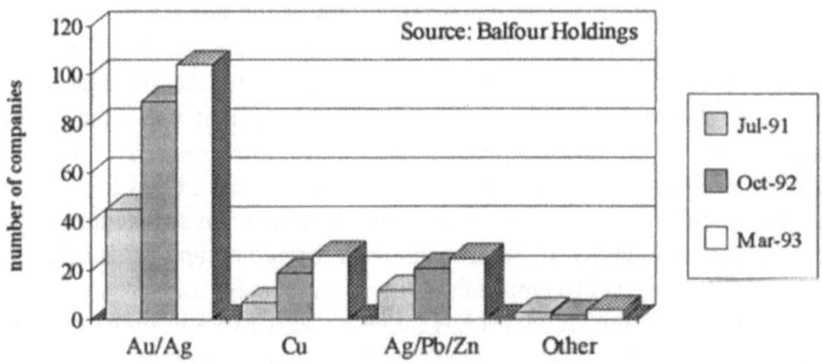

foreign mining companies, most of the expenditures are currently being made on exploration. Figure 8 shows the distribution of projects in Mexico managed by foreign firms. As this exploration results in new discoveries, the level of capital spent on feasibility and development is expected to increase. It is further estimated that each foreign mining company, on average, invests approximately US$500,000 per exploration property.[8] This would indicate that foreign mining companies are currently spending about 75 million dollars on exploration in Mexico, which is significant compared with the approximately 100 million dollars being spent by Mexican mining companies.

Figure 8. Foreign companies in Mexico by project status

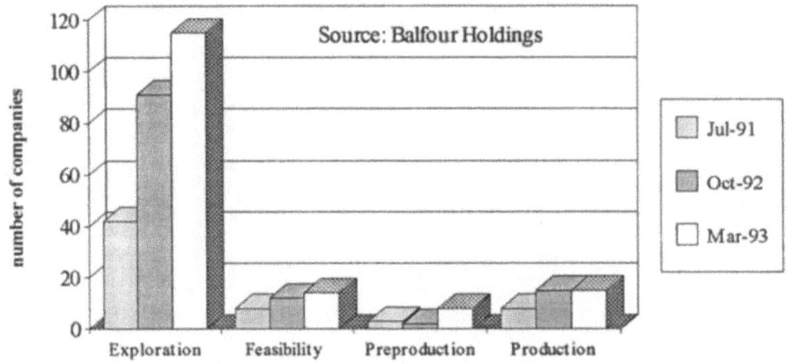

7 Environmental management

The question of the environment has become one of the key issues facing the mining industry. In Mexico's case, the environment has gone beyond being a mere concern and has taken centre stage. The debate leading up to the passage of the North American Free Trade Agreement made this very clear, as it was used by the treaty's opponents to try to embarrass Mexico's environmental record. The Mexican mining industry, aware of growing environmental opposition in the United States and Canada, has moved quickly to dispel the notion that Mexican environmental standards are subpar. Furthermore, given the sudden influx of foreign mining firms into the country, Mexican officials have had to emphasise that environmental standards will be enforced.

The Mexican Chamber of Mines has taken an active role in updating the country's environmental standards. The Chamber's committee on natural resources and ecology is charged with establishing the criteria and environmental standards necessary to regulate the industry, and is currently preparing new proposals for the design and construction of tailings dams. The Chamber is also reviewing the government's new water and forestry laws, and is actively lobbying against the fee currently being paid for water usage.[9]

The mining industry in Mexico will undoubtedly benefit from the country's pragmatic and flexible approach to mining regulation. For example, the level of detail required in an environmental impact statement depends on the size of the project and the environmental fragility of the project site. The *Instituto Nacional de Ecología* (National Institute of Ecology), an arm of the *Secretaría de Desarrollo Social* (Secretary of Social Development), has subdivided the country into ecological provinces which pre-determine the comprehensiveness and extent required of an

EIS.[10] Thus, a mining company has a clear picture of the time required and cost of preparing an environmental impact statement for a new project early in the project's development.

8 Challenges to the Mexican mining industry

While Mexico's mining industry has experienced dramatic, positive change in the past six years, it is still faced with a number of challenges. These include eliminating rules, regulations, and requirements that hamper the industry's competitiveness and improving the country's basic infrastructure. Mining companies in Mexico still require permits from the Secretary of National Defence in order to obtain and utilise explosives at the mine site. Industry has consistently lobbied to have the Federal Law of Firearms and Explosives modified in order to accommodate the specific needs of the mining industry. Industry is also trying to convince the government to eliminate the water fee which mining operations must pay for the use of sovereign water. In some cases, this fee is a hidden tax which adds as much as 3% to the costs of some mining operations.[9]

In the area of infrastructure, industry is concerned with the lack of roadways and the adequacy and maintenance of those roads already in service. Similar concern exists over the state of the country's rail system as well. The mining industry is also concerned with the lack of water and power lines and the costs associated with building such infrastructure. In addition, the Mexican mining industry is lobbying the government owned power company in order to obtain the preferential electric rate granted to large-scale power users, such as heavy industry and the agricultural sector.[9]

9 Conclusions

The importance of the restructuring of the Mexican mining industry cannot be underestimated. Although aggregate levels of mineral production did not fall as dramatically as they have in other countries with ill-conceived mineral policies, Mexico's mineral output came at an unacceptable economic price: the eventual destruction of the country's mineral production capacity. The relatively stable production levels masked a serious deterioration in the levels of capital reinvestment and greenfields exploration. Furthermore, the industry's structure eroded its ability to compete in an increasingly global market.

Mexico's economic miracle is already being hailed and emulated throughout the world. While the country's new minerals policies have not yet been in place long enough to garner the attention that the country's economic policies have, they should serve as a blueprint for minerals development in other developing countries. There is no question that every country has specific and unique cultural, geographic, political, and social considerations which might influence the development of a minerals policy.

However, this is often used as an excuse to pursue policies which make little economic sense. The beauty of Mexico's new minerals policy--and the reason why it should be so easy to apply in other countries--is its inherent simplicity and its reliance on basic economic incentives. The nature of the minerals industry is such that prices are set by the world market. By allowing competitive forces to determine the level of output the most efficient allocation of scarce resources is achieved. This simple economic tenet is the basis for Mexico's minerals policies and the key to its future success.

10 References

1. Unidad de Análisis Prospectivo El Financiero. (1993) **Sucesión Pactada**. Plaza y Valdés Editores, Mexico City.
2. Flores, Victor. (1986) **The Energy and Mineral Resources of Mexico: Strategies for Development**. Master's Thesis, The University of Texas at Austin.
3. Becerra Gonzalez, María. (1963) **Derecho Minero de México**. Editorial Limusa-Wiley, S.A., Mexico City.
4. Southworth, J.R. (1905) **The Mines of Mexico.** Blake and Mackenzie, London.
5. Bernstein, Marvin D. (1964) **The Mexican Mining Industry 1890-1950**. The State University of New York, Stony Brook, New York.
6. Secretaría de Energía, Minas e Industria Paraestatal. (1993) **Anuario Estadístico de la Minería Mexicana 1992**. Consejo de Recursos Minerales, Mexico City.
7. Secretaría de Energía, Minas e Industria Paraestatal. (1993) **New Mining Regulation**, English ed., Mexico City.
8. Silver, Douglas B. (1993) Mexican Mineral and Mining Activity, in **Mexico Mineral & Mining Directory 1993**. Howell Publishing Company, Denver.
9. Cámara Minera de México. (1993) **LVI Asamblea General Ordinaria 1993**. Mexico City.
10. Estrada Orihuela, Sergio. (1993) La protección ambiental y la industria minera mexicana. **Minero Noticias**, 159, 31-42.

Mining – Open-pit Technology

Application of advanced computer-based systems in open-pit evaluation

M. M. Randall
RTZ Technical Services, Ltd., Bristol, England

Abstract

The trend towards the exploitation of high tonnage low grade open pit deposits has meant that the level of capital investment exposes the investor to ever greater risk in a fluctuating market.

These risks are not normally addressed by the traditional methods of mine evaluation, which have tended to adopt a static approach in which costs capacities and revenue are considered constant.

This paper describes a powerful computer based system developed by RTZ Consultants that attempts to account for the variations in economic parameters and provides a platform for the rapid evaluation of 'what if' scenarios. The primary aim being to determine the most suitable exploitation strategy whilst attempting to maximise the net present value (NPV) of the project.

The practical application of this system is demonstrated by way of a number of examples drawn from recent mine evaluations undertaken by RTZ Consultants in Chile and Argentina. These examples also serve to highlight some of the problems that can be encountered when trying to apply these computerised planning techniques.
Keywords: Mine evaluation, open pit scheduling, pit optimisation

1 Introduction

The rapid development of faster and cheaper computer hardware over the last decade has led to a plethora of software packages that deal with all aspects of open pit planning. Whilst this would seem to be the answer to the mining engineer's dream it has in fact resulted in a relatively slow acceptance of the technology in the mining industry.

The main problem seems to lie with the diversity and lack of commonalty of the methods that all too often leaves the user staring at a garage floor strewn with tools and little idea where to start. The approach adopted by RTZ Consultants has been to utilise a combination of commercial and proprietary software packages that can be linked within a common framework. This system has been built up and developed over many years and is used extensively in the RTZ group for mine evaluation and long term planning.

The main components of the system are:

- Open pit optimisation using Whittle Programming Ltd's FOUR-D program.
- Detailed pit design with RTZ's Open Pit Design (OPD) program.
- Cut-off grade optimisation and long term pit planning with RTZ's open pit scheduler (OGREPlus).
- Medium and short term planning with RTZ's open pit scheduling system (OPSS).
- Financial evaluation spreadsheets (Lotus or Excel).
- Graphical presentation of results (AutoCAD).

These programs work within the common framework of regularised block models (grade, rock type, density etc) which can be used to pass information through the system with the aid of a number of utility programs, which, once set up, are largely transparent to the user. This process of evaluation will necessarily be iterative since the capital and operating costs determine the economic resource and are dependent on the mine schedule, whilst the mine schedule can only be determined having defined the resource. The process is summarised in Figure 1.

2 Pit optimisation

The objective of pit optimisation is generally to determine the pit outline which offers the highest profitability for a given set of economic and physical constraints. The largest economic pit is normally termed the ultimate pit and is used to determine the scale and life of the project.

Stemming from the work of Lerchs & Grossmann [1] this area of mine planning has attracted considerable attention from researchers and has led to the development of numerous methods of arriving at an optimal pit outline. Perhaps the best known are the 'floating cone' and Lerchs Grossmann graph algorithm techniques, which have become the standard methods of pit optimisation used throughout the mining industry. However, of these only the graph algorithm method provides a demonstrably rigorous means of arriving at the optimal pit shape [2].

The recent implementations of the graph algorithm technique by Whittle [3] have overcome the limitations of earlier versions by parameterisation of the block value equation and have led to the development of an extremely versatile tool that largely avoids the 'black box' syndrome and is easy to use. However there remain two important limitations:

- The model can only accept a single ore tonnage and recovered metal quantity for each ore parcel of a block. This means that an implied cut-off grade must be used and the effects of cut-off grade on selective mining cannot be taken into account.

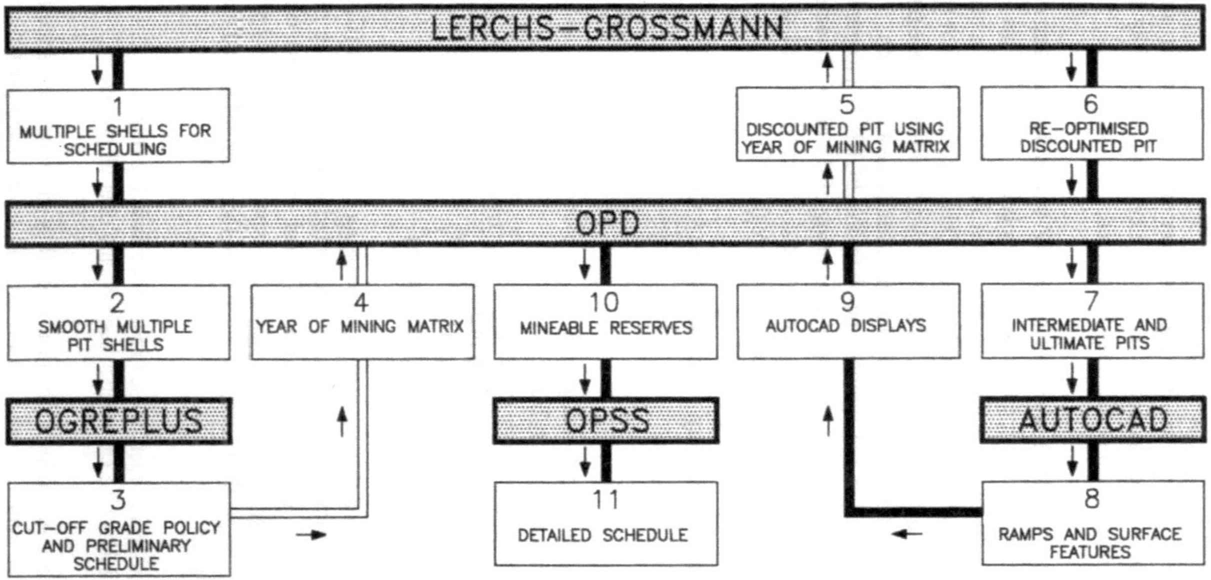

Fig.1 RTZ geological and mining software

3 Pit design

As with pit optimisation there exist a number of software packages that provided more or less the same basic tools for mine design. For the most part these mimic the traditional manual methods of pit design with the user laying out the pit perimeters for each pushback on a bench or level plan and then projecting the perimeters to successive levels. These tasks can be handled by the computer and provide the opportunity to consider complex projection decisions such as those provided by OPD.

- Smoothing of the perimeter according to the maximum and minimum length of its constituent chords and by the maximum and minimum internal angle defined by successive chords.
- Inclusion of berms at any level.
- Variation of the projection angle depending on rock type, sector of the mine or for a given range of levels.
- Automatic extension of the perimeter on each level in order to follow a particular rock code.

The basic input to OPD consists of block matrices for grade(s) and rock type which will have normally been generated by some geostatistical technique such as kriging and will have been used to create the FOUR-D model. These matrices can then be viewed level by level to explore the most suitable pattern of pushbacks. As mentioned previously, the results of the FOUR-D analysis can be used to guide the engineer during digitisation of the perimeters since a block model of the results can be produced like any other block model by reading the FOUR-D results file and writing the pit expansion number to a block model.

Besides providing a visual guide, the FOUR-D block matrix can be used as a control on projection in its own right. Using the facility in OPD for following a given rock code (referred to as footwall following) the pit perimeters can be made to roughly follow one or more of the FOUR-D pit outlines. This feature is particularly useful where a regular pushback geometry is not feasible as a result of ore pods or where the engineer wishes to ensure that the final pushback closely follows the ultimate pit outline. In either case the automatic perimeter extension will be subject to the perimeter controls discussed above (eg smoothness).

Having designed a pushback the contained material quantities can be viewed or printed for any specified cut-off grade and a 'dollar' value can be computed in terms of total profit and discounted cashflow. This type of incremental analysis is not strictly necessary if the OPD phases broadly follow the FOUR-D sequence but is normally used where:

- The assumption of overall pit slope angles cannot adequately account for the effect of including ramps.
- The ultimate pit limit contains small pockets of ore that cannot be accessed due to lack of ramp access or insufficient space in which to operate the machinery.

- Extension of the pit limit beyond the (FOUR-D) ultimate pit limit would result in an increase in the project valuation as a result of reduced mining costs.

These and other special conditions cannot be catered for in the pit optimisation other than by the manual insertion of additional arcs to force the required extraction geometry and are clearly best handled by the manumatic incremental (skin) analysis discussed above.

4 Cut-off grade optimisation

The cut-off grade is normally used in mining to discriminate between material that is ore and material that is waste. The point at which the revenue for the mineral equates to the overall costs of mining, processing and marketing is defined as the break-even cut-off. If additional value can be obtained by raising the cut-off grade above break-even then the cut-off grade can be used to maximise the present value of future profits. This concept is fundamental to the mine planning approach of RTZ and clearly states that principal planning objective is the maximisation of discounted cashflow.

It has been demonstrated [5] that the maximum present value can be found by maximising the sum of the present values for each mining period. Mathematically this application of the principal of optimality is known as dynamic programming and results in an optimal cut-off grade strategy for a given mining sequence.

Based on the work of Lane a computer program known as OGRE (Optimum Grades for Resource Exploitation) was developed by RTZ to provide a tool for rapidly evaluating the optimum cut-off grade strategy for any set of economic constraints. This program has since been modified and enhanced, as a result of its application to a wide range of projects, and has recently been incorporated with a long term scheduler (OGREPlus). Details of the scheduler are considered in the next section.

The main parameters used to determine the cut-off grade are:

- Mine, mill and market constraints.
- Unit costs.
- Recovery factors.
- Metal price.
- Stockpile grades, capacity and reclaim rates.

The above parameters may be fixed or varied with time. This flexibility is important when considering the sensitivity of the mine plan to that parameter (eg metal price) or when attempting to maximise the return by investigating alternative mining strategies (eg expansion of mine or mill capacity).

It should be noted that the calculated optimum cut-off strategy will only be valid for a given sequence of material throughput and that this will necessarily vary during subsequent re-scheduling. The integration of OGRE within OGREPlus handles this problem by allowing the user to iterate between the cut-off calculation and the semi-automatic scheduler until a stable solution is reached.

5 Scheduling

As with other areas of mine planning considerable advances in the application of computer methods have been made in recent years as a result of cheaper and faster computers. This is particularly true of the automatic scheduling systems, such as those based on linear programming [6, 7] which share the same kind of problems as that of the computerised chess program but without the rigorously defined moves.

This difficulty of generalising the rules has so far limited their application and these models have largely remained in the area of academic research at this time.

In view of the lack of suitable commercial packages RTZ decided to develop its own semi-automatic scheduler as an addition to the company's cut-off grade optimisation program (OGRE). The resultant program (OGREPlus) takes the optimum cut-off grade strategy of OGRE and applies the following constraints to the schedule:

- Required ore and waste tonnage per period.
- Maximum mining rate in any given area (phase) expressed in terms of tonnes or benches per period. The later defines the sinking rate and is dependent on access and drainage.
- Maximum equipment requirements (shovels, trucks and drills) by area. The effect of using mobile crushers can also be included by specifying their availability and location.
- Minimum ore exposure.
- Maximum production of deleterious minerals.

During the scheduling the discounted cashflow is automatically re-calculated so that any change to the schedule can be costed and alternatives ranked. A schematic animation is used to display the mining sequence, this allows the engineer to locate and remedy potential problems and is especially useful in the diagnosis of complex situations in which the schedule is constrained by a number of factors (eg sinking rate, mining rate and interaction of mining phases). Using this system, and allowing for two or three iterations between the cut-off grade calculation and the scheduler, a fairly complex mine schedule can be completed in a matter of hours. This plan is then exported to a spreadsheet for a more rigorous financial evaluation with the inclusion of capital costs and tax allowances.

Where more detail is required, RTZ's Open Pit Scheduling System (OPSS) can be used to plan the development of the OPD phases by scheduling each bench in logical mining units (blasts or blast composites) rather than on a bench by bench basis. By allocating the blasts to loading units a schedule can be produced by OPSS in much the same manner as that described for OGREPlus. This detail of planning is however not usually necessary at the preliminary mine evaluation stage and would only be undertaken during the subsequent period of the feasibility study. Further details of the application of OPSS to open pit and underground scheduling are given by Wooller [8] and are beyond the scope of this paper.

6 Presentation of results

A vital element of any planning tool lies in the presentation of data. This is particularly true of mine planning software where aids to the visualisation of the complex three dimensional interaction of the mining phases are essential in order to appreciate how and where modifications to the schedule can be made. In order to satisfy this need and provide maximum functionality, RTZ has adopted the policy of largely using commercially available third party programs for this function. With standard interface format exchange with AutoCAD (Autodesk Ltd, UK), DGM (L M Technical Services Ltd, UK) and QuickSurf (Schreiber Instruments Inc, USA) both functionality and portability are ensured over a wide range of platforms.

A similar approach has been adopted for reporting of numerical data. Adequate working screen reports are provided in the system whilst comma separated value (CSV) and fixed format files are exported to a spreadsheet for formatting, further manipulation (databases) and final reporting.

7 Case studies

A number of simple case examples are presented here in order to demonstrate some of the features of the system and to highlight areas of special interest. The data for these examples was drawn from recent mine evaluations conducted by RTZ in Chile and Argentina.

7.1 Cut-off grade optimisation

This first example deals with a medium sized copper deposit with a mineable reserve of 180 Mt at 0.93% Cu. The basic cost parameters are given below:

Mining Cost	$0.81/t rock
Processing Cost	$7.36/t ore
TC/RC	27 ¢/lb Cu

Using the software discussed previously, a series of mining phases were designed within the ultimate pit outline. The optimum cut-off strategy and mining schedules were then calculated for a range of milling rates.

In order to minimise the pre-production period the pre-strip was designed at 3 Mt. Whilst this reduced the initial capital it also limited the working area and access during the first few years. It was thus decided to use a stepped production rate that started at

20 000 tpd and rose to 30 000 tpd after 4 years. This then formed the base case for subsequent evaluations.

One part of this study included the comparison of a series of mine schedules based on fixed cut-off grade with the variable cut-off grade strategy of OGRE. For the fixed cut-off grade schedules it was found that the mine life varied between 11 and 20 years (Figure 4) with a maximum present value occurring at 0.6% Cu. However if the variable (declining) cut-off policy of OGREPlus is used then the present value increases by a further $12 M and contributed to a 10% increase in the maximum NPV of the project.

Applying a variable cut-off grade policy will necessarily mean varying the internal cut-off grade during the early years with the possible requirement to stockpile low grade material. The practicality of selective mining should thus be closely considered to see if it is workable and whether it is supported by the data. One approach to this problem is to apply factors derived from conditional simulation [4] as a correction within the tonnage grade curve analysis routines of OGREPlus. These factors cannot however be rigorously applied to the FOUR-D model other than as a fixed tonnage and grade correction by location. This approach is somewhat simplistic but can be used as a coarse guide as to its effects on selective mining.

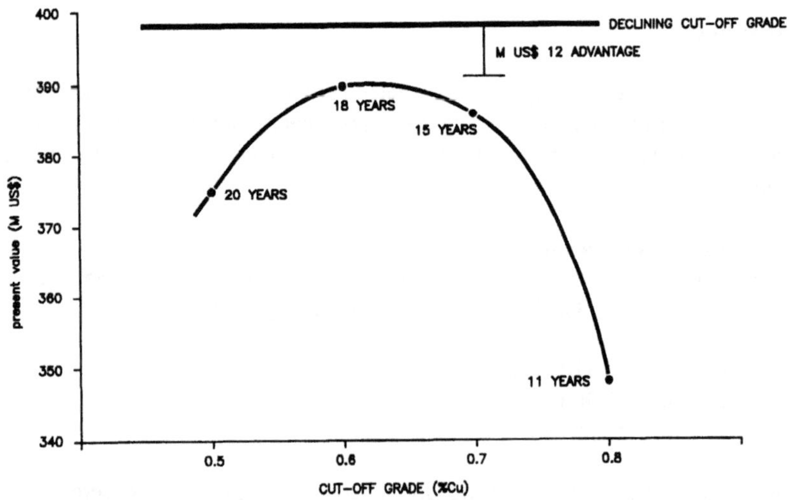

Fig. 4 Declining cut-off grade (DCOG) vs fixed cut-off (20/30 000 tpd case, pit A)

7.2 Modifying the ultimate pit limit

If the FOUR-D optimum pit outline is used as the limiting boundary for the design of the practical mine phases then it is likely that the final phase or a part of the final phase will return a negative discounted cash flow (DCF). This may occur due to the necessity to advance the waste stripping in order to avoid wild fluctuations in the required mine capacity and to provide an exposed ore reserve (typically 3 months ore production).

As discussed previously, the FOUR-D optimisation cannot take into account the effects of advance stripping and will thus result in the final shell or shells being uneconomic. This situation is demonstrated in Figure 5 in which the phases B and C are included within the FOUR-D pit limit under the following conditions:

- Minimum metal cost of mining (MCOSTM) of 0.00065.
- 'Best case mining schedule' in which the phases are mined sequentially.
- Ore production rate of 20 Mta.

However, if the phases are scheduled in OGREPlus, with the additional constraint that the limiting mine capacity is 80 Mta, then Phase C becomes uneconomic at some discount rate (DCF -$13.9 M) and the pit limit should be modified to exclude phase C or a portion of it. If a minimum mining cut width is now imposed, as in Figure 6, then the problem is accentuated and the inclusion or exclusion of the last cut can result in a difference of 50 to 100 Mt for a medium sized pit.

In practice the determination of the position of the pit limit in FOUR-D (version 2.05) is controlled by the range and increment of MCOSTM (a maximum of 50 pit shells can then be generated). The selection of the MCOSTM increment is thus critical to the (mathematical) precision with which the optimum pit limit can be determined and must be chosen such that the pit shells are commensurate with the proposed starter pit and subsequent mining phases. This is not always possible to arrange with a constant MCOSTM increment which will result in an increasing step size of the metal price as the pit limit is approached.

With the computer techniques discussed above the 'optimum pit outline' remains as a rather fuzzy concept due to the inherent flutter of the system. The variance can however be reduced by:

- Appropriate choice of the model and the cost parameters.
- Maintaining minimum cut widths for all mining phases. The assumption that the latter phases are unimportant can lead to significant over or under estimation of the mineable reserves **and** project value.
- Ensure that if an overall pit slope angle has been used, then the inclusion of the ramp will not significantly change the results.

The irregular step size for MCOSTM, offered in the latest release of FOUR-D (Whittle, 1993), also provides further scope for control and can be used to provide the detail where it is required.

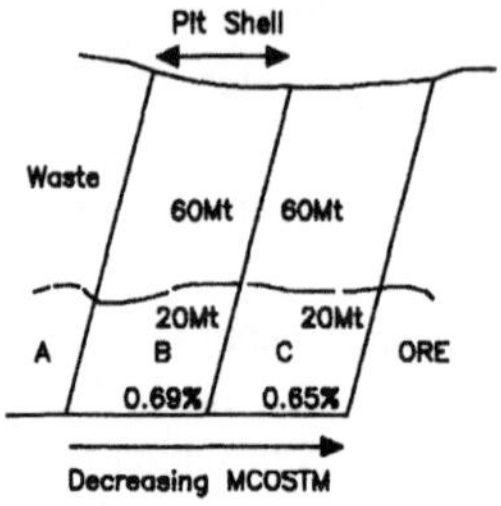

Input parameters:

	Ore	Waste
Limiting capacity (Mta)	20	60
Operating cost ($/t)	6	1
Recovery (%)	90	100
Nett revenue ($/t)	1543	0
Discount rate (%)		

Costs and Revenue (undiscounted)

Phase	Revenue	Costs
B	180.5	180.0
C	191.6	180.0

Mining of Phase B followed by Phase C
(FOURD best case schedule)

Period	Ore (Mt)	Waste (Mt)	Cash Flow (M$)	DCF (M$)	DCF Factor
N+1	0	0	0.0	0	1.0
N+2	0	0	0.0	0	0.9
N+3	20	60	11.6	9.6	0.8
N+4	20	60	0.5	45.5	0.8
Total	40	120	12.1	10.0	

Mining of Phase B followed by Phase C
(OGREPlus schedule)

Period	Ore (Mt)	Waste (Mt)	Cash Flow (M$)	DCF (M$)
N+1	0	0	0.0	0
N+2	0	60	−60.0	−54.5
N+3	20	60	11.6	9.6
N+4	20	0	60.5	45.5
Total	40	120	12.1	0.6

Fig.5 Evaluation of an Idealised pit optimisation sequence

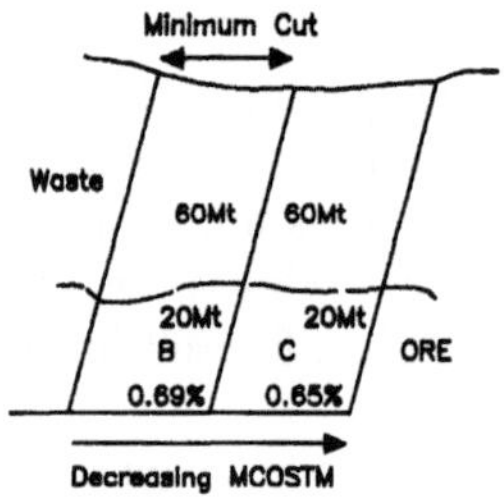

Input parameters:

	Ore	Waste
Limiting capacity (Mta)	20	60
Operating cost ($/t)	6	1
Recovery (%)	90	100
Nett revenue ($/t)	1543	0
Discount rate (%)	10	

Costs and Revenue (undiscounted)

Phase	Revenue	Costs
B	180.5	180.0
C	191.6	180.0

Mining of Phase B with C
(OGREPlus schedule)

Period	Ore (Mt)	Waste (Mt)	Cash Flow (M$)	DCF (M$)	DCF Factor
N+1	0	60	−60.0	−60	1.000
N+2	0	60	−60.0	−54.5	0.909
N+3	20	0	71.6	59.2	0.826
N+4	20	0	60.5	45.5	0.751
Total	40	120	12.1	−9.8	

Mining of Phase B with A
(OGREPlus schedule)

Period	Ore (Mt)	Waste (Mt)	Cash Flow (M$)	DCF (M$)
N+1	0	30	−30.0	−30
N+2	0	30	−30.0	−27.3
N+3	20	0	71.6	59.2
N+4	0	0	0.0	0
Total	20	60	11.6	1.9

Fig.6 Re-evaluation with a minimum mining cut width

7.3 Risk analysis

One great asset of a semi-automatic computerised scheduling system is the ability to rapidly evaluate a wide range of 'what if' scenarios. Using OGREPlus the complex interaction of unit costs, prices, capacities and recoveries can be investigated independently or as dependent variables. In either case the parameters can be varied with respect to time or tonnage.

One example of this type of analysis might be to consider the effect of varying metal price on the calculated optimum cut-off grade strategy. Figure 7 shows the results of an evaluation where the cut-off grades were calculated for a fixed metal price ($0.7/lb Cu) and for a variable metal price. In the second case the metal price was varied linearly from 0.7 to $1.2/lb Cu over the first five years of operation.

A better understanding of how the metal price affects the cut-off grade can be obtained by considering the formula for the mill limiting optimum cut-off grade (G_L) in which throughput is controlled by the mill:

$$G_L = \{h + (f + F)/H\} \,/\, (p-k)\, y \tag{2}$$

where

h = unit cost for processing ($/t)
f = fixed costs ($/y)
F = opportunity cost ($/y)
H = mill capacity (t/y)
p = metal price ($/t)
k = marketing cost ($/t)
y = overall recovery

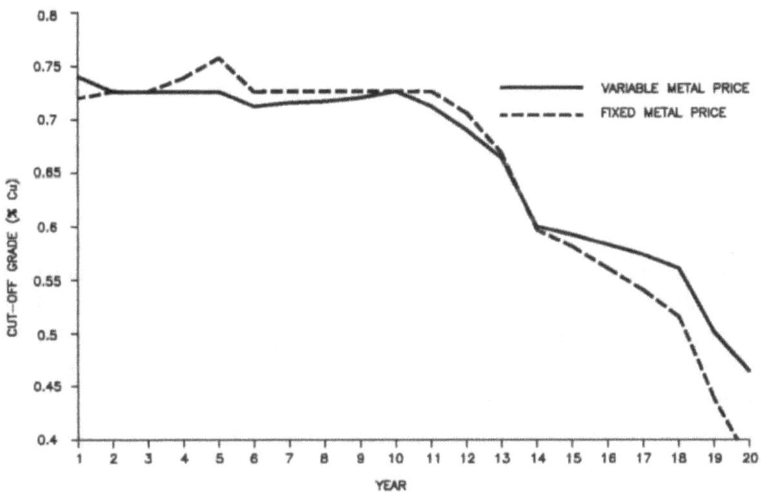

Fig. 7 Investigation of the influence of metal price on the cut-off grade strategy

The opportunity cost is defined (Lane, 1991) as:

$$F = V' - dV/dT$$

where

V'	=	discounted present value (PV)
dV/dT	=	rate of change of PV

An expected rise in price will thus reduce the opportunity cost and cause the mine to lower the cut-off grade now in expectation of higher profits later. Once the price is stabilised ($dV/dT = 0$) then the cut-off grade will decline in response to V' and approach the breakeven cut-off grade at $V' = 0$. For more complex cases the limiting cut-off grade may be controlled by the mine, mill and market and an equilibrium cut-off grade ($G_L = G_M = G_K$) is established for the limiting capacities of the mill, mine and market respectively.

This analysis demonstrates that the mine plan in this case is not particularly sensitive to the metal price and that the risk analysis associated with fluctuations in metal price could be treated as a separate issue in the financial model. There is of course no penalty in modelling the metal price changes in OGREPlus (provided the information is available) but is not part of the normal practice for the evaluations undertaken by RTZ Consultants.

8 Conclusions

Over a number of years RTZ has built up an integrated set of computer aided design tools for the evaluation of open pit mines. These programs allow the engineer to progress from pit optimisation to pit evaluation within a very short time frame and to consider many 'what if' scenarios as part of a risk analysis. The potential benefits and pitfalls of these tools have been demonstrated by way of their application to a number of case studies.

A key element in the choice of software has been RTZ's policy of avoidance of the black box syndrome. This has led to the use of a combination of commercial and proprietary software which taps the latest technology whilst maintaining a wide base of in-house expertise. This approach minimises the exposure to the all-too-common rise and fall in popularity of a certain mine design package whilst focusing on the development and incorporation of new techniques as they become available.

9 References

1 Lerchs H & Grossmann I.F (1965), **Optimum Design of Open-Pit Mines. Trans. C.I.M**, Vol LXVIII, pp17-24.

2 Dowd P.A & Onur A.H (1993), **Open-pit optimisation - part 1: optimal open-pit design**, Trans. Inst. Min. Metall. (Sect. A: Min. industry), Vol. 102, ppA95-A104.

3 Whittle J. (1989), **The facts and fallacies of open pit optimisation**. Whittle Programming Pty, Ltd., North Balwyn, Victoria, Australia.

4 Ravenscroft P.J. (1992), **Risk analysis for mine scheduling by conditional simulation**. Trans. Inst. Min. Metall. (Sect. A: Min. industry), Vol. 101, ppA104-A108.

5 Lane K.F. (1991), **The economic definition of ore, cut-off grades in theory and practice**. Mining Journal Books Ltd., London, p149.

6 Huang S. (1993), **Computer based optimisation of open-pit mining sequences**. Trans. Inst. Min. Metall. (Sect. A : Min. industry), Vol. 102, ppA125-A133.

7 Tolwinski B. & Underwood R. (1992), **An algorithm to estimate the optimal evolution of an open pit mine**. In 23rd Symposium on the application of computers and operations research in the minerals industries (APCOM) Littleton, Colorado: AIME, pp399-409.

8 Wooller R. (1992), **The Production Scheduling System**, Trans. Inst. Min. Metall. (Sect. A: Min. industry), Vol. 101, ppA47-A54.

9 Whittle J. (1993), **Whittle Programming Newsletter number 12**, Whittle Programming P/L North Balwyn, Victoria, Australia. p6.

Myth and reality: status report on computer open-pit optimization algorithms in the 1990s

D. M. François-Bongarçon
Mineral Resources Development, Inc., San Mateo, California, U.S.A.

Abstract

After years of fast development followed by a slowing down in the field of computerized open-pit ultimate outline optimization, the four most popular surviving groups of methods are reviewed, and described with their pros and cons. A particular group, the Reserve Parameterization method, is re-emphasized, its very nature and its advantages are described in more details, and the two existing families of algorithms commercially available for its practical implementation are discussed. The conclusion offers some updated guidelines to prospective users of open pit optimization programs.
Keywords: Open-pit outline optimization, parameterization method, 4-D algorithms.

1 Introduction

From the mid-sixties to the mid-seventies, with the advent of both large computers and large low-grade/high-tonnage deposits, the field of computer optimization of the ultimate open-pit outline has been the focus of a strong interest from the mining industry. Many individuals (engineers, mathematicians) and universities entered a frantic race for the ultimate solution to the problem of the ultimate pit, and numerous papers presenting a variety of algorithms were published during that period in mining conferences at large. This optimization fever then cooled down significantly, and by 1980, only a few algorithms had survived in the form of commercially available, widely used, proven software. Since then, research efforts have been mostly directed to improving these methods or their variants, and their programming. However, little has been published to help prospective users understand the differences between methods, and the usage of these methods usually still belongs very much to the infamous "black-box approach". To correct this situation, this paper intends to give a brief overlook of the most popular methods, and re-emphasize one particular method that has not yet received all the attention it deserves, although it revolutionized the field almost twenty years ago, namely the Reserve Parameterization method.

2 Available algorithms

Only four generic families of computerized methods are actively used in the industry today:

1. Incremental method, a computer assisted variant of the traditional manual push-back method.
2. Moving cone algorithms (also called floating cone, or cone miner).
3. Graph theory algorithms, popularly known as Lerchs & Grossman algorithms.
4. Reserve Parameterization method.

The four methods presuppose the availability of a computer block model containing, for each block, the estimated reserves and/or an algebraic quantity to be maximized, called "economic valuation" or "dollar value", and usually derived as a linear function of the estimated local reserves and relevant costs.

They are so extremely different that detailed direct comparisons would be meaningless. Instead, let us review their main characteristics and limitations separately, and explain how they function. The reader is referred to the abundant literature for more detailed presentations.

2.1 Incremental method

In the incremental method, the traditional pre-computer era's manual method of building the optimum outline from successive push-backs which are visually selected in turn from a series of level maps, is emulated in a very efficient and rapid way: the computer takes care in virtually no time of all the cumbersome 3-D arithmetics of identifying and summing up the blocks for each considered push-back, and of the tracking of the push-backs previously selected at any given time as material to be mined. Meanwhile, the mining engineer needs only to concentrate on designing and digitizing the next push-backs in plan view, and choosing in which order to explore them to best guarantee a reasonably optimal solution.

Sometimes one of the other, more automatic methods is used first, possibly in the form of a simplified (faster running variant) guide to this detailed engineering work, which is necessary anyway following any outline optimization, due to the inability of the available software packages to design the berms, ramps and detect/"clean up" unmineable geometries.

This method is highly recommendable against the moving cone methods (next section), as the engineer's eyes and brains, unlike any feasible algorithmic logic, are able to immediately spot on the level maps the most interesting ore pods, whatever their shape, connectivity or size. In Figure 1, a little hypothetical 2-dimensional example with 45° slopes illustrates a case where any moving cone algorithm will fail in finding any positive pit, unless a multiple block apex encompassing at least five ore blocks is used as first cone (which never happens in practical versions of the moving-cone method). On the figure, however, we can spot a large positive optimal outline at first glance.

Finally, the method should normally be used in several passes, descending as well as ascending, before reasonable optimality is achieved.

-1	-1	-1	-1	-1	-1	-1	-1	-1	-1	-1	-1	-1
-1	-1	-1	-1	-1	-1	-1	-1	-1	-1	-1	-1	-1
-1	-1	+6	+1	+3	+1	+7	+1	+3	+1	+6	-1	-1
-1	-1	-1	-1	-1	-1	-1	-1	-1	-1	-1	-1	-1
-1	-1	-1	-1	-1	-1	-1	-1	-1	-1	-1	-1	-1

Fig. 1: Shortcoming of the floating cone algorithm: No economic outline can be found blindly in this case, although one clearly exists.

2.2 Moving cone algorithms

In this most popular class of methods, the computer simulates the exploratory travel throughout the block model of an inverted cone representing the slope constraints and often called "the extraction cone". The apex of the cone is moved from ore block to ore block, and the current cumulative valuation of the cone calculated in each position explored. When this valuation is positive, the cone is said to contribute positively to the profit of the pit, and is recorded as material to be mined, along with the blocks of ore and waste contained in the cone. Altogether, this method tries to (blindly) mimic the manual incremental method of designing the ultimate pit as a series of successive push-backs. The example of Figure 1 indicates the limitations of such an emulation.

The solution, which depends largely upon the order in which the apex blocks are explored, is not rigorously optimal, and can even be far from a true optimum, depending on the degree of sophistication of the particular program used. Of course, the more sophisticated the programming, the lesser the risk to miscalculate a reasonably optimal solution, but the higher the computing time, which can be quite prohibitive, so that all the existing programmed versions of the method are trade-offs between these two concerns.

Often, the algorithm is run for a number of varying prices of the commodity of interest (hopefully at different corresponding cutoff grades), to provide a better insight into the sensitivity of the results and examine optimal pits of altogether different sizes. This recommendable usage originates in the often unconscious need for the features offered by the Reserve Parameterization method (see further), but represents an extremely imperfect way of achieving it.

Interestingly, the algorithm is also sometimes applied restricting the search down to each successive bench of the block model in turn, and the partial solutions obtained down to those benches are then used in an attempt to establish an optimal sequence of the exploitation of the optimal pit found. This, unfortunately, does not work, as demonstrated elsewhere [1] while it can be very successfully achieved within the frame of Reserve Parameterization.

2.3 Graph theory algorithms

These are the algorithms in which the extraction constraint is represented using a graph of which the nodes are the blocks of a reserve block model, and the arrows represent the precedence between blocks in the order of extraction imposed by the slope constraints. The graph theory allows the computation of the maximum closure of the graph, which happens to be the set of blocks which form a feasible pit outline while maximizing the algebraic values contained in the reserve model (usually economic valuations of each block).

When programmed with no short-cuts, the graph theory method guarantees the rigorous mathematical optimum (maximum of the algebraic block function), but the computing time is then always unpredictable, ranging from a few minutes to a few days on even a large computer, depending upon the particular solution being sought. With programming short-cuts, this drawback can be eliminated, but then the mathematical rigor of the solution is no longer guaranteed, and the algorithm, which is by no means among the most flexible to use, then loses most of its interest.

A few programmed version of this method are available on the market, mostly in its rigorous form. Ever since the first program [2], research has been going on to find faster variants and incorporate any simplifications allowing to cut the computing time down without losing the rigor of the solution. Every 5 years or so a new, usually better performing, variant makes its appearance. The most recent known to the author as worth of interest dates back to a few years ago [3]. More will be said further about a possible use of graph theory algorithms as tools for implementing the more useful Reserve Parameterization method.

2.4 Reserve parameterization method

2.4.1 History and basic principles

If the graph algorithm by Lerchs and Grossman was the first "scientific" breakthrough in the field, the second and only significant other methodological breakthrough came with the idea of Reserves Parameterization applied to ultimate pit design [1][4][5][6].

In that method, the only quantities ever manipulated in the optimization are grades, tonnes and metal quantities. Profit formulae are purposely eliminated. As a consequence, the optimization is a purely technical optimization, to be used in the subsequent true economic optimization, and the output of the method is not, as usual, a single optimal pit outline, but rather a series of pit outlines (usually about 30) representing the range of truly optimum projects for the full range of possible economic conditions and management policies. As pointed out in the literature on the subject, the end result, after the subsequent economic optimization has been performed, is closer to the true optimum than with any other technique, more suitable for sensitivity analyses, and easier to adapt to changes in the economic context.

The fundamental concepts are not new. They can be traced back to Lasky's pioneering idea [7] that the reserves (even geological) should be represented as a function of a parameter such as a cutoff grade. This productive concept of "grade-tonnage curves" then entered the field of open pit optimization with Lerchs and Grossman's strong concern, in their initial paper [2], that varying parameters controlling the reserves should always be tried in turn, and the maximization algorithm re-run each time. Finally, a consistent theory was put forth by Matheron [8] and specific applications in the case of open pit mines developed [9] [10] with advances described in many papers [1][4][5][6].

In this theory, any technically feasible project outline can be characterized by a triplet (Q,V,T) where T the tonnage of recoverable ore to be processed, Q is the corresponding recoverable quantity of metal of the project, and V the total tonnage to be extracted. For each possible value of the pair (V,T), we define the technically optimal projects as those existing feasible projects (Q_{max},V',T') that maximize Q with both tonnages (V',T') inferior or equal respectively to the values (V,T). Performing the parameterization consists of finding those technically optimal projects (pit outlines) under the reasonable assumption that the true economic optimum, whatever its definition, is to be found among them. Variations of V correspond to variations in the overall geometrical shape of the final project outline, while variations of T are related to a selection of the ore to be treated within V. In the general theory, these variations can equivalently be replaced by mathematical duality with the variations of two parameters λ and Θ, coefficients of the equation of a generic plane in the space (Q, V, T) as suggested in the continuation. The quantity to maximize is then $Q-\lambda.V-\Theta.T$, and Q, V, T are cumulative sums over each project of block quantities q, v, t.

In an open-pit operation, the selection of the ore tonnage T often makes explicit reference to an internal cutoff grade z_c, which therefore happens to be one of the fundamental parameters (in this case, Θ). Indeed, for a given project geometry of given total size V, the only variations of the definition of the ore (tonnage T) are those due to changes in the cutoff grade z_c. Under such conditions, independent variations of z_c and V generate all admissible variations of T. The parameterization is therefore obtained by maximizing $Q(z_c)-z_c.T(z_c)-\lambda.V$. For each likely value of the internal cutoff grade z_c, we can therefore simplify the search by replacing the quantity of recovered metal in each individual block, q, which depends on z_c, with its useful portion $q^+(z_c)=q(z_c)-z_c.t(z_c)$ (metal on top of process costs), and focus only on projects maximizing the cumulative useful metal quantity $Q^+(z_c)$ of the project for each possible value of total tonnage V. Some of these technically optimal projects are of more particular interest: those with particular values of V corresponding to points on the convex hull of the graph $Q^+_{max}(V)$, i.e. the projects maximizing function $Q^+ - \lambda.V$ for all possible values of a parameter λ, and it can be shown that the other projects can safely be ignored. Thanks to this convenient resort to convex analysis, each "critical" project thus retained is in fact attached to a full, continuous range of values of λ, while the set of projects achieving the maximum of function $Q^+- \lambda.V$ for the transition values of λ between two successive ranges includes a smallest and a largest pit, which are precisely those retained. The smaller pit retained for a transition value of λ is also the larger one for the next critical value of λ, and it follows that the set of all the retained critical projects corresponding to a given internal cutoff grade z_c is finite and limited, and they can also easily be shown to be nested. λ and z_c are the actual parameters of the parameterization. V has therefore conveniently been replaced by mathematical duality with an equivalent dynamic "local" parameter λ that also has the dimension of a cutoff grade to be applied individually to a certain function of each block mineralized content. This considerably simplifies the task of determining the critical projects (detailed description, example and diagrams in [5]). In practice, this technical optimization needs to be performed only for a few values of the cutoff grade z_c, i.e. parameter Θ (in rare cases, though, it is always possible to include "external", continuous variations of T within V in the optimization process, by performing additional full Θ-parameterizations for a number of fixed values of λ, using as value of the parameter Θ a variable value of z_c; as an example, this would handle the case when processing costs for a production unit would drastically depend not only on the tonnage of contained ore, but also on the grade of this ore).

Once these technically optimal critical projects have been found, a full scale economic evaluation of a selection of them can be launched, and the best rated project can be selected as the true optimum. Indeed one can demonstrate that, in particular, the smallest or largest feasible pit maximizing any linear function of Q, V, and T corresponding to a given internal cutoff grade (and thus any classical profit formula) almost surely is to be found among the critical projects retained and found.

The advantages offered by the method are clear, and include the following: 1.- non-numerical choice criteria may be used at the economic evaluation stage (e.g. a project may be too small or too large or have too poor a geometry to be included in the short list); 2.- the economic impact of usually unquantifiable factors such as the sub-optimality introduced by the design of access ramps, or the influence of the plant size on the project present value, can be fully accounted for, as the exact geometry of each project is known before the purely economic optimization takes place; 3.- any kind of sensitivity analysis can be performed within the full or the short list of technically optimal projects, with no re-run of the outline optimization program whatsoever (the costly part).

Finally, and above all, the outline optimization method deals only with tons and grades. As a consequence, for each block of the block model, the method can accept as standard inputs estimates of the recoverable ore tons and metal content at various internal cutoff grades, and therefore is able to more fully account for any (even unspecified) variations of the global economic context.

2.4.2 Algorithms

2.4.2.1 MULTIPIT

The major difficulty in the development and marketing of the technique of Open Pit Reserve Parameterization, is its programming complexity, aimed at overcoming the tremendous challenge of calculating a large number of alternative optimal pit outlines in a reasonable computing time. In spite of several re-programming attempts by individuals, companies and universities, the original algorithm by Francois-Bongarcon (program MULTIPIT), based on a functional rather than combinatory approach originally due to G. Matheron, has remained until recently the only practical application program of its kind to be commercially available for the parameterization for open-pit reserves, but it has not been marketed very actively to date, and a real effort of promotion of the method itself is still needed.

Yet, the program is in routine use in several major mining companies around the world. It has been largely described in the specialized literature, and has been the object of on-going research ever since its first version in 1975, including regular benchmark testing against rigorous Lerchs and Grossman optimization runs. Further developments [1] made the optimization of the global mining sequence a mere by-product of the program runs, another important feature that deserves more exposure than it has received. Finally this program finds the entire series of pit outlines of the parameterization in λ in about the same computing time as any other commercial algorithm to calculate one single optimal pit outline.

2.4.2.2 Use of graph based algorithms: Whittle's FOUR-D

In 1987 , however, a new algorithm was put on the market by J. Whittle. Under the name FOUR-D, Whittle proposes a commercial program for the application of Matheron's theory of Reserve Parameterization using Lerchs and Grossman's graph theory approach to pit outline optimization, as an alternative to the functional approach developed for MULTIPIT. Thanks to Whittle's innovative ideas in the use of the graph theory, the Parameterization Method has finally received more attention in the mining world, and, at last, an alternative way of programming it has been proposed.

Unfortunately, in addition to our previous remarks on the limitation and pitfalls of graph theory based algorithms, a detailed examination of Whittle's formula for the value of a block, and of the modus operandi of his program as of 1990, show intrinsic weaknesses that tend to cancel out the advantages of the reserve parameterization technique itself. These weaknesses it is hoped, may have been corrected since, but are interesting as they illustrate the type of features one should carefully examine before using/buying a software package.

First of all, the only quantities the program can handle as input in a block of the model are one single original ore tonnage and the corresponding recoverable quantity of metal, from which an original input ore grade can be derived and which implies the choice of some kind of an initial cutoff grade (hopefully equal to 0.) prior to the program run. FOUR-D has been programmed to find critical projects when both fundamental parameters ($\Theta = z_c$ and λ in previous section) vary but remain in a constant, predetermined ratio. Within the algorithm run, when the value of the λ parameter ("the metal cost of mining") changes, the internal cutoff grade z_c therefore changes accordingly, but the new cutoff grade can only be applied each time to the original input grade of the block. In other words, a standard geostatistical input in the form of a local grade tonnage curve in each block, i.e. a series of recoverable ore tonnages and metal quantities corresponding to various cutoff grades, cannot be handled by the program, which means the critical impact of the mining selectivity on the mineable reserves cannot be properly taken into account, even though it is now a standard requirement in ore reserve modelling.

It is easy to verify and illustrate the negative implications of this particular pitfall. Suffices imagining the hypothetical case of a vertical, outcropping pipe of constant grade ore surrounded by barren rock. A correct and complete parameterization of the optimal pit reserves should be expected to provide us with a series of pits ranging from very small to very large, corresponding to a general economic context going from very bad to very good, or to a variety of management constraints. A parameterization (i.e. 4-D) algorithm with the weakness described above WILL NOT be able to provide the smaller pits because its varying internal cutoff grade will, in due course, necessarily reach the constant ore grade value, and the ore will therefore all vanish, before the smaller pits have been added to the solution. In other words, a number of the smaller pits will be missed. This is highly undesirable, as these smaller projects, in a real case, although they may tend to skim the deposit, give us very valuable information about the real extent of the risk factor in the case, say, of a drastic, unexpected metal price drop during production. They can also just be justifiable projects within the frame of certain management decisions.

Secondly, before any practical simplification, the two technical cutoff parameters found relevant by the theory of reserve parameterization are in fact sorts of theoretical counterparts, in a very broad sense, to the metal cost of mining and the metal cost of processing the ore. We have proposed to reduce them to one per run, by equating the one corresponding to the ore tonnage T to the internal cutoff grade (or metal cost of processing the ore) and running the program for two or three of its values chosen across and slightly outside the realistic range of actual internal cutoff grades. Whittle, instead, decided to fix the ratio between the two parameters. The idea seems to make sense at first glance, since the counterpart of that ratio in the real world is not very variable.

It was nevertheless found by the author and collaborators, over the 15 years the parameterization method has been used, that linking the two "true" optimization parameters too directly to their real world counterparts considerably diminishes the advantages of the method: as already mentioned, in the general theory these parameters are initially introduced naturally as director coefficients of a family of planes in the 3-D space defined by the metal quantity Q, the ore tonnage T, the total rock tonnage V; they must be let free to take any possible value whether or not it corresponds to a realistic value of the physical counterpart; they are NOT originally introduced for their similarity to real world breakeven grades, and their physical interpretation is at most an aid to a better understanding of the work performed by the algorithm, and a hint for reasonable practical simplifications to the search. As to Θ and λ, they have such a direct interpretation only in the case of an actual profit formula strictly of the linear form a.Q-b.T-c.V (a very restrictive and simplified case, which became a standard only due to former algorithmic limitations), under the additional condition that the tonnage to be processed be controlled only by direct processing costs . One of the main advantages of the method of Reserve Parameterization is to give access to all the other cases, among which the real one is usually to be found. In fact, it is the very decision to purposely ignore any such direct interpretation that is the essence and the strength of the parameterization method: with no direct (calculated) interpretation, the two dynamic parameters manage to represent a lot more than just operating costs and commodity prices; they allow us to break the methodological vicious circle until now inherent to open pit optimization. By taking certain possibly "unrealistic" values, they become able to also account for the impact of non quantifiable factors such as management decisions, maximum or minimum acceptable mine life span, minimum mill size, etc.... This quantification is only possible if the relation between the two parameters corresponds to a large enough range of ratios. In final analysis, again, it is therefore a fundamental methodological error to fix the ratio between the two parameters.

Finally, theory and experience have shown that the correct objective of reserve parameterization is to identify as many critical outlines as possible corresponding to the critical transition values of the parameters, so as to not miss any major critical outline. This can only be achieved efficiently by a dichotomic variation of the λ parameter, and even then implies a large number of iterations (i.e. in this case re-runs), in the order of hundreds. It has been found that fixing at most ten or so values of a parameter, as would necessarily be the case in any graph theory based algorithm because of obvious computer time considerations, allows the determination of only a few of the critical projects, making the result totally inappropriate for the objective it intended to pursue in the first place.

2.4.2.3 Use of other algorithms

From the above considerations, it becomes obvious that if time, and more especially computer time, is not an issue, careful use of any classical algorithm (including Lerchs and Grossman's) should in the principle allow one to perform a full reserve parameterization optimization study, albeit a very tedious one. Suffices using the algorithm a large number of times, in a dichotomy mode, to determine the critical contours generated by the maximization of block model function $q'(z_c)- \lambda.v$ taken as a profit formula, when λ varies from 0 to a value high enough for no optimum outline to subsist . Although such an implementation is not recommended as a routine procedure, it certainly constitutes a good start to test and evaluate the amazing increase in optimization capabilities the method has to offer.

3 Conclusions

In spite of the critical impact the ultimate outline may have on the economic value and the technical feasibility of an open pit project, there is a lack of consensus on how and where open-pit optimization methods may be applied. Most algorithms are used as single-run black-boxes by their users, and reputable firms even market programs with a concerning lack of awareness for their limitations, pitfalls, possible lack of performance, flexibility and correctness.

Two methods, however, emerge as being less propitious to the black-box syndrome, and more flexible, and should therefore be preferred: the Incremental Method, for the human intervention it implies, and the Parameterization Method because a large range of possible projects can be examined. Other methods should be used only as preliminary guides to the Incremental Method, or, in the case of the rigorous Lerchs and Grossman's algorithms, as a benchmark tool to test the performance of other programs.

Finally, a few basic guidelines and recommendations can be proposed to the prospective users of such programs:

- One can test one's available program by running it against hypothetical block models for which the correct optimization results are known. The synthetic example mentioned above in section 2.4.2.2. would be a good starting point.
- A correct implementation of the Parameterization Method seems to be the most efficient choice, because of the many advantages not offered by other methods.
- One should be cautioned against all the disadvantages and limitations of the Moving Cone and the Graph Theory methods, and use them with prudence.
- A manual or computerized incremental optimization should be performed in parallel with any fully automatic algorithm, as a safety check, very much for the same reasons why a hand calculated grade block model should always accompany a detailed computerized block model.

Finally and more generally, other factors that can play a significant role in the final results should be handled more carefully than can unfortunately usually be observed: unlike for the underground case, the cutoff grade used to report open pit reserves should always be the internal cutoff grade, NEVER a breakeven cutoff grade including mining costs as is still too often being seen; the specific gravity of the rock should always be given careful consideration, as it often varies and always directly affects both the tonnages of ore and waste; bench heights can have a critical impact on the selectivity, therefore on the ability to achieve predicted reserves; last, but not least, the type of technique used to build the grade block model should not only be as precise as possible, and account for the selectivity of the mining operation to be, but also be of a nature to fit the optimization objective. This is generally best achieved by the proper use of geostatistical methods for recoverable selective reserves.

4 Bibliographical references

1. Laille, J.-P., Francois-Bongarcon, D. 1984. **Parameterization of the Reserves of an Open-Pit: A Complete Case Study**. in Geostatistics for Natural Resources Characterization. Nato ASI series. Reidel publisher, Dordrecht, Holland. Vol. 2, pp 1045-1056.
2. Lerchs, H., Grossman, I.F. 1965. **Optimum Design of Open-Pit Mines**. CIM Bulletin, vol. 58, January 1965, pp 47-54.
3. Zhao, Y., Kim, Y.C. 1990. **A New Graph Theory Algorithm for Optimal Ultimate Pit Design**. SME Annual Meeting, Salt Lake City, February 1990.

4. Dagdelen, K., Francois-Bongarcon, D. 1982. **Towards the Complete Double Parameterization of Recovered Reserves in Open-Pit Mining.** 17th International APCOM Symposium., Golden, Colorado. pp 288-296.

5. Francois-Bongarcon, D., Guibal, D. 1982. **Algorithms for Parameterizing Reserves Under Different Geometrical Constraints.** 17th International APCOM Symposium, Golden, Colorado. pp 297-309.

6. Francois-Bongarcon, D., Guibal, D. 1981. **Parameterization of Optimal Designs of an Open-Pit: Beginning of a New Phase of Research.** Proceedings of the A.I.M.E. Annual Meeting, Chicago, 1981, and in Transactions of the A.I.M.M., bound vol. 274, 1984, pp 1801-1805.

7. Lasky, S.G. 1950. **How Tonnage and Grade Relations Help Predict Ore Reserves.** Engineering and Mining Journal, April 1950. Vol. 151. No 4. pp 81-85.

8. Matheron, G. 1975a. **Le Paramétrage Technique des Réserves.** Technical Note 453. Centre de Géostatistique de l'Ecole des Mines de Paris, Fontainebleau, France.

9. Matheron, G. 1975b. **Paramétrage des Contours Optimaux.** Technical Notes 401 and 403. Centre de Géostatistique de l'Ecole des Mines de Paris, Fontainebleau, France.

10. Francois-Bongarcon, D. 1978. **Le Paramétrage des Contours Optimaux d'une Exploitation à Ciel Ouvert.** Docteur-Ingénieur Thesis. Institut National Polytechnique de Lorraine. Nancy, France.

Mineral Processing

Construction of the Kori Kollo gold plant, Bolivia

G. May
Minproc Engineers, Inc., Denver, Colorado, U.S.A.
R. Brough
Minproc Engineers, Inc., Denver, Colorado, U.S.A.
J. Cederberg
Empresa Minera Inti Raymi, S.A., Bolivia

Abstract

In February 1993 Minproc Engineers Inc. successfully completed a lump sum contract to design, procure and construct a process facility for Empresa Minera Inti Raymi, S.A., at Kori Kollo, 42 kilometers from Oruro in central Bolivia. The operation treats 14,000 tpd of ore to produce 240,000 ounces of gold, and 1.4 million ounces of silver annually. The flowsheet consists of crushing, SAG/ball mill grinding, with carbon-in-leach for precious metal recovery.

The facilities were largely constructed with Bolivian based contractors with a minimum of field engineers from Minproc's home office in the United States. Other subcontractor supervisors came from Chile for the mill installations, and Australia for the electrical work. Most of the mechanical equipment was sourced from North America. The structural steel and platework came exclusively from Chile. The logistics of coordinating the transportation of equipment and materials over rugged terrain and assembling the plant are discussed.

Few significant problems arose during construction. Local shortages of construction materials were supplemented by supplies from other regions. Minor problems with the erection of some of the vendor-supplied equipment were quickly rectified on the site or at local machine shops with no adverse impact on the overall schedule.

Construction started in November 1991, and the plant performance trials were successfully completed in mid-February 1992, some 76 days ahead of the original schedule. The safety record was excellent, with only 1624 lost time hours occurring in a total of 1.5 million hours worked.

1 Introduction

1.1 General

The Kori Kollo operation is owned by Empresa Minera Inti Raymi S.A., a Bolivian mining company. Inti Raymi is an eighty-five percent owned subsidiary of Battle Mountain Gold Company (BMGC). The remainder of Inti Raymi is held by Zeland Mines S.A., a Bolivian mining company. The project is located near the village of La Joya, approximately 220 km south of La Paz, and 40 km north west of Oruro on the Bolivian altiplano. Inti Raymi started a heap leach operation in 1985. By 1988, following several expansions, 4,000 tonnes per day of oxide ore were being treated. At this time drilling was started to delineate the larger sulfide orebody. A feasibility study for the development of this deposit was completed in 1990, and following project approval, the engineering design was started later that year. During this period orders were placed for the major items of mechanical equipment. In July 1991, Minproc Engineers Inc. was awarded a firm price contract to design and construct the 14,000 tonne per day sulfide ore process facility.

1.2 Process Selection and Design

The precious metal mineralization at Kori Kollo is associated with pyrite veins, veinlets of disseminated pyrite and minor zones of tectonic and/or hydrothermal rocks. Sixty to 65 percent of the gold is associated with the veins, and the remainder with the disseminated pyrite. Petrographic studies have shown that the gold particles range from sub-micron to plus 100 microns in size, with approximately 30% smaller than 5 microns. An important feature of the ore is its high silver content, typically 8-10 times that of the gold.

Although the Kori Kollo ore is partially refractory, BMGC decided to treat it by conventional grinding and cyanidation, offsetting the value of gold lost against the savings in capital and operating costs realized by not including an oxidation process in the flowsheet. Primary crushing of run-of-mine ore is followed by a coarse ore stockpile, grinding and classification in an open circuit SAG mill and two parallel, closed circuit ball mills, cyanide leaching and precious metals adsorption onto carbon, carbon desorption, leach tailings thickening for water recovery, and disposal in an impoundment. Figure 1 is a simplified flowsheet representing the process.

The physical and operating parameters presented to Minproc as the basis for the plant design are outlined in Table 1. Table 2 summarizes the design criteria developed for the carbon circuit.

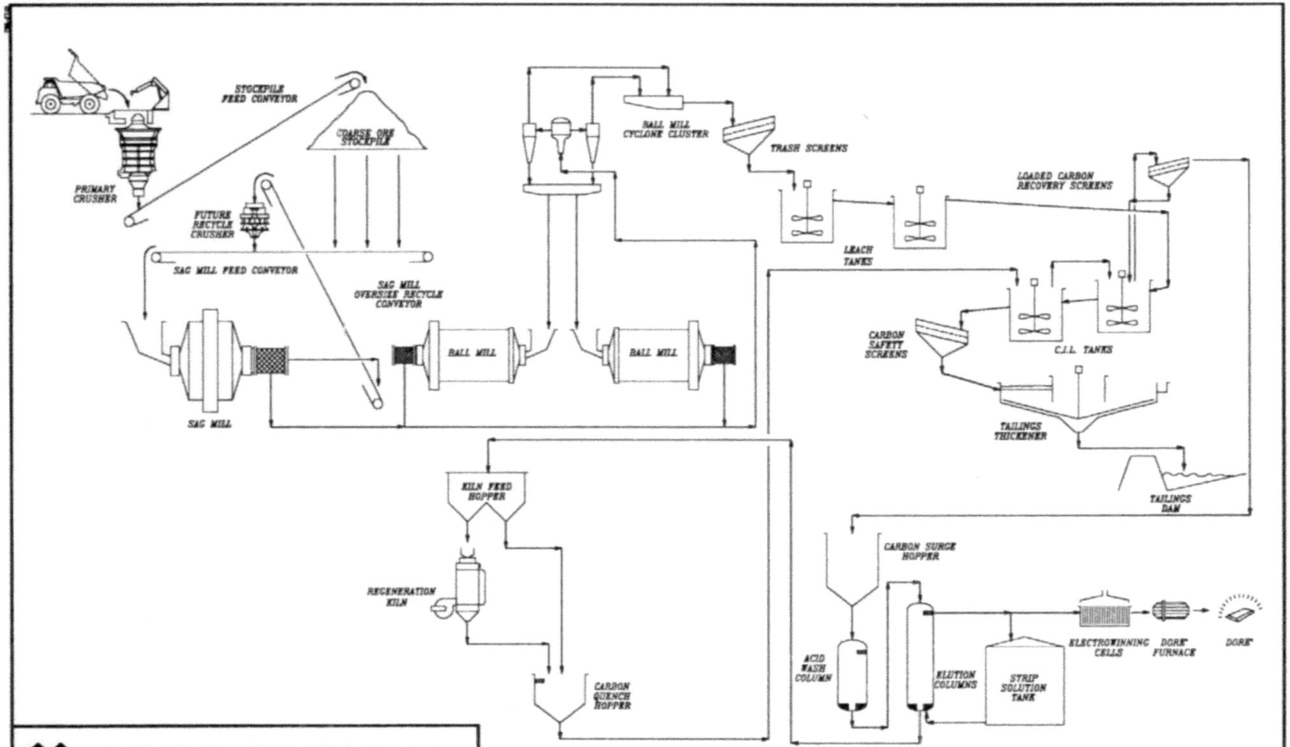

MINPROC
MINPROC ENGINEERS INC.
SUITE 8300 5800 SOUTH QUEBEC ST., ENGLEWOOD, COLORADO 80111
TELEPHONE: (303) 721-8111
FAX: (303) 721-8008

FIG.1

Table 1. Design Requirements

General		
Plant altitude		3,715 m ASL
Circuit throughput		641 t/h
Plant availability		91 %
Feed grade,	Au	2.3 g/t
	Ag	20.3 g/t
Recovery,	Au	70 %
	Ag	65 %
Leach and Adsorption		
Nominal retention time		32 hours
Carbon		
Design loaded carbon grades		
Au		802 g/t
Ag		6,352 g/t
Desorption		
Elution type		Pressure Zadra
Metal recovery		Electrowinning

Table 2. Desorption Design Criteria

Adsorption	
Metal to be loaded,	
Gold	24.8 kg/day
Silver	203 kg/day
Design loaded carbon grades	
Gold	802 g/t
Silver	6,352 g/t
Design barren carbon grades	
Gold	50 g/t
Silver	200 g/t
Carbon movement required	33 t/d
	1,375 kg/h
Desorption	
Carbon treated	231 t/week
Elution temperature	135 °C
Solution flow for elution	
Total flow	50 Bed Volume
Flow rate	2 Bed Volume/hour

The crushing and grinding circuits are reasonably conventional, as are the leaching and adsorption circuits. However, the individual equipment items are large by any standards, and very large slurry flows are handled throughout the circuit.

Figure 2 is a simplified overall site plan of the facility.

The silver-to-gold ratio in the ore is 8-10:1. The relatively large quantity of precious metals leached requires the movement of 33 tonnes of carbon per day at the design carbon loadings. The Zadra desorption process was selected, and BMGC experience at their Fortitude operation provided the basis for the desorption sequencing and the design of downstream operations.

The two, 21.5-tonne elution columns are believed to be the largest operating anywhere. The use of two units, each capable of treating two-thirds of the daily carbon stripping requirement, provides more operational flexibility than would a single 33-tonne column. This scheme allows the individual process steps in the elution sequence to be completed well within the available time, and permits the equipment sizes and solution flow rates to be kept to reasonable levels.

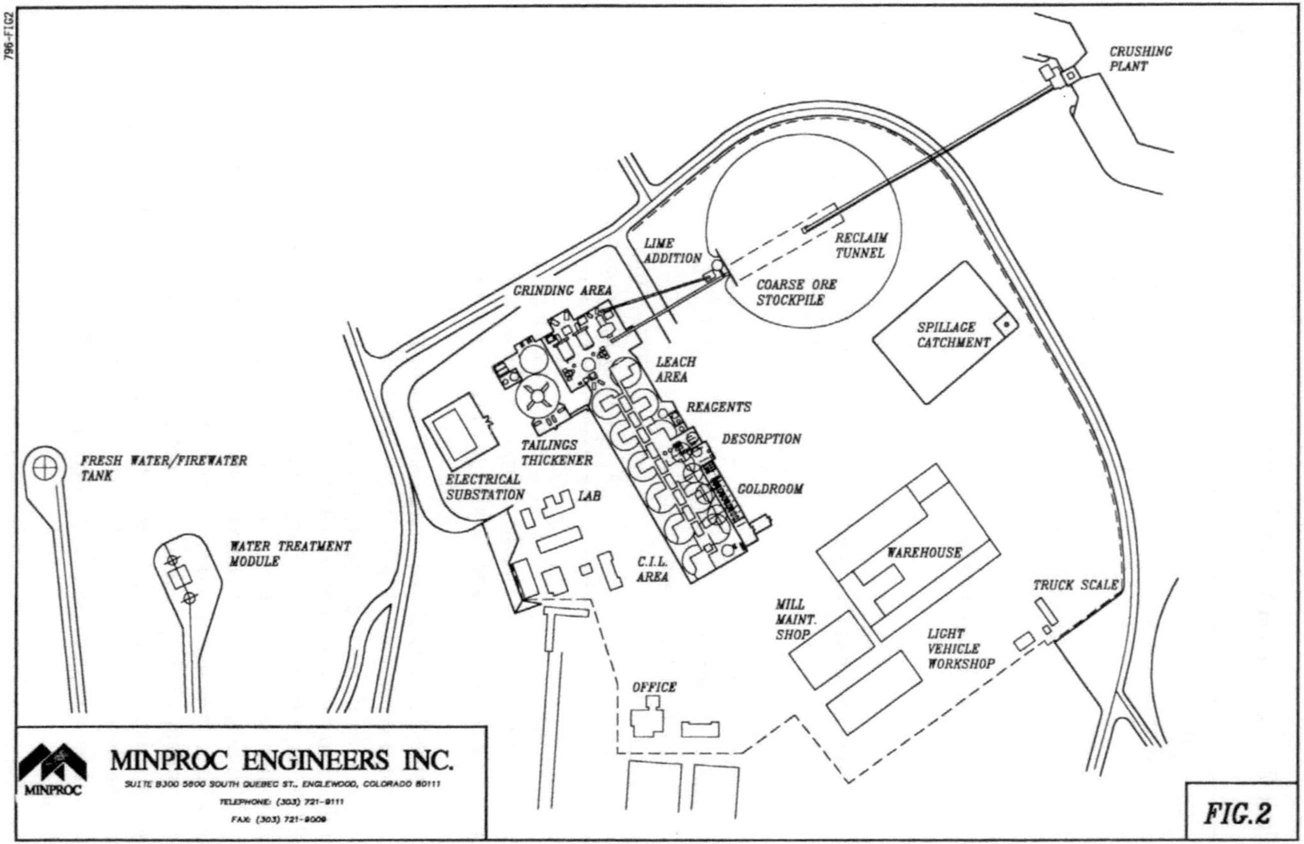
796-FIG2
CRUSHING PLANT
RECLAIM TUNNEL
LIME ADDITION
COARSE ORE STOCKPILE
GRINDING AREA
SPILLAGE CATCHMENT
LEACH AREA
REAGENTS
DESORPTION
TAILINGS THICKENER
FRESH WATER/FIREWATER TANK
ELECTRICAL SUBSTATION
LAB
GOLDROOM
WATER TREATMENT MODULE
WAREHOUSE
C.I.L. AREA
TRUCK SCALE
MILL MAINT. SHOP
LIGHT VEHICLE WORKSHOP
OFFICE
MINPROC
MINPROC ENGINEERS INC.
SUITE 8300 5800 SOUTH QUEBEC ST., ENGLEWOOD, COLORADO 80111
TELEPHONE: (303) 721-9111
FAX: (303) 721-9009

FIG.2

1.3 Project Schedule

Table 3 shows the significant project dates. The period from award to completion of the performance trial was 18½ months. Field mobilization to the introduction of ore to the grinding circuit took 14 months.

Table 3. Project Milestones

Contract award to Minproc	Jul. 17, 1991
Review of process concepts	Aug.-Sep. 1991
Minproc site mobilization	Oct. 25, 1991
First concrete poured	Jan. 13, 1992
Power connected to substation	Nov. 5, 1992
First ore through grinding circuit	Dec. 22, 1992
First gold/silver pour	Jan. 30, 1993
Performance trial completed	Feb. 14, 1993
Design daily throughput achieved	Mar. 1993
Design availability achieved	Apr. 1983
Contracted completion date	May 1, 1993

2 Flowsheet Description

2.1 Crushing and Grinding

Run-of-mine ore is delivered to the primary crusher in 50 and 75 tonne trucks, which dump directly into the 1.07 by 1.73 m (42 × 68″) gyratory crusher. The crushed ore discharges to a surge pocket and is conveyed by variable speed apron feeder and conveyor onto the coarse ore stockpile. Ore is withdrawn from the stockpile by combinations of three variable speed apron feeders discharging onto the SAG mill feed conveyor. The ore is ground in open circuit in a 8.3 m dia. by 3.8 m long, 7500 hp SAG mill, and two 5 m dia. by 9 m long, 5500 hp ball mills in closed circuit with twenty-three, 0.38 m dia. cyclones. The SAG mill product discharges over a trommel screen, with the plus 10-12 mm screen oversize recycled to the mill via the main feed conveyor.

The screen undersize is pumped to the ball mill discharge pump box, and the combined mill discharges form the feed to the cyclones. The cyclones are arrayed radially around a single distributor, with nine underflows directed via sloped feed launders to each of the two ball mills. The underflow from the remaining five cyclones can be directed to either of the ball mills, or recycled to the SAG mill feed. The grinding circuit product, the cyclone overflow, is distributed to three 1.52 m wide by 2.44 m long vibrating trash screens. Screen undersize gravitates via a cross-stream sampler into a sump, from where it is pumped to the first of the leach tanks.

2.2 Leach and Adsorption

The six leach and six carbon-in-leach (CIL) tanks are 15.6 m dia. by 16.2 m high, and are stepped in the classical staggered arrangement to enable bypassing any tank. The slurry is suspended by a 150 hp, dual propeller agitator in each tank . Each CIL tank is equipped with three 1.1 m dia. by 1.7 m high North Kalgoorlie Mines style intertank screens, the third unit being an operating spare to facilitate screen cleaning. Leached slurry flows from the last CIL tank to a distributor and over three 1.52 m wide by 2.44 m long vibrating carbon safety screens. The screen undersize gravitates via a cross-stream sampler to the thickener feed box, where plant water is added.

2.3 Tailings Pumping and Disposal

The slurry is further diluted in the feed well of a 22.9 m dia. high rate thickener. The overflow gravitates to the process water tank, from where it is distributed throughout the process. The solids are discharged from the cone at 50% by weight.

A two-stage pumping system transports the tailings slurry 2.5 km to the disposal area. The circular, 2.5 km diameter impoundment is enclosed within a 3 m high berm, constructed of local materials. The tailings slurry is carried around the perimeter embankment in two 3.8 km lines, each serving half of the impoundment. Each line is fitted with 336 valved take-off spigots, of which 14-16 are operating at any time. Deposition is advanced around the impoundment, with time allowed for evaporation of water and consolidation of the tailings before the next layer is deposited. There is no provision for recycle of water to the plant. The tailings impoundment and distribution system were designed by Knight Piésold and Co., of Denver.

2.4 Carbon Handling

Carbon is advanced through the six tanks in the CIL train by submersible pumps, and is washed over two 1.22 m wide by 3.04 m long recovery screens. The cleaned carbon gravitates into a surge hopper, sized to contain a 30 tonne batch. When ready to advance, the batch gravitates into the acid wash column, a rubber-lined vessel 2.74 m dia. by 10.16 m high. After acid washing, the carbon is pressure transferred to one of the two, 1.98 m dia. by 18.37 m high elution columns. A single tank provides strip solution for the two elution circuits, which can be operated either independently or in parallel. Each system consists of a primary heat exchanger, a heat exchanger for recovery of heat from the strip solution leaving the column, the column, eluate filters, and four 8.5 m^3 electrowinning cells. Barren solution from both circuits is collected in a tank, and returned to the strip solution tank for recycling through the process. Heating for each of the systems is provided by a dedicated 150 hp steam boiler, with independent condensate recovery systems.

Stripped carbon is pressure transferred from the columns, and dewatered on a 1.22 m wide by 2.44 m long screen. Approximately one third of the carbon gravitates to a vertical regeneration kiln, with the remainder by-passing into the kiln quench hopper. Carbon is returned to a two-deck 1.22 m long by 2.44 m wide sizing screen at the end of the CIL circuit by a combination eductor and pump.

2.5 Reagents and Services

Facilities are provided for the handling, mixing and distribution of reagents.

Separate compressor systems provide air for leach aeration, instrument service and general plant requirements. Water drawn from the nearby Rio Desaguadero is treated by coagulation and sand filtration for general plant use. Water for use in the elution system and for reagent make-up is further treated in an ion exchange system.

2.6 Operations Control

All plant sections are controlled via a state-of-the-art Programmable Logic Control system, and can be run in automatic mode from the control room, or in manual mode, local to each piece of equipment.

The main control room is located above the grinding area and the leach tanks, with panoramic views of the important wet plant areas. An overview of plant operations is provided on a series of screens displayed on color graphic monitors. The control room operator can start or stop equipment, or change process variables, from the pertinent control screen.

The crushing plant is provided with a separate control room which overlooks the dump pocket. Crushing operations are also monitored in the main control room.

Pre-commissioning of the plant control software was undertaken using a simulation package. This provided inputs into the PLC to simulate plant operations, providing confirmation of correct control response. The important control loops were tested under a range of operating conditions, allowing optimization of instrument parameters before wet commissioning. The simulations were also used for operator training, allowing the control room users to become familiar with the screens and control functions before start-up on ore.

3 Construction

3.1 Contractors

The facilities were largely constructed with Bolivian based contractors, with overall supervision provided by field engineers from Minproc's home office in Denver. Other subcontractor supervisors came from Chile for the mill installations, and Australia for the electrical work. The civil contractor was the first mobilized to site in November, 1991. Initial delays with his materials' supply and completion of the batch plant were overcome with active support by Minproc home office and local employees.

This additional supervisory effort assisted throughout all phases of construction, and ensured the problems that arose were quickly solved. Local shortages of construction materials were supplemented by supplies from other regions, and minor problems with the erection of some of the vendor supplied equipment were quickly rectified on site or at local machine shops. Four major sub-contractors completed construction, and approximately 1.52 million manhours were expended to complete the project. Less than 10% were for Minproc personnel.

3.2 Safety

A concerted effort to make the issue of safety a prime priority of the contractors was initiated form the onset of the project. This was imposed at all levels by each contractor in the form of weekly meetings and awards (both individual and site wide) presentations, and in general this proved to be extremely successful. The project was completed with an outstanding safety record throughout, with the most serious accident being a broken leg. This is considered a major achievement in the light of the relative inexperience of many of the contractors' personnel.

3.3 Logistics

Most of the mechanical equipment was sourced from North America, with the structural steel and platework fabricated in Chile. Equipment and materials were imported through Arica and Antofagasta, Chile, and transported by road and railroad to site. The planning and implementation of all the logistical aspects proved to be crucial in achieving early project completion. The first shipments were transported by road from Arica, but by June 1992, the unpaved portion of the road had deteriorated to the extent that unnecessary damage was occurring to some of the cargo. After this time shipments were imported to Antofagasta, transported by rail to a marshalling yard set up in Oruro, from where they were moved to site by road.

4 Conclusions

The successful completion of the Kori Kollo sulfide project, which is Bolivia's first modern gold producing operation, represents a significant achievement by all parties concerned. Project highlights include:

The owner's acceptance of a firm price contract for completion of the project facilities.
Successful integration of the owner's technical representatives in the decision making process of the firm price approach, without growth in costs.
Use of modern, cost effective technology throughout the plant.
Introducing Bolivian subcontractors to more efficient methods than those traditionally used without compromising the project schedule, allowing a very high local contribution to the overall construction effort (87% of the total hours expended).
Rapid achievement of design throughput and gold production well ahead of the contracted end date.

Performance to date has proved that the operation has achieved the design parameters and metal production, at low production costs. BMGC pursued an aggressive approach in choosing a firm price contract instead of the more usual cost reimbursable contract for construction of the plant facilities.

This strategy minimized the risk to the owner as Minproc took total responsibility for contract completion, and resulted in handover to plant operations 2½ months earlier than had been contracted.

5 Acknowledgements

The authors thank their respective Managements for permission to publish this paper. The contributions of BMGC technical people throughout the design and commissioning process; Minproc's project management and designers, and field engineers throughout the project; and the Inti Raymi operators are all gratefully acknowledged.

La Coipa modified grinding circuit, Chile

Julio Binvignat T.
Jorge Omori V.
Compañía Minera Mantos de Oro, Copiapó, Chile

Abstract

Grinding circuit modifications to increase throughput normally involve high capital costs and relatively long construction periods, which in some cases can interfere significantly with production targets. The proposed modification of the La Coipa grinding circuit will result in an increase in mill throughput of 10% with a relatively low investment cost (approximately US$1 million), will not affect the normal operation of the plant during its installation and will keep the actual flowsheet as an alternative. The modification involves prescreening of the SAG mill feed to maximize power utilization of the pebble crusher in the SABC circuit. Simulation studies and in-plant testing have confirmed that during periods of "hard ore", significant increases in throughput and mill efficiency can be obtained.

Resumen

La modificación de un circuito de molienda para aumentar el procesamiento de una planta en operación generalmente involucra altos costos y períodos de realización largos. Se estima que el circuito propuesto para la planta La Coipa permitirá aumentar el procesamiento horario en un 10% con un costo de inversión relativamente bajo (US$1 millón) sin afectar la operación normal de la planta, permitiendo mantener el flowsheet actual como alternativa. El tiempo para completar el proyecto se ha estimado en 6 meses.

Introducción

Compañia Minera Mantos de Oro, subsidiaria de Placer Dome y TVX Gold, tiene su faena La Coipa ubicada en la III región, Chile, 140 km al noreste de la ciudad de Copiapó y aproximademente a 4,000 m de altura sobre el nivel del mar, la cual entró en operaciones en julio de 1991 con una instalación para procesar 15,000 tpd de los yacimientos Ladera, Farellón y Coipa Norte. El proceso consiste en chancado, molienda SAG, lixiviación agitada, Merrill-Crowe fusión, y depositación de relaves compactados. El producto final es barra de metal doré, obteniendose como sub-producto mercurio.

La capacidad de procesamiento se ha visto afectada por la presencia de minerales volcánicos y sedimentarios. Los primeros presentan volúmenes significativos de mineral con baja moliendabilidad autógena por su baja densidad y alta dureza, por lo que su aporte como medio de molienda es nulo y permite la formación de tamaños críticos, los que se acumulan en el interior del molino produciendo problemas de transporte. Por otro lado el material sedimentario, con altos contenidos de arcilla, genera un alto porcentaje de producto fino que afecta las tasa de filtrado de la pulpa de relaves. Estas dos situaciones no han permitido alcanzar consistemente los parámetros de diseño de la planta. Para alcanzar y exceder el criterio de diseño en la molienda, se ha evaluado una modificación al circuito actual.

Circuito Actual Planta de Molienda

El criterio de diseño para la planta actual de molienda de La Coipa estimó un procesamiento de 695 tph. El circuito actual considera un molino SAG de 28' x 14' con 5,200 kW y dos circuitos inversos de molienda clasificación con molinos de bolas de 16' x 24.5' con 3,150 kW cada uno. La clasificación se realiza con dos baterías de ciclones D26 con seis unidades cada una. Completa el circuito un chancador Omnicone 1560 y 185 kW. Esta unidad está sub-utilizada, con un tonelaje actual no superior a 45 tph y una demanda de 50 kW, teniendo una capacidad de 280 tph con setting de 19 mm y 163 tph con setting de 10 mm.

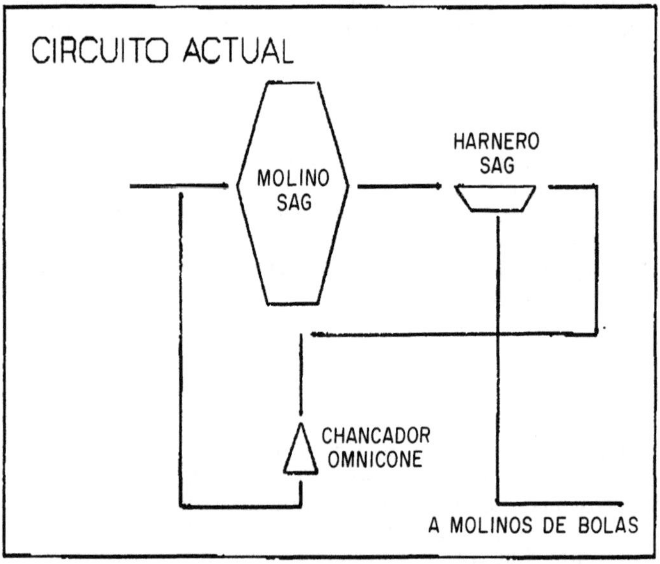

Figura 1: Flowsheet Circuito Actual

Circuito Modificado Propuesto

El diseño actual de La Coipa (SABC) es usado comúnmente para evitar la acumulación de tamaños críticos en el molino SAG, que pueden disminuir el procesamiento horario. El circuito actual de La Coipa presentas problemas de transporte dentro del molino SAG y un aumento en el área de evacuación de las parrillas no ha mejorado esta situación. Para aumentar la capacidad de molienda se ha considerado modificar la granulometría de alimentación al SAG, chancado la fracción que contiene tamaños críticos, para lo cual la alimentación consistirá esencialmente en dos productos: producto -175 mm + 75 mm y producto -19 mm. Para alcanzar esta distribución de alimentación, se considera usar la capacidad ociosa del chancador Omnicone, traspasando el material de tamaño critico de baja moliendabilidad desde el molino SAG al chancador.

El circuito propuesto consiste en la instalación de un harnero con doble deck de 8' x 16 ' antes del molino SAG. El oversize del primer deck (75 mm) alimentará directamente al molino SAG. El oversize del segundo deck (-75 mm +13 mm) alimentará al chancador de cono. Este producto se enviará al molino SAG, al igual que el undersize del segundo deck, según se ilustra en la figura 2.

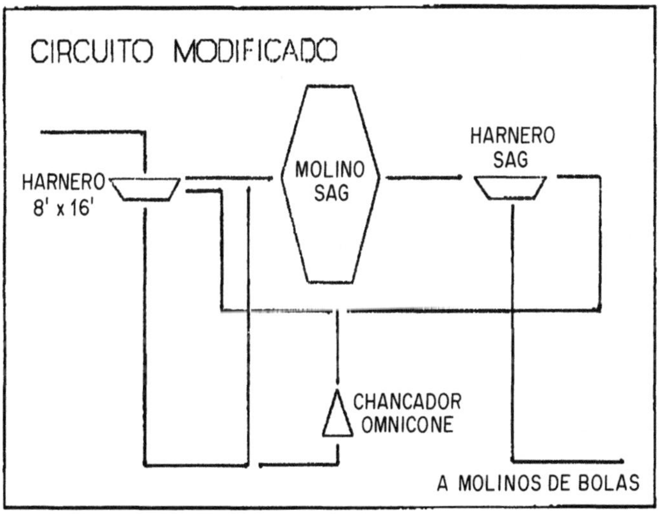

Figura 2: Flowsheet Circuito Modificado

Evaluación del Proyecto

Para evaluar el circuito propuesto se decidió seguir dos rutas paralelas:

- La instalación de una correa auxiliar que alimenta solamente pebbles al chancador Omnicone para evaluar el comportamiento de este circuito a nivel industrial.

- La contratación de una empresa consultora para simular el circuito actual y el propuesto con el fin de determinar el beneficio adicional y las posibles restricciones que pudieran presentar los equipos existentes.

- Esta estrategia ha podido estimar los tonelajes a procesar por los dos circuitos para un mineral relativamente duro. La simulación indica que es posible aumentar el tonelaje horario en un 12% con un consumo máximo de potencia de los molinos de bolas (Anexo I).

- Finalmente, con este resultado favorable se ha contratado los servicios de una empresa de ingeniería nacional para evaluar el costo de la inversión requerida (Anexo II).

Conclusiones

El circuito propuesto traerá los siguientes beneficios:

- Permitirá alcanzar consistemente tonelajes que exceden los valores de diseño al eliminar los problemas de transporte al interior del molino SAG, por reducción de los tamaños críticos presentes en la alimentación fresca al molino.

- Favorecerá una mayor tasa de filtración de la pulpa de relaves al aumentar la granulometría del producto final de molienda, mediante la disminución del proceso de molienda por atrición.

- Disminuirá el consumo de medios de molienda al favorecer la molienda autógena y minimizar el proceso de abrasión.

- Reducción del consumo de reactivos aglomerantes (floculantes y ayuda filtrante) en las etapas de espesamiento y filtrado al favorecer el engrosamiento de la granulometría final.

- Disminución del costo fijo (10%).

ANEXO 1

SIMULACION DE LOS CIRCUITOS DE MOLIENDA

TABLA 1: Simulacion de Circuito Actual

Condición: Mineral relativamente duro	W_i = 22.8 kWh/t
Carga de Alimentación Fresca	F_{80} = 190,300 um
Alimentación Molinos de Bolas	F_{80} = 1,850 um
Producto Molinos de Bolas	P_{80} = 230 um

ALIMENTACION FRESCA 600 tph	Tonelaje tph	Energía Específica kWh/t	Energía Consumida kW
Alimentación SAG	600	8.00	4,800
Alimentación Omnicone	36	1.17	42
Alimentación Molino de Bolas	600	10.50	6,300
TOTAL	600	18.57	11,142

ALIMENTACION FRESCA 650 tph			
Alimentación SAG	650	8.00	5,200
Alimentación Omnicone	39	1.17	46
Alimentación Molino de Bolas	650	9.69	6,300
TOTAL	650	17.76	11,546

TABLA 2: Simulacion Circuito con Pre-Harneado

Condición: Mineral relativamente duro	W_i = 22.8 kWh/t
Carga de Alimentación Fresca	F_{80} = 190,300 um
Alimentación Molinos de Bolas	F_{80} = 3,118 um
Producto Molinos de Bolas	P_{80} = 230 um

ALIMENTACION FRESCA 650 tph	Tonelaje tph	Energía Específica kWh/t	Energía Consumida kW
Alimentación SAG	650	6.93	4,505
Alimentación Omnicone	280	0.36	101
Alimentación Molino de Bolas	650	9.69	6,300
TOTAL	650	16.78	10,905
ALIMENTACION FRESCA 700 tph			
Alimentación SAG	700	6.93	4,851
Alimentación Omnicone	280	0.36	101
Alimentación Molino de Bolas	700	9.00	6,300
TOTAL	700	16.07	11,252
ALIMENTACION FRESCA 730 tph			
Alimentación SAG	730	6.93	5,059
Alimentación Omnicone	280	0.36	101
Alimentación Molino de Bolas	730	8.64	6,307
TOTAL	730	15.71	11,467

ANEXO II

Estimación de Costos de Inversión

La estimación de Costos de Inversión de Capital considera todas las nuevas obras e instalaciones y las modificaciones a las existentes, y se resumen como sigue:

1.	Costos Directos	'000 US$
	- Suministro de Equipos	211.91
	- Suministro de Materiales	291.60
	- Contratos de Construcción y Montaje	121.38
Total	Costo Directo	624.89
2.	Gastos Generales y Utilidades	48.19
3.	SUB TOTAL SIN IMPREVISTOS (1 + 2)	673.08
4.	Imprevistos	89.99
5.	Inspección Técnica de Obra	39.00
6.	COSTO DE INSTALACIONES (3 + 4 + 5)	802.07
7.	Ingeniería del Proyecto	80.21
8.	Administración La Coipa	48.12
	TOTAL INVERSION (6 + 7 + 8)	930.40

Tiempo de implementación del proyecto.

El plazo de ejecución del proyecto se ha estimado en seis meses sin interrupción de circuito actual.

Optimization of control systems for mineral processing

Carlos Leal
ABB Industrial Systems AB, Västerås, Sweden

Introduction

Data based control systems have today increased in all kinds of industry. The mineral processing industry has followed this development and various control systems have been installed in existent and new plants.

The goal is to improve product quality, raise productivity and lower production costs. To reach these goals improvement of the different units of mineral processing have been applied. This leads to an increasing of information for the operators. To help the operators and for a better overview of the process a more total concept of process control system is normally required.

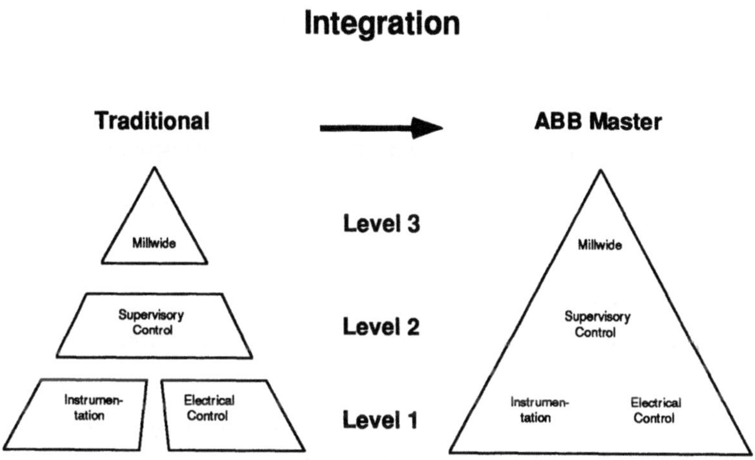

Figure 1

The triangle above shows the different levels/categories of automation in a plant. All these levels/categories correspond in normal cases to different departments at the plant. All these departments have in a traditional automation system worked with different systems supplied by different vendors. In a non-integrated system all common data has to be transferred through gateways.

The ABB Master in Mineral Processing concept covers all levels/categories in the same system:

Level 1 - Integrated instrumentation and electrical control functions in one Distributed Control System (DCS), ABB Master. The electrical and the instrumentation departments can share and co-operate in the same system.

Level 2 - The computer/process control departments deal daily with supervisory control functions from crushing to dewatering measurements and controls, all can be implemented in ABB Master. In many cases the systems have been installed as packaged black box systems combining mini-computers, software, sensors and production guaranties. Model based control strategies can also be implemented on this level.

Level 3 - Plant wide production management and co-ordination system for mineral processing plants implemented in ABB Master. This is a big advantage for the computer/process control departments to be able to share all data in the system and no special hardware is needed.

With an integrated system the operator is working in a uniform environment where all objects like analog measurements, PID-loops, sensors, valve controllers, motors, measurement frames etc., are treated in the same way.

OPTIMIZATION STRATEGIES FOR MINERAL PROCESSING

In the following section ABB's control strategies are presented for the different process areas.

CRUSHER CONTROL

The main advantages with automatic control of crushers are minimized downtime and maximized throughput.

No control of the crushers is usually performed in processes where autogenous and/or semi autogenous grinding mills are used. Only supervision and monitoring of the process is applied. In the case where other types of grinding mills are used, the particle size must be lower and more accurate. In these cases automatic control will give many advantages such as minimized downtime.

Hydrocone Crusher Control

Here the closed-side-setting control, CSS, is applied. CSS means the smallest distance between mantle and cone. The parameters measured and controlled are power and hydraulic pressure. The philosophy is to minimize the CSS. This is performed by following the calibration curve for each ore type. The CSS varies according to the operating power and the hydraulic pressure.

Microwave sensors are used for point-level detection in the crusher. The CSS measurement is accurate to 0.05% of full scale.

The crusher control has resulted in increased crusher capacity, decreased antel wear and decreased downtime. CSS has been reduced from 25 mm to 2-9 mm thus improving the next step, rod mill capacity, by 8-10%.

CONCENTRATOR CONTROL

The most common process steps in a concentrator are as follows:

- Grinding
- Flotation
- Dewatering

Some of the main control objectives are as follows:

- Maximizing recovery while the grade is kept above a minimum level
- Maximizing the grade
- Maximizing the throughput
- Minimizing reagent costs

The control may be divided in three different stages:

- Stabilizing control
- Supervisory control
- Optimization

The stabilizing control is performed to achieve stable conditions in the process. The supervisory control is defined as adjustments of the setpoints for the stabilizing control, and finally the optimization is performed to calculate setpoints for the supervisory control.

The control of a concentrator can be illustrated as in Figure 2.

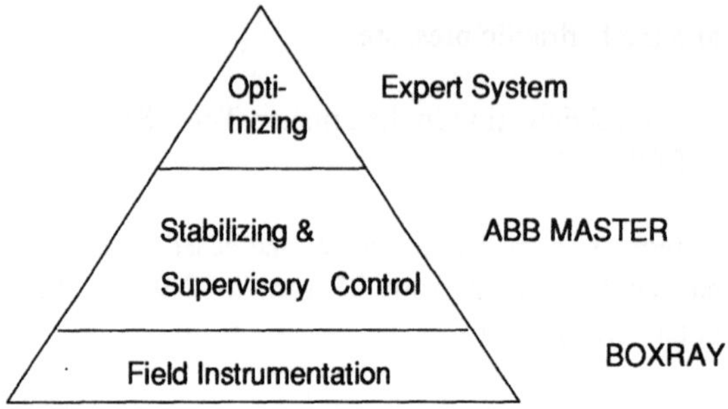

Figure 2: Concentrator control and attached products

GRINDING CIRCUIT CONTROL

Automatic control of grinding circuits is usually the first step to total plant automation. The details for the control are always specified by the customer.
In the following will be described some features of grinding control:

Stabilizing Grinding Control

To achieve stable conditions in the grinding circuit, the most common variables to control and manipulate in the stabilizing ABB/Boliden grinding control strategy are as follows:

Controlled Variable	**Manipulated Varible**
- Mill Feed Rate	- Feeder Speed
- Mill Power Draw	- Grinding Media Addition
- Mill Pulp Density	- Water Flow Rate Ratio
- Cyclone Sump Level	- Cyclone Pump Speed
- Cyclone Pressure/Density	- Sump Pump Water Flow Rate

Mill feed rate control; The ore feed setpoint is manually set. The feed rate is monitored by a weightometer and a controller generates a signal to the feeder.

Power control; A set point for the mill power controls the mill feed. The set point is choosen by the operator in order not to overfill the mill. This control is common for pebble (or ball) mills when automatic pebble (ball) supply is essential.

Mill pulp density; The water flowrate is kept at a constant preset ratio to the ore feed in order to provide a constant slurry density in the mill. This control is often combined with ore feed rate control.

Cyclone sump level control; The level in the sump box is kept at a constant level to assure as constant conditions of the flow to the cyclone as possible.

Cyclone pulp density control: Water addition to the pump sump in order to adjust pulp density to the cyclone.

Cyclone pressure control; The cyclone pressure is controlled by a variable speed pump in order to keep the pressure at its optimum where the cyclone efficiency is best.

Studies show that stabilizing control with PLCs and DCS systems will give up to 15% higher production in the grinding circuit.

Supervisory Grinding Control

For supervisory grinding control there are three different strategies, or control objectives:
* Maximum throughput
* Maximum throughput and constant size of grind
* Constant throughput and constant size of grind

The particle size is measured by a particle size analyzer or estimated by experience.

Our strategy for maximum throughput, comprises the following functions:

Mill power control finds the power level which gives the optimum throughput at different grindabilities. This control is combined with peak power supervision which detects the mill peak power in order to prevent mill overfill. Also included is a feed interruption compensation, which compensates for shorter ore feed interruptions in order to avoid mill power drop.

To achieve constant size of grind ABB the following method is used:

Ore feed is set to a given setpoint. Flow fluctuations depending on ore feed are avoided giving stable conditions for the flotation. This control improves the process metallurgy and makes the flotation control easier.

At constant grinding energy (kWh/ton ore) the size of the grind is relatively constant for a certain ore type. Metallurgical tests determine the optimum size of grind. Since every ore type has its own characteristics the grinding energy needed can vary significantly.

Density meters based on radio-isotopes are very expensive and could also be dangerous to handle. An alternative way, which we have used for some installations, is to utilize two pressure probes mounted in the pump box with a fixed distance. The pressure difference indicates the density and the same signals are used for level calculation.

Both signals from the sensors are connected to the control system where the density and level is calculated. The calibration is easy to perform.

A 4-20mA signal is connected to the control system. The operator sets the set point in a PID controller and the output is connected to a frequency converter. Higher level gives higher speed.

A supervisory/optimizing control on top, in the DCS and in a level 3 computer, will give another 5-7% increase of the production.

Field Instrumentation

Extensive instrumentaion is required to achieve proper grinding control. Minimum variables to be measured are as follows:

Ore throughput, power draw, water flowrate, bearing temperature, hydrostatic bearing pressure (primary autogenous mills), motor torque (mills with variable speed), mill speed (mills with variable speed), charge volume (primary autogenous mills), pulp density, pulp flow of discharge, circulating load (pulp flowrate to cyclone) and cyclone pressure.

Advantages and benefits

The major process benefits by the ABB's automatic control of grinding circuits are increased capacity, availability and better metallurgical quality.

The capacity is kept at an optimum level through the power load control, where the power is kept at a level where maximum throughput is achieved. Power peak supervision will minimize the risk for mill overfill causing production losses and the compensation for feed interruptions will also avoid productions losses and mill power drops for AG mills.

Radio control improves grinding efficiency by keeping the pulp density in the mill at its optimum.

The availability is increased through power optimization and power peak supervision, which eliminates the risk for mill overload.

The metallurgical properties are improved through constant feed, eliminating pulp flow fluctuations and by control to get optimal size of grind and cyclone efficiency.

FLOTATION CONTROL

The flotation process is the most important process step to control the grades, the recovery and the reagent consumption. The control can be divided into stabilizing, supervisory and optimizing control.

Definitions of some variables:

G=Grade; Concentration on valuable mineral.
R= Recovery; The amount (in %) of the valuable mineral, which is recovered.

Stabilizing Flotation Control

The main control features for stabilizing flotation control are as follows:

Automatic control of:

- pulp level in the flotation cells
- air flow to the flotation cells
- the pH to preset set point

Pulp Level

The pulp level (or froth depth) in the flotation cell is an essential flotation control variable. Pulp level variation depends on pulp flow fluctuations or blower air flow. The pulp flow variations can depend on the grinding or recirculation in the process. If the pulp level is kept constant the conditions for a steady concentrate grade are improved drastically.

The froth depth can be used to control concentrate or tailings grade from a particular bank. Increase of the froth depth produces a higher concentrate grade, but at the expense of a loss of recovery.

The pulp level in the flotation cell is measured by a level sensor (ultrasonic, floating device, bubble tube etc.) which is connected to a PI-element.

The control system gives an output signal via an I/P-converter to a pneumatic pinch or dart valve. Usually the control system is a part of an operator terminal close to the floation cell where the operator can see how different setpoints affect the flotation.

Air Flow

Blower air is a basic variable for flotation. The control system keeps the air flow constant at a given setpoint. Blower airflow variations or counter pressure variations in the flotation cell are compensated.

Decrease in air flowrate increases concentrate grade and decreases recovery of the valuable minerals, because the air affects the flowrate of water to the concentrate and this results in a proportionally greater reduction of the flowrate of gangue to the concentrate than that of the valuable mineral.

The flotation process responds much faster to changes in aeration rate than to those in froth depth (pulp level), and aeration rate is often more effective than froth depth in maintaining stable circuit behaviour.

It is important to install the flowmeter and control valve according to the recommendations. The airflow meter could be of venturi or fluidistor type.

A 4-20 mA signal is connected to the control system where the operator can change the setpoint. The control valve is usually a butterfly valve.

pH

The pH is one of the most difficult variables to control and requires some special arrangements: the dosage equipment especially requires correct design. Our long experience has proved that an on/off valve which flushes dirt and particles from the pipes is preferred. Normal control valves will be plugged in in just a few hours.

The pH or in some cases conductivity is normally measured in the flotation cell inlet. For maintenance and calibration the transmitter is placed close to the measuring point. It is preferable to have as short a cable as possible between the pH probe and the transmitter to avoid disturbance.

A 4-20mA signal is connected to the control system. The operator sets the set point and the output is converted in the process control program to a pulsed output.

A pneumatic cylinder squeezes a rubber tube in closed position or opens it fully. Also other types of full area valves can be used.

Supervisory Flotation Control

The supervisory flotation control can be divided in two levels, feed forward and feed back control.

Feed forward flotation control	Feed back flotation control
1. Determine your current status	1. Determine the result of the actions
2. Predict your needs	2. If not satisfactory, determine the cause of the problem
3. Take action	3. Take action to correct it

Feed forward flotation control

Feed forward flotation control can be done in the DCS. Included in the control is reagent dosage rate control.

The reagent dosage rate control is based on the ore feed rate and the assays from the on-stream analyzer.

A description of the control is illustrated.

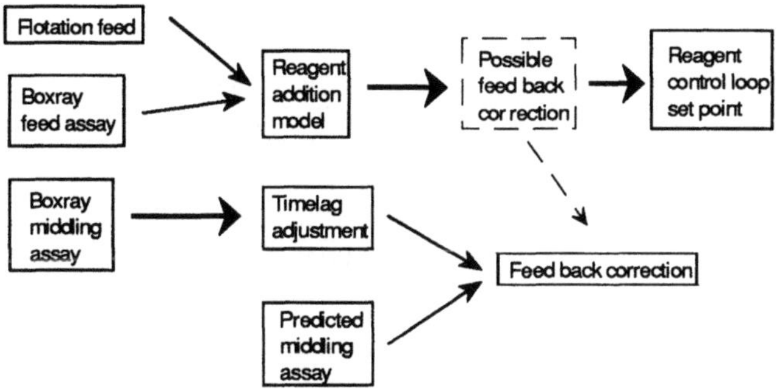

Figure 3: Supervisory flotation control

The figure also explains the correlation between feed forward and feed back control.

The reagent addition model is a standardized package, which can be customized for each project.
The reagents can be fed by gravity flow giving a nonpulsating flow which is feasible for the most accurate flowmeters.

The reagent flow is measured by a magnetic inductive flowmeter.

The reagent flow is controlled by an adjustable flow coefficient control valve specially designed for small flowrates. Usually the control system is a part of an operator terminal close to the floation cell where the operator can see how different setpoints affect the flotation.

Feed back flotation control

Feed back flotation control is more difficult to perform compared to feed forward flotation control.

The first feed back control was installed at one of Boliden's concentrators in 1975.

The feed forward flotation control is usually expert system based. The decision structure is illustrated in Figure 4.

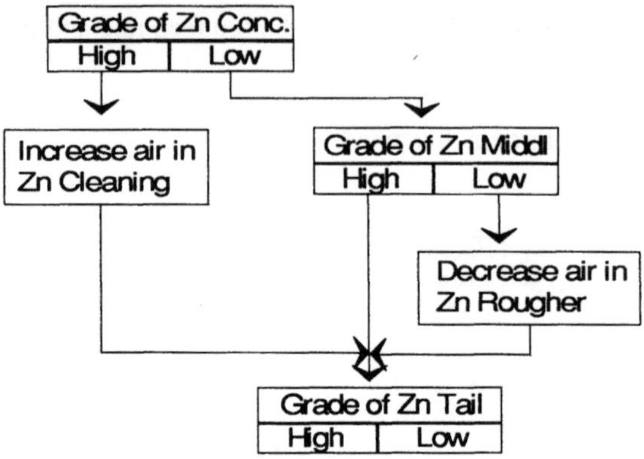

Figure 4: Decision structure for feed back flotation control

DEWATERING CONTROL

Normally thickening, filtering and drying are the processes included in the dewatering section.

Thickening Control

The solids level in the thickener is kept constant by a variable speed slurry pump. The pump speed is controlled by the solids level in the thickener. This control eliminates the risk for solids in the clarified water.

The machinery is protected by a torque control.
The level is measured by a level sensor which senses the counter pressure from the bottom of the thickener. This pressure is transmitted to a control signal for the slurry pump.

The power consumption is measured which is proportional to the torque load of the machinery.

When the solids level in the thickener increases the slurry pump speed increases.
When the machinery torque increases the scraper blades are lifted.

The level control minimizes the risk for overfilling of the thickener and solids running with the overflow water.

The torque control eliminates machinery damage and clogging of the rakes.

Filtering Control

Our control strategies for vacuum and pressure filters are presented in this section.

Vacuum filters; For vacuum filters, control of the vacuum pressure and revolution speed are the main features. We also monitor pumps and the hydraulic system.

Pressure filters; Pressure filters require an extensive control to minimize maintenance and optimize production.

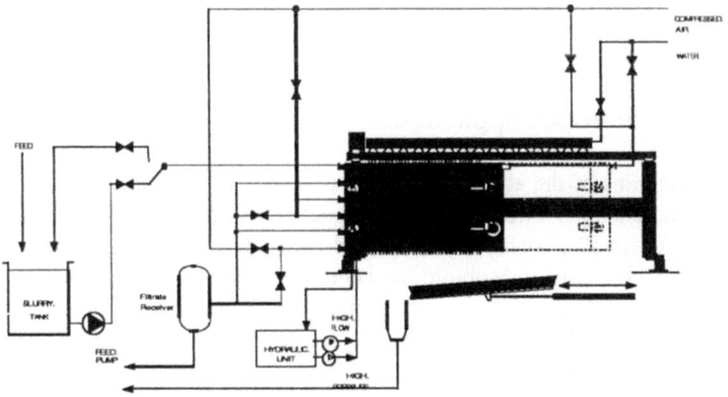

Figure 5: Pressure filter, type Sala.

The concept for pressure filters includes:

- Automatic calculation of pause sequence time
- Automatic filling of slurry buffer tank
- Automatic calculation of slurry pumping time
- Automatic calculation of air dewatering time

Installations where our control concepts have been used show much higher capacity of the filter. The cycle time decreases because the air is dewatering to a fixed moisture, the slurry pumping time is calculated which will lead to optimally filled chambers. The strategy also maximizes the density of the slurry i.e the weight of the cake per cycle will be higher. The weighing system is integrated in the control system.

The references also show that availability of the filter drastically increases. Extensive statistics like trends and correlations give important information to operators and maintenance people. The cycle time and the number of cycles are minimized, i.e wear is minimal and there will be time over for planned maintenance. The system tells when it is time to change filter cloths.

The air consumption is minimized because the cake moisture is controlled. The limitation of air flow during the end of the air blowing cycle will save more than 5% air.

The filter cake has a controlled and uniform moisture content.

Drying

Our dryer control includes monitoring of the exit gas. An alarm is sounded if the temperature exceeds or falls below a set temperature range. The signal from the temperature device is connected to the burner controls and the fuel rate to the burner is regulated automatically.

Some reference applications

Applications of the above mentioned strategies for process control have been made in some plants in Sweden: Aitik´s Copper Ore Concentrator, Boliden´s Concentrator plant, Laiswall´s concentrator of Lead-Zinc , in Canada: Val´Dor´s Precious Metal plant, Chicoutime´s Niobium concentrator. Specific application for flotation and filtration supervision in Spanien: Andalucia (Apirsa) concentrator of complex ores (copper, zinc, lead) in Portugal: Aljustrel (Pirites) concentrator for complex ores (copper, zinc, lead), in Saudi Arabia: Sukhaybarat´s Precious Metal Plant.

Deliveries of ABB Master System have been also made for supervision and monitoring in Chile: Lince´s copper recovering by leaching and SX/EW and recently to Louvicourt´s Aur Resources in Canada.

OPTIMIZATION AT LEVEL 2

The SuperView package is totally integrated in the ABB Master control system. This feature gives the opportunity to optimize the process. In the user data bases user defined files can be stored. These files can be accessed from the user displays, calculation module and from the historical database.

Recently a new tool based on a multivariable process overview in real time for predicting the behaviour of a plant has been installed. This new tool, IPI, Integrated Process Intelligence is capable of handling a large number of variables of the process, quantification of the relationships, creation of models from process data and predict quality and laboratory tests.

Modelling of process is based on measured data from the plant where a multidimensional space is obtained and projected in different direction. The technique is based in the so called multivariate data analysis.

Multivariate methods. Projection technique

Tables are a natural way to organize data. They summarize our observations and they are suitable for displaying, storing and retrieving data. Analysis of large tables may be made in a multidimensional space and then projecting down the points swarming in a plane. By this way it is possible to analyse all the variables at the same time. Multivariate analysis is a mathematical method by which it is possible to analyse, optimize and control a process when the variables change simultaneously.

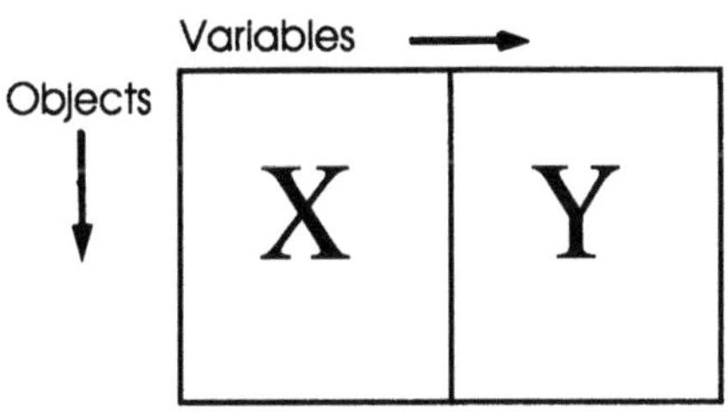

Object: Experiment, material, i.e. all measurements
X-variables: Variables parameters
Y-variable/responses: Measured parameter as a quality of the process

Figure 6: Organization of data in Multivariate Data Analysis

With the help of projection methods it is possible to make windows in the space to project down all the data points. With projection methods it is possible to make a model from reality and thus in a simple way to make conclusions on the observed data.

The more important projection methods for data analysis are Principal Component Analysis (PCA) and Projections to Latent Structures (PLS). PCA and PLS are based on fast and stable algorithms.

Principal component analysis, PCA

PCA is the base for multivariate data analysis. PCA gives a mathematical tool to interpretate the variance in a data table, X, which is normally the process data. PCA approximates a data matrix, Y, in terms of two small matrices T and P. These two new matrices contain the most important information on the pattern in the original matrix, Y.

Various principal components can be calculated based on the mathematical algorithms. The first principal component is the linear combination of all variables after scaling and centering that explains the majority of variance in data. The geometrical representation of the first principal component is a line in the K-space which best approximates the data. The line then goes through the mean value and between the points lying longest away from each other.

The second principal component is represented by a line in the K space which is orthogonal to the first principal component and goes then through the mean value.

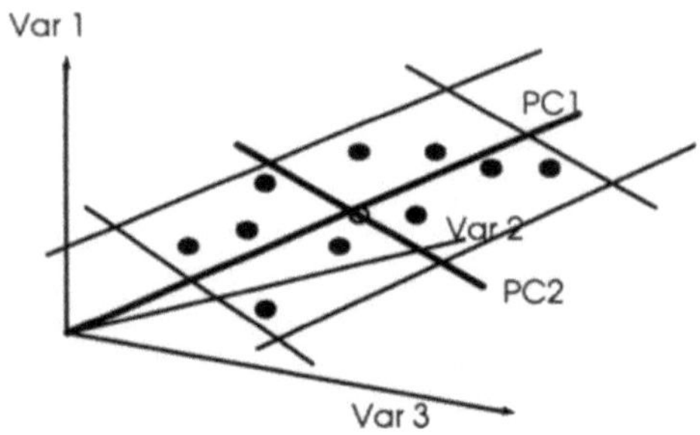

Figure 7: PC1 and PC2 in the K-space

From a PCA are obtained scores, which summarize all the variables entering the analyses. It is normal to calculate two or three scores, and then plot them against each other (tt-plots). This results in a picture that represents the best summary of the process behavior during time. We can see information like trends, unusual behavior etc, in this plot. One will recognize in this plot the area corresponding to normal process behavior and the area where the process goes out of control.

The PCA also gives results of values called loadings, where it is shown how the variables are combined to form the scores. In this plot of loadings the more important variables are pointed out as well as the direction of them.

The PCA also shows residuals, deviations between the data and the principal component model. When the residuals are significant an abornmal situation is detected.

The graphical presentation of PCA is shown in Figure 8 where the score and load are plotted.

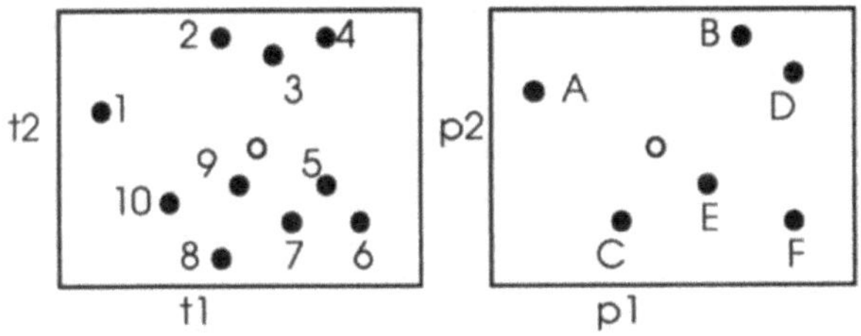

Figure 8: The score and load plot

PLS, the partial to latent squares regression

As mentioned above, the PLS is also a projection method based on mathematical algorithms. PLS is an extention of PCA, which is used as a tool for process diagnostics, showing which input and process variables, X, are responsible for the change of output variables, Y.PLS is used in situations where there are many colinear input and process variables and many result variables.

The objective of PLS-modelling is to make a model of X (data block) in a way that Y (block data for qualities properties) may be predicted.

Row vectors P and C are so called loading vectors and they express the contribution from each X and Y variable to the model component.

The PLS analysis also results in model coefficients for the variables, called PLS-weights. Then a PLS-weight vector, W_{ak}, is calculated for each model projection. The weights for the X-variables, W_{ak}, show how much they participate in the modelling of Y.

The coefficients PLS-weight can be plotted in a so called CW plot. These plots show the relationships between X and Y, which X-variables that are important, which Y-variables that are related to which X, etc.

The total number of PLS projections are formed by cross-validation which minimize the errors of prediction in the PLS-model.

Applications of PCA and PLS

A great number of applications in the chemical area have been made off-line by using the software developed by UMETRI AB, Umeå, Sweden.
Application of multivariate data analysis have been made in some mineral and metallurgical processes:

- Determination of optimal zinc content in a concentrate in relation to the proper amount of collector need during flotation has been applied in a swedish mine.
-The study of parameters like the coke quality, SiO_2 content in pellets, additions of briquettes, different iron ores, carbon powder, oxygen gas and oil addition which influence the reduction process in the blast furnace have been applied in a Swedish ironwork. Important conclusions have been deduced from these studies for the performance of the furnace.
- A multivariate pilot study to guarantee an uniform cathode quality has been made in a swedish copper work. Seventeen variables have been studied at the same time in order of further improve the cathode quality.

Application in ABB Master System

LKAB, the Swedish Iron Mine in Malmberget is using the ABB Master system for the ore handling, sorting plant, concentrator and pellets plant since 1987. In this ABB Master system there is also a VAX computer including a SuperView 900 management information system, which provides an overview of the processing operation for long-term storage of trend studies, operation reports and measurements. In the last two years the application to the sorting plant for modeling and testing by using multivariate data analysis has been performed. The PLS-modelling was made in the SuperView environment. At first modelling was performed using an IBM-PC containing the software SIMCA-P developed by UMETRI AB, and the resulted PLS-model was transferred back to the VAX-computer for on-line presentation and executing in SuperView 900 environment.

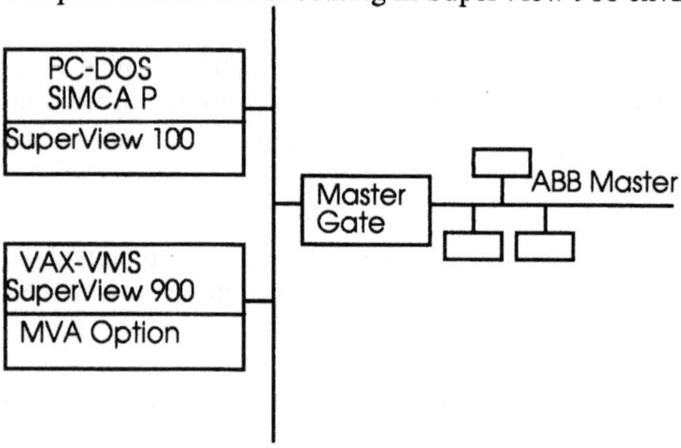

Figure 9:Configuration of ABB Master system for collection of data and modelling

The IPI system procedures as follows: first is collected a reference dataset, training set, for modelling from an historical data base in SuperView 900 (with the option of VT340 support). The training set is then transferred via the remote file system (RFS) within the SuperView 100 in the PC. Data are then converted to models with the SIMCA-P Software in the PC-environment. The final and completed PLS-model is sent then back to the SuperView 900.

The model can now be used for presentation and can execute on-line for historical as well as for new process data.

The sorting plant in Malmberget

From Malmberget underground mine the iron ore is transported via conveyors to the crushers of the sorting plant. After screening, further crushing and dry magnetic separation, two products are delivered to the beneficiation plant. One product consists of raw material for producing sinter fines and the other product is used to produce pellet feed. Waste material is also produced in this process. A description of the process is shown in figure 10.

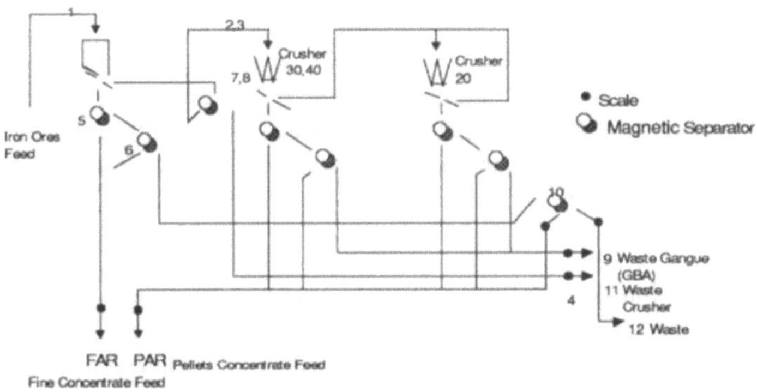

Figure 10

The multivariate study for the sorting plant

In the following table arc shown the twelve variables studied from the process.

Table 1: Process Variables (X)

1.-Total feed tonnage (ton in)
2.-Loading crusher 30 (kr 30 in)
3.-Loading crusher 40 (kr 40 in)
4.-Pellet feed concentrate earth (PAR m)
5.-Speed magnetic separator 1 (HS1)
6.-Speed magnetic separator 2 (HS2)
7.-Power crusher 30 (Pkr 30)
8.-Power crusher 40 (Pkr 40)
9.-Gangue waste (GBA)
10.- Loading magnetic separator 3 (ton S3)
11.- Crusher waste (kr Avf)
12.- Total waste (tot avf)

Variation of the total feed tonnage (1), the speed of separator 1 (5) and magnetic separator 2 (6) were studied in a systematic way according to a central composite design (CCD). The CCD is an experimental design to calculate a model containing quadratic terms and correlated effects.

The six used quality variables are shown in table 2.

Table 2. Quality variables (response Y)
1.-Pellets feed concentrate (PAR)
2.-Fines sinter concentrate (FAR)
3.-Relative Fines Sinter concentrates (rel F)
4.-Iron in Fines Sinter concentrate (%Fe FAR)
5.-Phosphorus in Fines Sinter Concentrate (%P FAR)
6.-Iron in ore

The first two responses, PAR and FAR, are the quantities of preconcentrates which go to pellets and fines sinter. FAR and PAR are measured on-line with conveyor weight and represent the distribution of products from the sorting plant. Distribution of products can also be represented by a relative expression according to FAR/(FAR + PAR)

The response of iron content in FAR with respect to phosphorous content in FAR were analysed through measurement in the laboratory. The last response, iron in ore is calculated from analysis in the subsequent process together with the material balance of the sorting plant. The last three responses iron in FAR, phosphorus in FAR and iron in ore are produced off-line.

Results of IPI application

Three PLS models were described. In the first PLS-model the original 12 variables and quadratical and crossterms making a total of 18 variables were used as X-block data with pellet feed concentrates (PAR), fines sinter concentrates (FAR) and relative FAR used as Y-block data.

The predictions of amounts PAR, FAR and relative FAR were made each minute under a time period of 20 hours. These are shown in figures 11, 12 and 13.

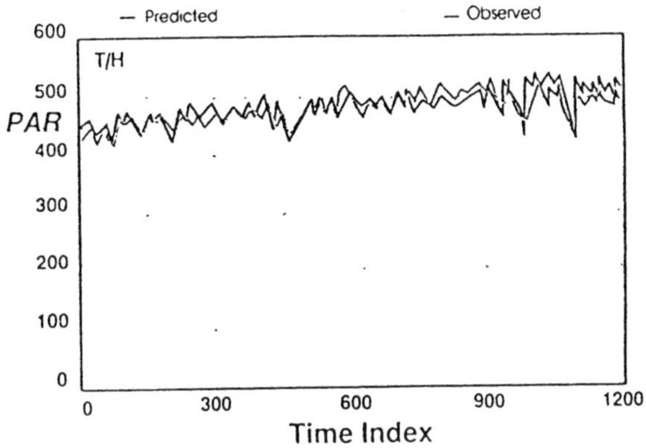

Figure 11: Observed (grey curve) and predicted (black curve) of amount PAR

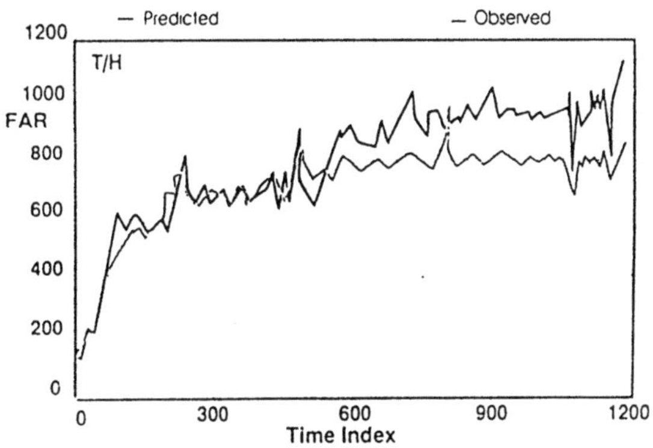

Figure 12: Observed (grey curve) and predicted (black curve) of amount FAR

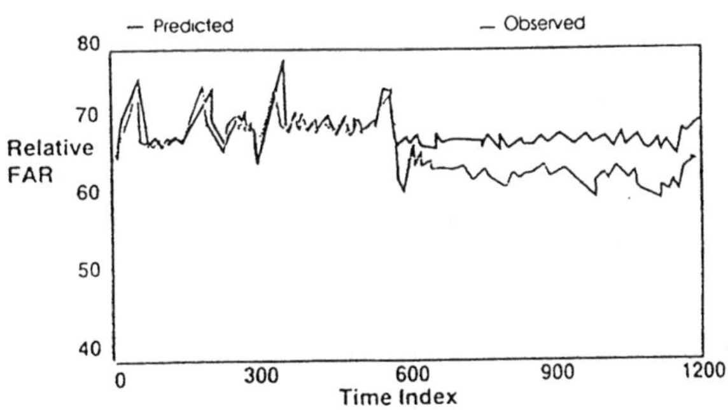

Figure 13: Observed (grey curve) and predicted (black curve) of amount relative FAR.

The predictions agree with actual operation data for PAR. Some deviations are shown for FAR and relative FAR after time index > 600. Deviations can be explained from the shifting in the tonnage loading in crusher 40 which has an important significance in the model responding FAR product. At that time it was show that some obstruction in the screens to this crusher occurred caused by moisture in the feed ore.

The other model had as goal to model the chemical conditions in FAR where the responses were iron and phosphorus contents.

In figure 14 and 15 are shown the observed and predicted curves for percentages of iron respective phosphorus during time.

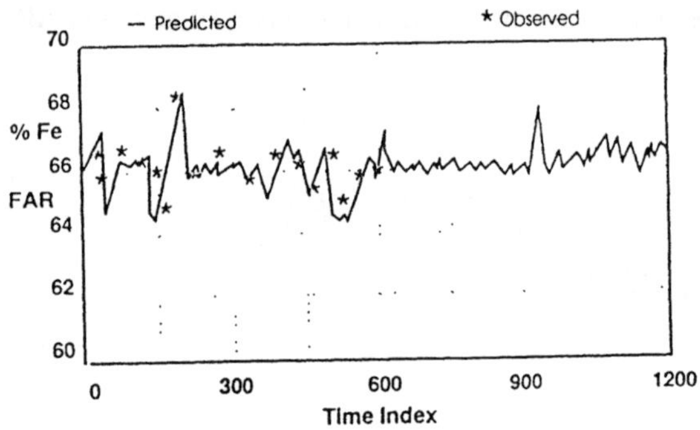

Figure 14: Observed (grey points) and predicted (black curve) iron content in FAR

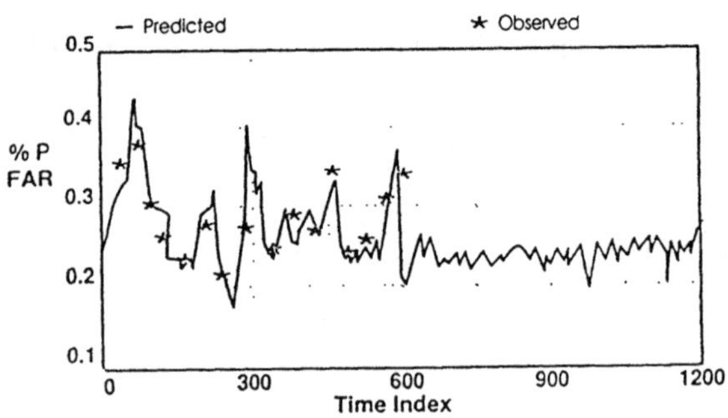

Figure 15: Observed (grey points) and predicted (black curve) of phosphorus content in FAR

The third model was the iron content in the iron ore raw material feeding the sorting plant correlated against the same X-block data as showed before in the previous models.

The observed and predicted results are shown in figure 16.

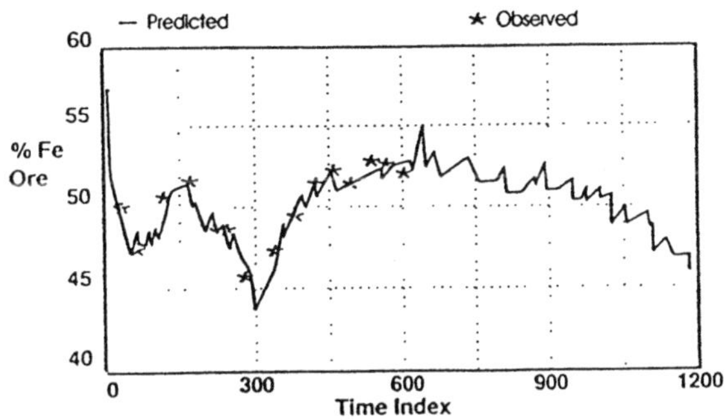

Figure 16: Observed (grey points) and predicted (black curve) of iron content in raw ore.

A relation between loading tonnage and speed of magnetic separator 1 were shown. They indicate that both variables can be handled as a quote regulation. This leads to that various loading tonnages can be maintained as a good control of the process through manipulating the speed of separator 2. By this way it is possible to control the amount of produced fines sinter feed concentrates, FAR.

Good predictions of iron content in the iron raw material can also be used to build a regulation loop. Such a loop gives a possibility to maximize the tonnage with help of the actual raw material quality.

Since data handling are continously feeding the system, it is possible to update the model parameters and perform on-line simulation.

IPI in connection to other techniques like expert system, fuzzy logic and/or neural network might provide a powerful tool for optimization, simulation and trimming of a control system.

Practical implementation of the bacterial oxidation process of refractory gold ores

D. P. Ryan
Signet Engineering Pty., Ltd., Perth, Western Australia

Abstract

The effective treatment of refractory arsenopyritic gold ore is of growing significance, especially in countries such as Australia where many of the more readily accessible oxide deposits are approaching depletion. Traditionally these ores were subjected to roasting to liberate the occluded gold, often with preconcentration by flotation.

Bacterial oxidation is gaining increasing application as an important viable treatment process for refractory ores, providing benefits in cost, performance, and environmental aspects. Design of current large scale bacterial oxidation plants has presented new challenges to the engineer, particularly in materials selections and construction techniques for handling the corrosive pulps at low pH levels and elevated temperatures.

Reference is made in this paper to the experience gained from the recently completed Wiluna sulphide project in Western Australia, which when commissioned was the world's largest bacterial gold treatment plant with full sulphide oxidation.
Keywords: Arsenopyrite, bacterial leaching, refractory gold ores, sulphide oxidation.

1. Introduction

Refractory gold ores historically have produced only 30%-40% recovery when treated with cyanide for a typical grind of 70% to 80% minus 75 microns. Some recoveries have been considerably lower.

The commonly used pretreatment of refractory ores is by roasting and this process has been in operation since the turn of the century.

Roasting of concentrates or ore containing arsenopyrites requires a stage of roasting which leads to increased capital and operating costs.

In recent times, SO_2 producing processes such as roasting have become more and more environmentally unacceptable worldwide. Alternatives to roasting have been investigated for the past 20 years. These alternatives include the bacterial oxidation of refractory ores.

Against this background Asarco investigated the available processes to treat their refractory type ore reserves at Wiluna in Western Australia.

2. Wiluna gold mine : brief history

Gold was first found in the Wiluna area in 1896, but the main Wiluna orebodies were not discovered until 1903. Oxidised ore was mined from 1904 until 1911 when attempts to treat arsenic ores were unsuccessful.

Mining recommenced in 1931 with refractory underground ores being roasted.

Operations ceased in 1947.

The area was again explored by Asarco from 1982 and a 3 million tonne oxide ore reserve was established by 1985.

In 1984 a tails retreatment plant was commissioned and in 1986 a CIP plant was constructed. In 1989 the CIP plant was expanded from 400,000t/a to 1,000,000t/a and the tails retreatment plant was expanded to 300,000t/a.

From 1904 to 1992 over 16 million tonnes of gold ore were mined at Wiluna producing almost 3 million ounces.

3. Wiluna sulphide project

3.1 Process Selection

In the late 1980's with the depletion of their oxide reserves, Asarco investigated alternative processes to treat their proven sulphide reserves of 2,000,000 tonnes.

Testwork was commissioned for the following process routes:

- fine grinding
- whole-ore and concentrate roasting
- bio-oxidation
- pressure oxidation

Asarco established the following criteria for determining the most suitable process route for their Wiluna sulphide ores:

- amenability of the ore to the selected process route
- high gold recoveries
- a proven operating process
- operating costs consistent with the reserve head grade

- capital cost consistent with an estimated 5 year project life.

Various studies completed in the period 1989 to 1991 reduced the process route selection to the following processes:

- two-stage concentrate roasting
- bio-oxidation of the ore using the Genmin Biox process from South Africa.

Roasting of concentrates or ores containing arsenopyrites presents the following problems:

- Often a two stage roasting operation is required to drive off arsenic from the mineral matrix as arsenic trioxide during "reducing roast", followed by an oxidizing roast to drive off the remaining sulphur. This 2 stage roast leads to increased capital costs.

- Gas scrubbing is required to collect all the arsenic trioxide driven off, whether or not SO_2 requires scrubbing. Generally this As_2O_3 will have a purity of 95% or less and will be difficult to sell, given the oversupply in todays arsenic market. If not neutralised and safely disposed of (which is costly) the arsenic can be drummed and stored; but this will only create a larger problem at the end of the mine life. From a strategic and safety point of view, roasting of ores containing arsenopyrites entails difficult handling problems.

After a visit by Asarco to Fairview in RSA to view an operating Biox plant and an additional detailed study of the two final options confirmed bio-oxidation of the Wiluna sulphide ores as the selected process route.

The following factors had the greatest influence in Asarco's selection of the Bio-oxidation process:

1. Amenability of Wiluna ores to leaching.

2. The timely emergence of bio-oxidation as a feasible process technology.

3. Availability of sufficient supplies of water of acceptable quality.

4. Decision by a nearby mining company to adopt the Biox® process.

5. The cost of arsenic disposal, permitting requirements and onerous ongoing environmental monitoring requirements mitigated against the roasting option. Permitting for a Biox plant was estimated at 6 to 9 weeks while for a roaster 6 to 9 months.

6. The high cost of Biox® effluent neutralisation was reduced by the availability of an extensive local calcrete resource.

The feasibility study by Asarco concluded the following in favour of Biox over two stage roasting:

- **CAPITAL COST : less by 20%**
- **OPERATING COST : less by 8-10%**
- **AU RECOVERY : greater by 2-3%**
- **CONSTRUCTION TIME: less by 25%**
- **ENVIRONMENTAL :**

  (a) Permitting considerably shorter than roasting
  (b) Arsenic is neutralised and stabilised instead of being produced as an impure by-product
  (c) Less onerous monitoring and reporting.

4. Project Implementation

4.1 General

Signet Engineering were commissioned to design and construct a 400,000 t/a sulphide ore treatment plant with a concentrate treatment rate of 41,500 t/a by the Biox process. The mineralogy of the ore is primarily pyrite and arsenopyrite, grading 6g/t Au, 1.5% to 2% sulphur and 0.7% arsenic. The concentrate treatment plant was designed for a concentrate grading 24% S, 11% As and 90 to 120 g/t Au.

Figure 1 is a simplified isometric view of the Wiluna plant.

4.2 Major Design Considerations

- The major consideration was the detail design, specification and construction of the largest bio-oxidation plant in the world. The next biggest plant was the Harbour Lights project at Leonora in Western Australia treating 20 tonnes/day of concentrate, one fifth the capacity of the Wiluna plant.

- Other major considerations included expanding the existing crushing and milling plants with minimal downtime.

- Construction of the complete bio-oxidation plant, CCD, neutralisation on circuit and flotation plant within the confines of an operating plant and ensure continued plant production.

- Expanding most plant services including the power station and water supply without loss of plant operating time.

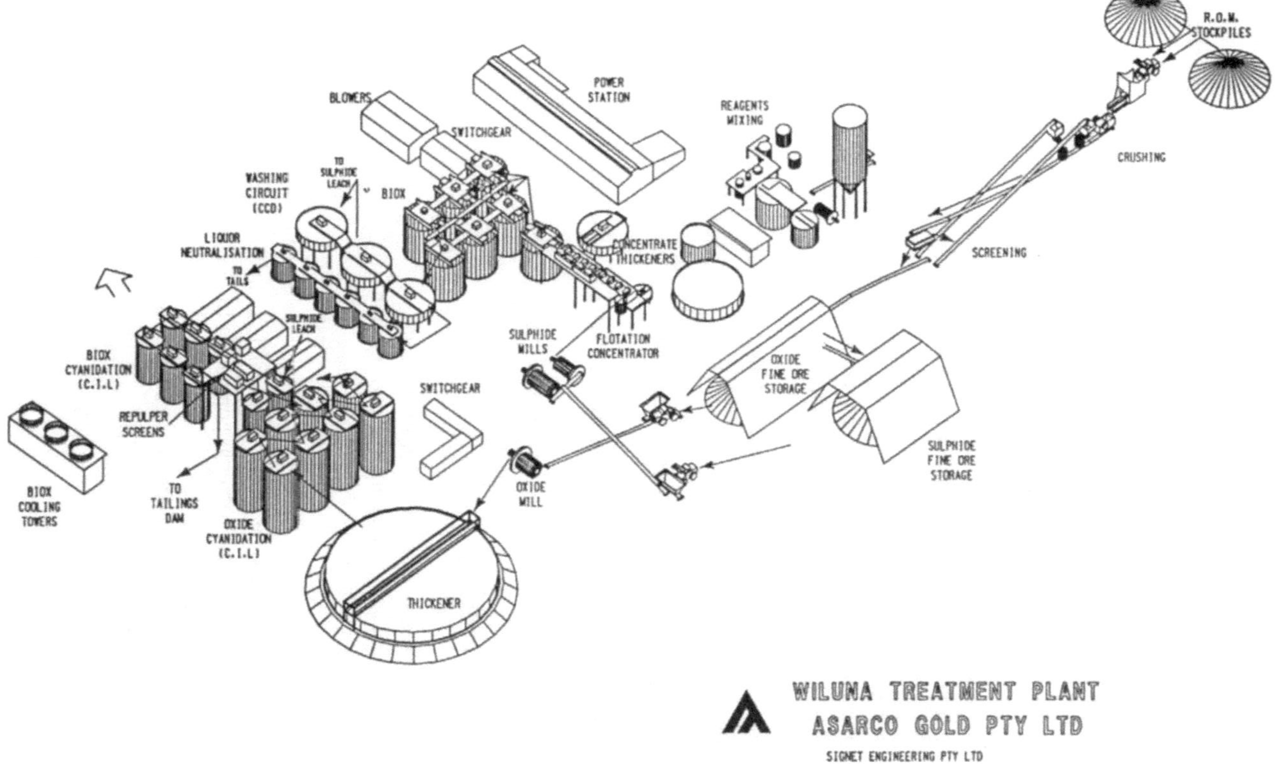

Figure 1 Simplified isometric view

- Converting the leach circuit to handle oxide ore, expanded tails retreatment throughput, and sulphide concentrate again without downtime.

- Available area within the existing plant was extremely limited, however the final layout achieved a highly operable plant, compactly designed which has optimised available space while maintaining access for maintenance by heavy lift equipment.

- The plant was laid out to minimise the number of operators required, by providing easy access and maintaining similar operating levels in adjacent areas.

4.3 Bacterial growth and tolerance

The bacteria survive and grow in an aerobic environment, at 40°C ± 2°C and in a pH range of 1.4 to 1.8. Among the bacteria of interest include:

- Thiobacillus ferrooxidans oxidises the sulphides to sulphates and ferrous iron to ferric iron.

- Thiobacillus thiooxidans oxides sulphate sulphur.

- Leptospirillum ferrooxidans preferentially oxidises ferrous iron in solution.

The Factors that influence bío-oxidation reaction rates are:

1. Nature of the mineral
2. Solubility of the mineral
3. Type of semiconductor normal on "positive" hole
4. Electrode potential between the mineral and solution
5. Ionisation energy of the mineral

4.3.1 Bacterial Growth

Bacterial growth takes place by binary fusion and is characterised by a doubling time which is the time taken for the bacterial population to double in the exponential phase of growth.

Doubling time measurements are not accurate but the following ranges have been reported in various literature:

T. Ferrooxidans	3.6-12 hours
T. Thiooxidans	10-24 hours
L. Ferrooxidans	9-32 hours

The time varies with mineral species which seems to be a major rate determining factor.

The overall Biox reactor retention time of 5 days is considerably greater than the doubling time of the bacteria.

4.3.2 Toxic Effects of Soluble Components

The above bacteria species have a high tolerance to the toxic effects of metal ions in solution. The more specific tolerance levels relevant to Wiluna are noted below.

Lead	200ppm
Antimony	250ppm
Bismuth	200ppm
Mercury	50ppm
Tellurium	150ppm

Table 1. Chloride Toxicity

Chloride (ppm)	Inhibition
60	None
600	5%
15.000	100%

The levels of all of these components in the solution phase of the Biox® reactors must be maintained below these levels. Chloride levels are normally 300-400ppm.

4.3.3 Effect of Temperature

The optimum temperature in the Biox® reactors is 35°C to 45°C. Variations to this range reduce bacterial activity and hence oxidation rates.

4.3.4 Effect of pH

The optimum pH range is 1.4 to 1.8, but growth occurs over a pH range of 1 - 5.5 but over 2.4, bacterial growth is reduced and is inhibited by the precipitation of various compounds on the sulphide mineral surfaces.

4.3.5 Population of Bacteria in Solution Phase

There should be a high enough population of bacteria in solution to enable the secondary oxidation reactions of Fe^{2+} to Fe^{3+} and As^{3+} to As^{5+} to proceed.

The optimum pulp density reflects the balance between mineral surface for solid phase bacterial attack versus the bacteria required in solution to effect the secondary reactions. This balance for a continuously operating Biox® plant is 20% w/w; above this density the rate of oxidation is reduced.

4.4 Brief Description of the Crushing, Grinding and Flotation Area

4.4.1 Crushing

The 2 stage crushing plant was expanded from 120t/h of -12mm ore to 175t/h of sulphide ore and 230t/h ore of oxide ore, each producing a -12mm product by the addition of a 54" El Jay tertiary crusher. Oxide and sulphide ores are crushed on alternate shifts and stored separately.

4.4.2 Grinding Circuit

Fine crushed oxide and sulphide ores are ground in parallel. Oxide ore is processed through an existing Marcy Mill.

Sulphide ore is treated through a ball mill/rod mill circuit. The rod mill is an Allis Chalmers 550 kW unit. The ball mill is an existing ANI-RUWOLT mill of 1250 kW.

The rod mill, new cyclones and a relocated trash screen (Delkor, Linear screen) were installed and commissioned prior to completion of the Biox® circuit. This circuit was operating on oxide ore for some weeks prior to changeover to the sulphide ore.

A three day shutdown was required to tie in the new conveyor feeding the rod mill, relocate the cyclone feed pumps, mill discharge pump box, and the Delkor screen, erect a new mill structure and install the new cyclone pack.

The mill throughput rate is currently 60t/h against a design of 48.6t/h albeit on the softer transition ores currently being treated.

4.4.3 Flotation Circuit

A conventional flotation circuit comprising roughers, scavengers, cleaners and scavenger cleaners has been installed.

The flotation circuit is designed to utilise gravity as much as possible for all major pulp flows.

The only operational pumps in this circuit are the concentrate and tailings thickener underflow pumps, the cleaner circuit feed pump and the scavenger cleaner flotation feed pump.

The roughers and scavengers are naturally aspirated 8 Wemco 120 Cells in 2 x 4 banks with a residence time of 20 minutes. The cleaners and scavenger cleaners are 8 agitair 48 cells in a bank of 2 x 4 cells with a residence time of 12 minutes.

The design mass recovery to concentrate is 8.7% at a 90% sulphur recovery level - a large proportion of this, 60 - 70% of total concentrate weight, is recovered as rougher concentrate, this reduces the load on the cleaner circuit.

Concentrate thickener	6 m dia
Tailings thickener	10.5 m dia

The rougher cons gravitates directly to the cons thickener (6 m dia) and flotation tails gravitates to the tails thickener (10.5 m dia).

Current concentrate pull on transition ores is 3-5%, at a concentrate grade of 20-27%S, 4-10% As, 80-127 g/t Au.

It should be noted that some of the concentrate grades achieved from the transition ores would have caused roaster difficulties, not being autothermal, had that circuit been employed.

Flotation concentrate is pumped to a stock tank which has a 48 hour residence time at design concentrate pull. This residence time was chosen to ensure uninterrupted and stable cons feed flow rates to the Biox® circuit.

4.5. Biox Circuit

4.5.1 General

The bacteria in the Biox circuit live in the following environment:

Pulp pH	1.2 to 1.8
Pulp Temperature	40 to 45°C
Oxygen Concentration	±2ppm O_2
Pulp Concentration	10-20%

The following strains of bacteria are specifically adapted for the process:

- Thiobacillus ferrooxidans
- Thiobacillus thiooxidans
- Leptospirillum ferrooxidans

4.5.2 Biox Plant

Figure 2 is the simplified Biox® area flowsheet.

Detailed criteria for the Wiluna Biox plant to provide the optimum environment for the bacteria to survive and grow are as follows:

Flowrate	115t/d at 10-20% solids slurry
6 Reactors	3 Primary in parallel 3 Secondary in series
Oxidation Level (Design)	93%
Cooling Water	1200m³/h
Air Supply	31,300m³/h
Agitation	Primaries and first secondary; 135kW axial flow impellers Final two secondaries; 90kW axial flow impellers

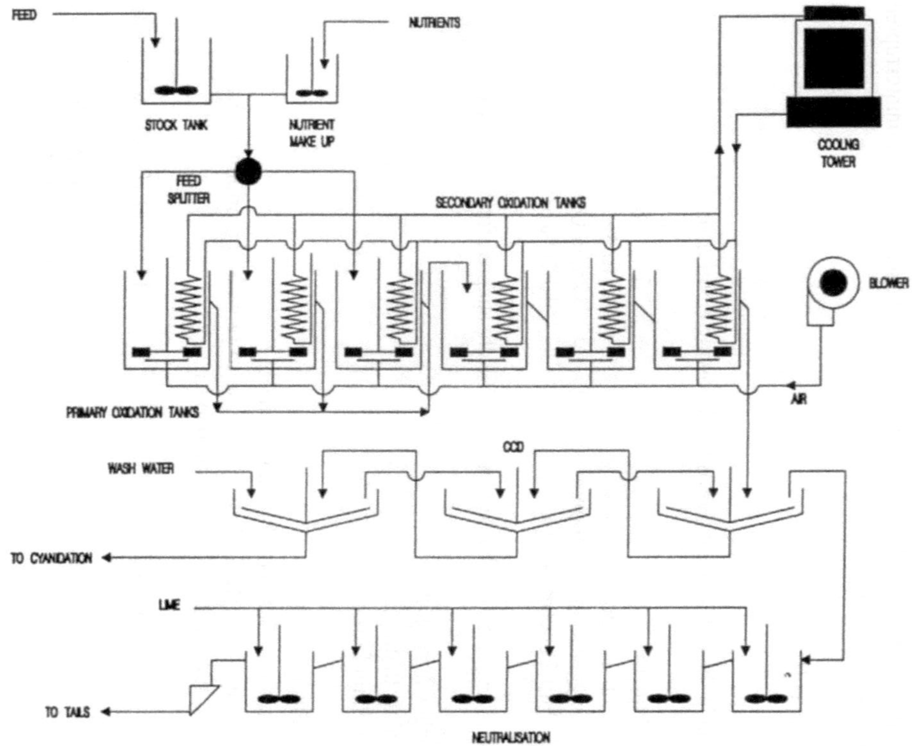

Figure 2 Simplified Biox area flowsheet

Cooling Tower	Design 12MW/hr of low grade heat at 37°C
Temperature	40°C ±2°C
pH	1.4 to 1.8
Retention Time	5 days

The three reactors were installed in parallel to ensure sufficient residence time in the first stage for the reproduction growth of bacteria to be greater than the flow of bacteria from that stage, ie. to prevent a washout of bacteria.

It was essential to design a circuit using materials of construction suitable to withstand the extremely harsh environment and that were non-toxic to the bacteria.

Also the design had to be sufficiently robust and secure to ensure continuity of air supply, withstand the extremely harsh Wiluna environment, and provide for a long life design.

All aspects of the design required considerable attention to detail.
The following outlines the various details considered:

- Water Supply

  The Wiluna water supply was split into low chloride water, up to 300ppm and medium chloride water up to 600ppm. This necessitated the selection of stainless steel with a tolerance to chloride attack and strict separation of the water services. Bacteria are very susceptible to high chloride levels.

- Air Supply

  Air supply was considered to be one of the critical project selections. To ensure survival and growth of the bacteria, it is essential to maintain an air supply to the bacteria. When transition ores would be encountered for the first 2 years it was essential to select a system with a large turndown ratio.

  Blowers were selected to provide the supply of oxygen required for the reaction. 31,300m³/h is required by the reactor bank at full load; this low pressure air is added to the high shear zone below the impeller.

  Three blowers from Aquatec Maxcon were chosen each capable of supplying 12,000m³/h. The blowers are of a high speed (16,400 rpm) turbine type with variable inlet and outlet vanes which give a turndown capability to 40% of rated max capacity of each blower. On this basis air output can be varied from 12% to 100%.

Each unit is fully instrumented and the operation is controlled by the air demand of the reactors. The facility to shutdown reactors and vary output loads has considerable impact on operating costs as the blowers represent a total connected electrical load of 1.2 MW. The Biox® blowers provide 3tph oxygen for the reaction at maximum output.

Each blower is fitted with an aftercooler to reduce air temperature to 50°C to protect the bacteria by avoiding high temperature zones in the reactors (where bacteria would be destroyed).

Materials of construction of the aftercoolers was important as acidic slurries could be syphoned into the system under some conditions.

- Materials of Construction

  Considerable time and effort was spent in the correct selection of the materials of construction in Biox, CCD and Neutralisation areas.

  Selection criteria were based on the following:

  - suitability for the environment, not only the bacterial environment but also Wiluna
  - location of the source of the selected materials
  - delivery time including shipping
  - cost of the materials
  - availability of sufficient skilled labour to fabricate and install the selected materials
  - ability to repair in a timely and cost effective manner.

All of the above were considered against a tight schedule and budget.

A linatrite N50 rubber was selected as the liner to protect the reactors.

The Biox area was constructed free from any galvanised materials in cable trays, pipe, pipe fittings and light fittings. Generally any contact of the acidic slurry with reducing metals (brass, copper, zinc, aluminium) can generate arsine gas, which is considerably more toxic than hydrogen cyanide.

4.5.3 Cooling Water Circuit

The cooling circuit selected was an evaporative cooling tower designed and constructed by Marley Australia.

Design criteria in the selection of a suitable cooling tower considered the following:

- At design capacity there is an expected heat load of 12MW/h at 37°C.

- There was an expected wide variation in heat load due to treating transition ores initially.

- At design the airflow is 1,000,000Nm³/h through the cooling towers.

- The Wiluna environment; dust loads, insect control, scale and microbiological control all had to be addressed in the equipment selection.

In the Biox reactors 8km of stainless steel cooling coils were installed; therefore, it was essential for the correct selection of a suitable anti-scalent system. This scale inhibitor also had to be non-toxic to the bacteria. It was also necessary to design the coils for the ease of removal and replacement in case of failure.

Selection of the correct fill for the towers, insect screens, microbiological control of the cooling water which was non-toxic to the Biox® reactor bacteria should a cooling coil failure occur and the selection of suitable side stream filtration unit, all took considerable detail design time and effort.

4.5.4 Concrete

In light of the aggressive nature of the pulp, the detail concrete design was varied from a normal tank farm design to minimise the potential leaching for low pH by the pulp into the concrete which would then destroy the reinforcing steel. This required minimising any possible cracks, elimination provision of construction and expansion joints, and the application of protective coating to the surface.

4.6 CCD and Neutralisation Circuit

Biox® residue gravitates to 3 x 10.5m diameter high rate thickeners for removal of soluble iron and arsenic species together with dissolved sulphate. A wash ratio of 4:1 is used.

A prime consideration is the production of a stable basic ferric arsenate precipitate; thus a 2 stage neutralisation circuit is required.

CCD overflow is neutralised in a 6 stage tank bank with calcrete and lime additions to remove soluble iron and arsenic by precipitation as ferric arsenate and sulphate as gypsum. The majority of these reactions are completed in the first two stages and there is considerable scale build up in these tanks which will require routine descaling.

Soluble arsenic levels in the discharge liquors to tailings is maintained below 0.4 mg/l As.

The CCD thickeners operate in an acid duty (at pH2-3). The units are lined with natural rubber, and the rake and launder components are fabricated in SAF2205 for corrosion protection. Underflow pumps on all thickeners are Bredel hose pumps.

Final thickener underflow is pumped to a neutralisation circuit and then to the first leach tank in the Biox® residue CIP circuit.

4.7 Concentrate Leaching

Biox® residue leaching is carried out in one tank of 250m³ capacity and adsorption in 6 tanks of 250m³ capacity. Total residence time is 64.5 hours. The concentrate circuit was developed by utilising the existing adsorption circuit, hence the extended leaching/adsorption times which are considerably longer than normal requirements.

Concentrate leach recoveries in excess of 90% are achieved for reagent consumptions of 25.85 kg/t of cyanide and 30-40 kg/t of lime.

4.8 Oxide Leaching

Eight existing leach tanks were converted to 2 leach, 6 CIL to treat oxide ores and the reclaimed tailings.

5. Project information

Location:	Wiluna WA
Environment:	Semi Arid
Temperature Range:	8 °C to 50 °C
Project Schedule:	10 months
Peak construction labour force:	128
Average construction labour force:	75
Planned Plant Shutdowns:	There were approximately 6 major shutdowns to tie in the new plant to the existing plant, ranging from 4 hours to 84 hours. (All proceeded on schedule).
Expanded Power Station:	4.8 MW (additional)
Concrete:	1300m³
Plate work:	480 tonnes
Structural steel:	375 tonnes
Industrial Relations:	There was no lost time due to any industrial relations problems
Inoculum Buildup:	July 92 to February 1993

Inoculum Buildup: July 92 to February 1993

6 Conclusions

The Wiluna project has shown that bacterial leaching for gold recovery can be successfully applied to medium sized refractory gold deposits. From a capital, operating and environmental standpoint the process is superior to other alternatives.

Signet have paid considerable attention to design details in the many critical areas of the process and this has contributed to the success of the project.

7 Acknowledgements

Signet wish to acknowledge the courage of Asarco senior management Cay Williamson and Paul Odd in taking the decision to go with a new technology and their approach to engineering design decisions to ensure the most functional rather than lowest cost approach was taken.

8 References

1. Baxter, K.G. and Odd, P.A.R., (1993) "Design of the Wiluna Biox Plant", **Lecture to AUSIMM meeting**, September 20, 1993.

2. Odd, P.A.R., Craven, B., and Irvine, W. (1993) "Bioleaching - A feasible process for Wiluna Refractory Gold Ores", **Biomine Conference** 22-23 March 1993, Adelaide

3. Van Aswagen, P.C. (1993) "Bio-oxidation of Refractory Gold Ores; the Genmin Experience", **Biomine Conference** 22-23 March 1993 at Adelaide .

Exploration

The Geoscan Airborne Multispectral Scanner as an exploration tool applied to El Halcón prospect, Chile

R. A. Agar
GeoScan Pty., Ltd., Perth, Western Australia
N. R. Fraser
GeoScan Pty., Ltd., Perth, Western Australia
N. H. Lockett
Ashton Mining, Ltd., Perth, Western Australia

Abstract

The Geoscan Airborne Multispectral Scanner (AMSS) is a narrow band remote sensing device which can record up to 24 channels from 46 available spectral bands within the range of 0.49 microns to 12.0 microns. These spectral bands comprise 10 in the visible/near infrared (VNIR), 8 in the shortwave infrared (SWIR) and 6 in the thermal infrared (TIR) portions of the electromagnetic spectrum.

The remotely sensed data are processed by the Geoscan developed Geoscan Image Processing System (GIPSy). This system facilitates rapid manipulation and interpretation of the data. Of particular value as an exploration tool is the ability of the system to aid the user in the discrimination of contrasting rock types and alteration styles by using multi-band treatments.

This paper describes this system as applied to the El Halcon Prospect, near Copiapó, Chile, where it has aided the delineation of hydrothermal alteration zones. The known alteration comprises propylitic, argillic and sericitic zones plus areas of silicification. The presence of adjacent low grade copper mineralisation at depth has been confirmed and economic Cu and Au mineralisation is present at the periphery of the alteration zone. The prospect is interpreted as having significant potential to host a substantial porphyry copper deposit.

Keywords: Geoscan Airborne Multispectral Scanner, El Halcon prospect, porphyry copper.

1 Introduction

The use of narrow band remote sensing to distinguish surface mineralogy is a powerful exploration tool in areas with little vegetation cover. Identification of mineralogical groupings may be achieved by the use of a multi-channel, narrow band, airborne imaging system such as the Geoscan Airborne Multispectral Scanner. This system routinely samples data over 24 spectral bands at a spatial resolution of 10m. The system therefore provides very good ground resolution and at the same time can detect alteration zones related to mineralised systems such as epithermal gold or porphyry copper deposits. This paper describes the Geoscan system and outlines how it has been applied to porphyry copper exploration at the El Halcon prospect, which is located 45km north east of Copiapó, in the III Region of Chile (See Fig.1). Copiapó is located approximately 800km north of Santiago.

2 System design and instrumentation

Remote sensing systems are a trade-off between signal-to-noise ratio and band width. If there are many, narrow bands, the signal-to-noise ratio is lower than if there are fewer, wider bands. However, if the bands are too wide, then the ability to distinguish individual mineral species is compromised [3], as is the case with the space-borne systems.

Another limiting factor is the rate of data capture. An increase in spatial and spectral resolution will greatly increase the amount of data captured and so both must be limited to a certain extent to maintain a manageable amount of data.

The Geoscan Airborne Multispectral Scanner can simultaneously record data from 24 spectral bands out of an available 46 bands. The choice of bands to be recorded depends upon the nature of the study to be undertaken. For mineral exploration work, the bands chosen are as presented in Table 1. The system is an imaging spectrometer which uses grating dispersive optics with three sets of linear array detectors. These detectors are sensitive in the visible/near infrared (VNIR), shortwave infrared (SWIR) and the thermal infrared (TIR).

Prior to data acquisition, the target area is flown once, during which time the operator adjusts the gains and offsets in each channel to best suit the conditions, thus providing maximum surface contrast. The data are then collected using these settings. They are first written to hard disk and then transferred to optical disk. During data acquisition, the operator can monitor the input data in real time. The optical disk recordings require no processing beyond backscatter and panoramic distortion corrections which are routinely made at the commencement of each image processing session. Processing takes place on the Geoscan developed Geoscan Image Processing System (GIPSy).

3 Geology

3.1 Porphyry copper model

The geological setting and identified alteration at the El Halcon Prospect is interpreted as being compatible with that found at porphyry copper deposits. Porphyry copper deposits are formed as a result of magmatic hydrothermal

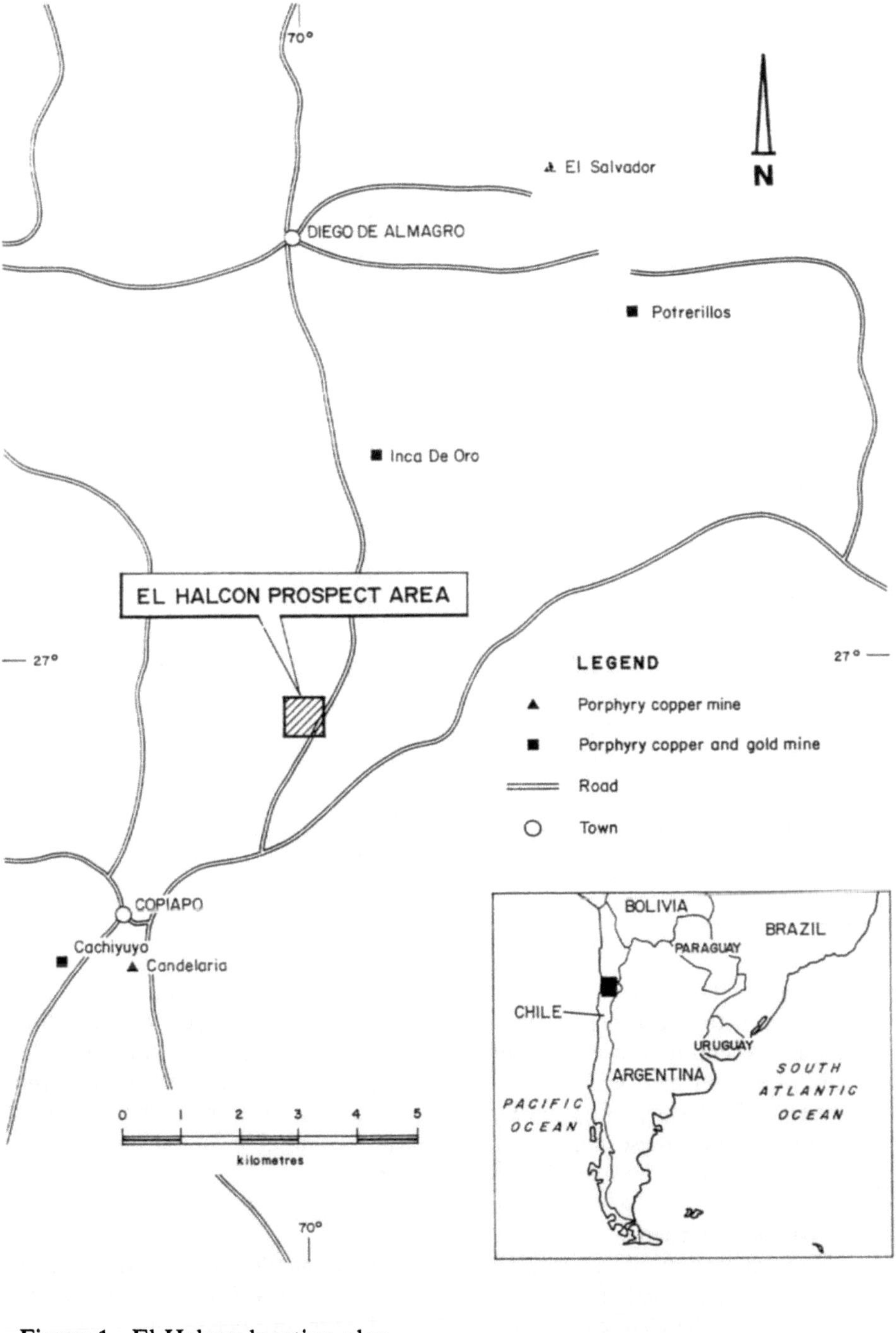

Figure 1. El Halcon location plan

Table 1. Geoscan band specifications

	Band Number	Central wavelength (μm)	Bandwidth (μm)
	1	0.522	0.042
	2	0.583	0.067
	3	0.645	0.071
V	4	0.693	0.024
N	5	0.717	0.024
I	6	0.740	0.023
R	7	0.830	0.022
	8	0.873	0.022
	9	0.915	0.021
	10	0.955	0.020
	11	2.044	0.044
	12	2.088	0.044
S	13	2.136	0.044
W	14	2.176	0.044
I	15	2.220	0.044
R	16	2.264	0.044
	17	2.308	0.044
	18	2.352	0.044
	19	8.64	0.530
T	20	9.17	0.530
I	21	9.70	0.530
R	22	10.22	0.533
	23	10.75	0.533
	24	11.28	0.533

processes. Typically, they are large, roughly equidimensional and are associated with intrusions of intermediate composition, typically adamellite, granodiorite and quartz monzonite. [1],[2] They exhibit a large, concentrically zoned alteration halo which may extend for hundreds of metres. Lowell and Guilbert (1974) (in [1],[2]) developed a model for porphyry copper deposits which shows that these deposits have clearly recognisable lateral and vertical alteration zoning. The four zones identified are:-

1. **The potassic zone:**
   This alteration is a result of potassium metasomatism which may occur in conjunction with leaching of Ca and Na from aluminosilicate minerals. The characteristic minerals are biotite, orthoclase and quartz.

2. **The phyllic or sericitic zone:**
   The sericite zone results from the leaching of Mg, Na and Ca from aluminosilicate rocks. This assemblage is characterised by quartz, sericite and pyrite. Pyrite may constitute up to 10% by volume of this zone. In the process of silicate sericitisation, quartz is generated in large quantities, giving a prominent silicification component in this zone.
3. **The argillic zone:**
   This style of alteration is characterised by the formation of new clay minerals. Argillic alteration is represented by plagioclase altering to kaolinite, montmorillonite, or illite, with accessory biotite and chlorite. Pyrite is the major sulphide present, although it is less abundant than in the sericite zone.
4. **The propylitic zone:**
   Characterised by chlorite, epidote and calcite, this zone is often quite extensive. The chloritization of biotite releases potassium, resulting in accessory sericite.

3.2 El Halcon geology

The El Halcon prospect is located 45km north east of Copiapó at an altitude of approximately 1 750m in a desert region. Consequently, vegetation coverage is minimal. It is situated near the centre of the Carrera Pinto mountains which form a circular feature up to 20km in diameter, representing a volcanic centre of Lower to Middle Tertiary age, which has been faulted and truncated. [5] See Fig. 2.

The flanks of the mountains are formed by an andesitic sequence of Paleocene age known as the Venado Formation. Situated inside the flanking Venado Formation is the intrusive Carrera Pinto Batholith which covers an area of approximately 10km to 12km in diameter and is composed of granodiorite. Intrusive quartz monzonite bodies are related to the hydrothermal alteration present at the prospect. [4]

Alteration styles identified comprise propylitic, argillic and sericitic. The alteration zones comprise two main areas; the northern area is approximately 2.5km by 2km and the southern area is approximately 2km by 1.5km. The alteration styles detected are compatible with the Porphyry Copper model as outlined above. The propylitic zone forms the outer halo of the two alteration zones, with argillic and sericitic zones situated within these two zones.

4 Image processing

The data used in this study were collected in May 1992 at an elevation of 15 670 feet above ground level which resulted in a ground resolution of 10m and a swath width of 9.9 km. Processing of the data with the GIPSy normally consists of using a background treatment to simulate a colour image and then

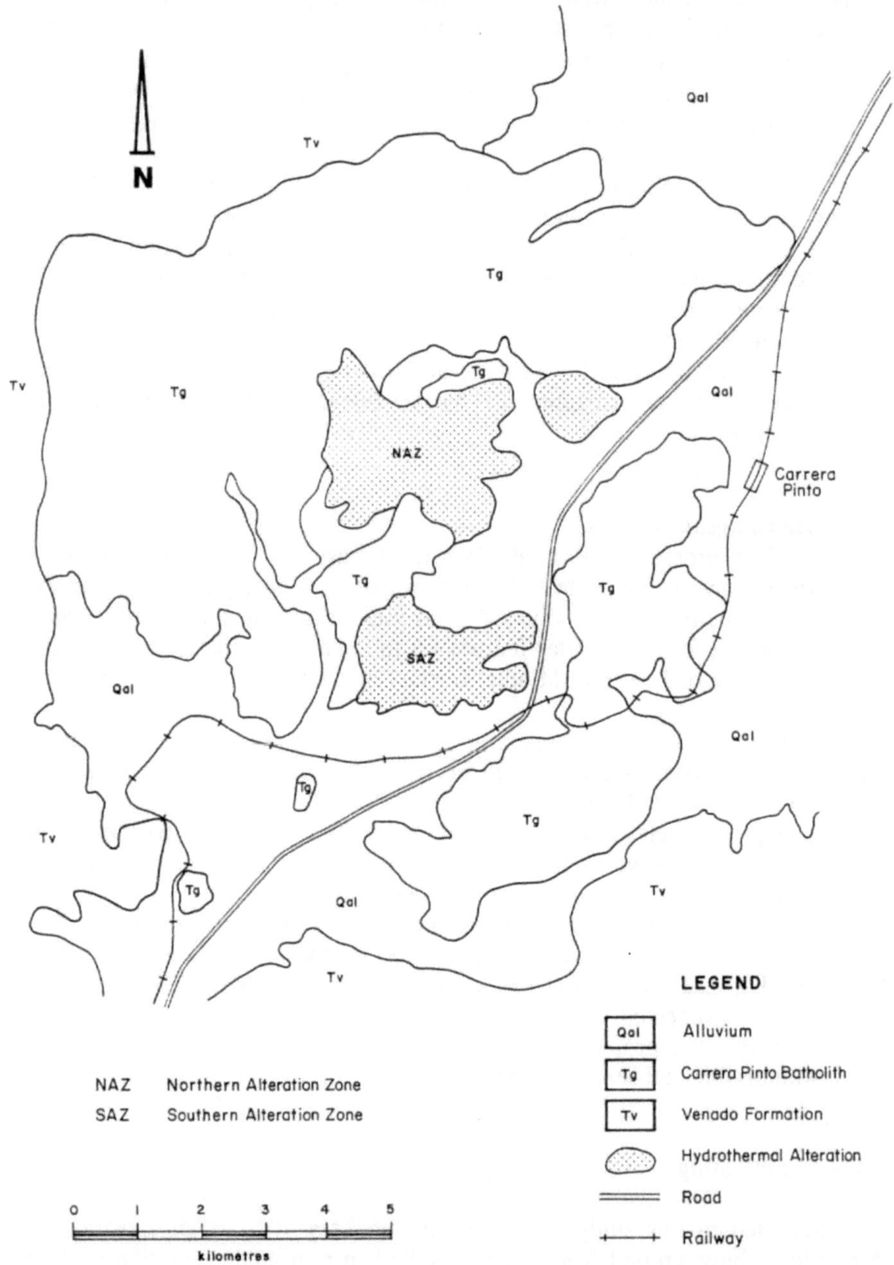

Figure 2. El Halcon geological plan

overlaying various discriminatory treatments over this, with interpretation being recorded on base maps or hardcopy images. The overlay treatments employed usually consist of simple band combinations or combinations of band differences.

Figures 3, 4, and 5 show typical laboratory responses for a variety of minerals in the VNIR, SWIR and TIR respectively. A large number of combinations are obviously possible, but a range of standard treatments have been developed. Figures 6-11 are reproductions of the imagery covering the Northern Alteration Zone. See Figure 2.

Figure 6 is a simulation of an air photo, reproduced in black and white, with Band 3 set to red, Band 2 set to green and Band 1 set to blue. Figure 7 is a monochrome band difference treatment using Bands 13-18, which is designed to show propylitic alteration. The propylitic alteration zones are outlined in this image, with the most strongly altered zones showing a strong white response. There is also a weaker response from the colluvial and alluvial material derived from these zones. Examination of Figure 4 shows that epidote and to a lesser extent, chlorite, will display quite strong positive values for this particular band difference due to the strong MgOH absorption feature centred on Band 18. AlOH bearing minerals will not provide the same response due to their shorter wavelength absorption feature in bands 13/14. It is worth noting that carbonates can exhibit a similar spectral response to that of MgOH minerals at these wavelengths.

Figure 8 is a monochrome band difference treatment using Bands 11-14 which is designed to highlight areas of argillic alteration. This works on the same principle as the propylitic treatment, but utilises the shorter wavelength absorption feature of the AlOH minerals which characterise this alteration style. i.e. montmorillonite, kaolinite and illite. MgOH bearing minerals would produce a poor response from this treatment as may be seen by examining Figure 4.

The treatment displayed in Figure 9 is from the TIR region of the spectrum and is a monochrome band difference treatment using Bands 22-20 which is designed to highlight areas of silicification. Examination of Figure 5 shows that quartz would produce a positive response for this difference treatment and so siliceous zones give a white response on the image, which can be seen in the centre-right portion of Figure 9. Another treatment which may be employed is 19(red), 20(green), 23(blue) which can also distinguish silica and in addition is useful in enhancing structural features. A black and white copy of this treatment is presented as Figure 10 with siliceous zones showing as dark patches in the centre-right portion of the image.

Figure 11 is a monochrome band difference treatment using Bands 6-8. This treatment is designed to highlight areas of iron oxide staining, which show as white on the image. Examination of Figure 3 shows that iron bearing minerals such as haematite and goethite would produce a positive response with this treatment.

It can thus be seen that with careful selection of treatments produced by

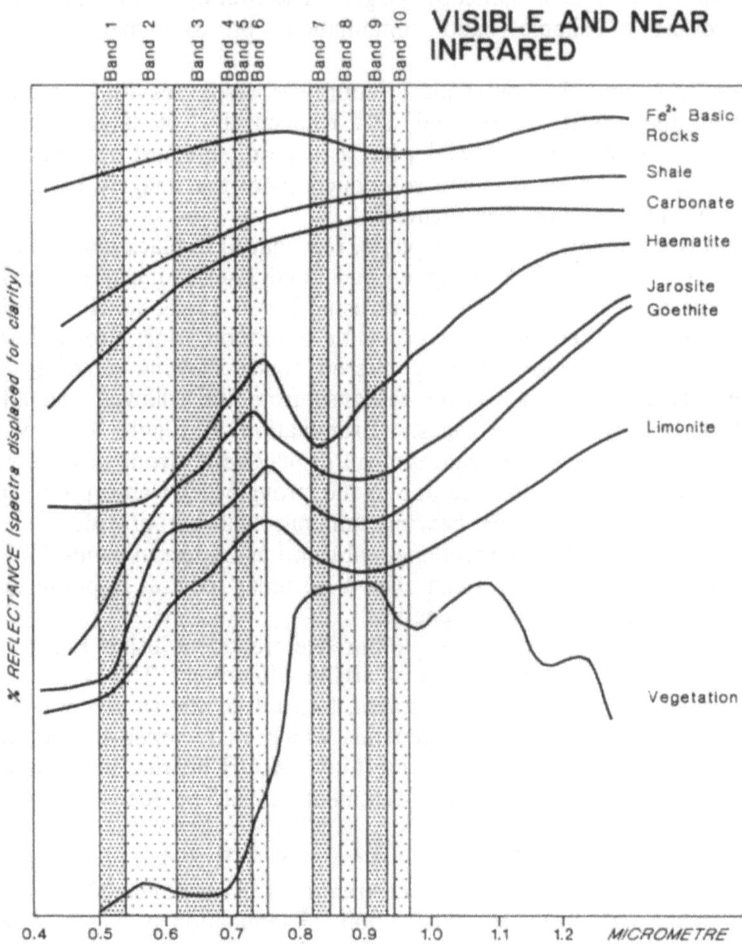

Figure 3. VNIR selected mineral spectra

examination of representative spectra, it is possible to extract a significant amount of mineralogical information from the imagery. Further improvements to the interpretation may be achieved by using spectral data obtained from samples from the prospect. This will provide "ground truth" data. These laboratory derived spectra will be used to devise refined treatments to better delineate the distribution of the alteration styles present at the prospect.

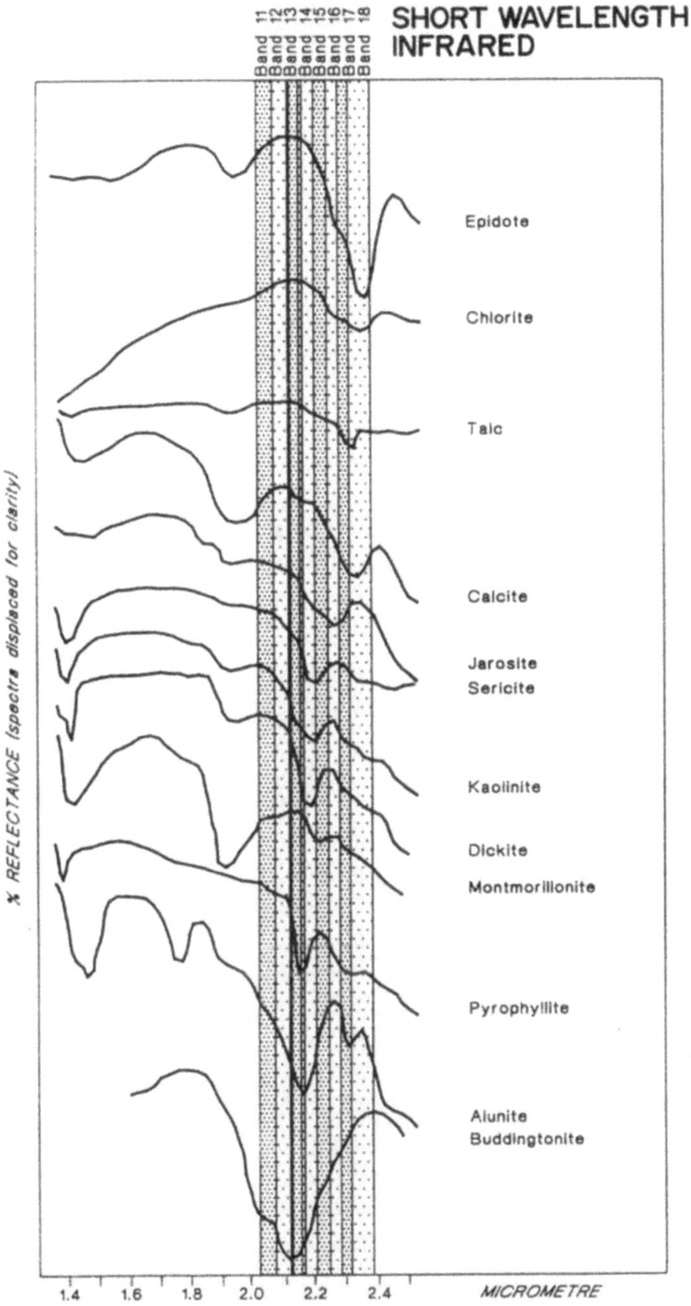

Figure 4. SWIR selected mineral spectra

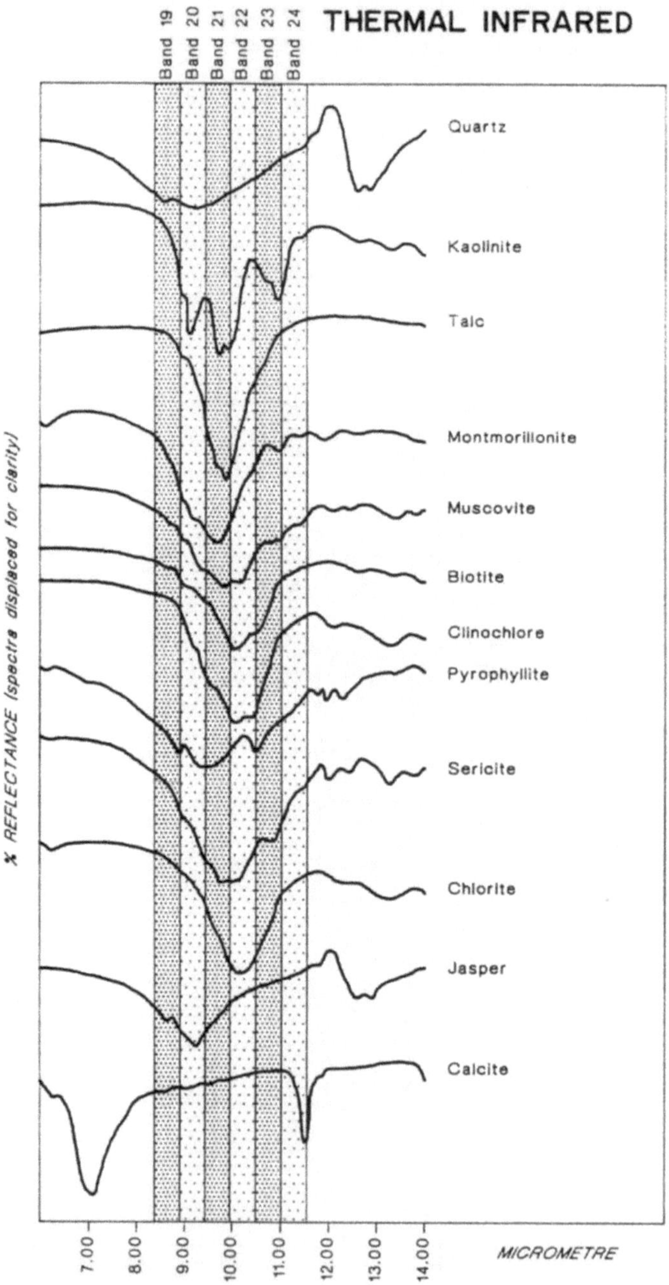

Figure 5. TIR selected mineral spectra

Figure 6. Bands 3 2 1 composite image

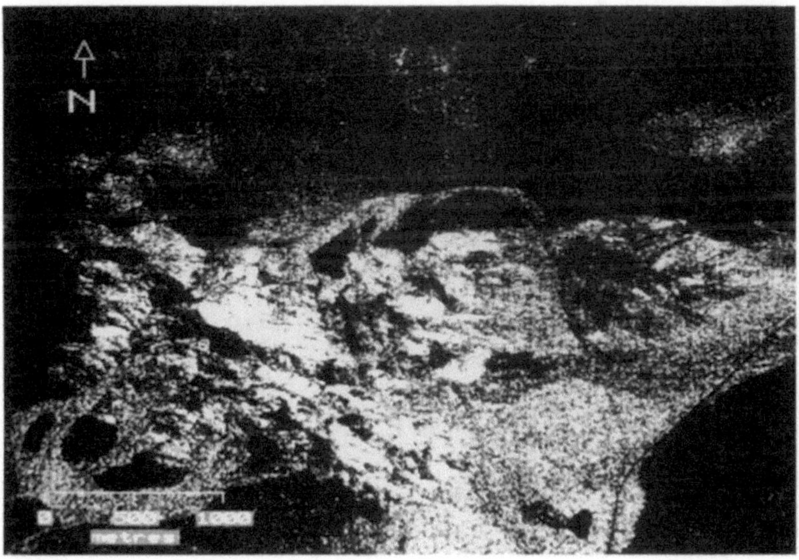

Figure 7. 13-18 band difference image

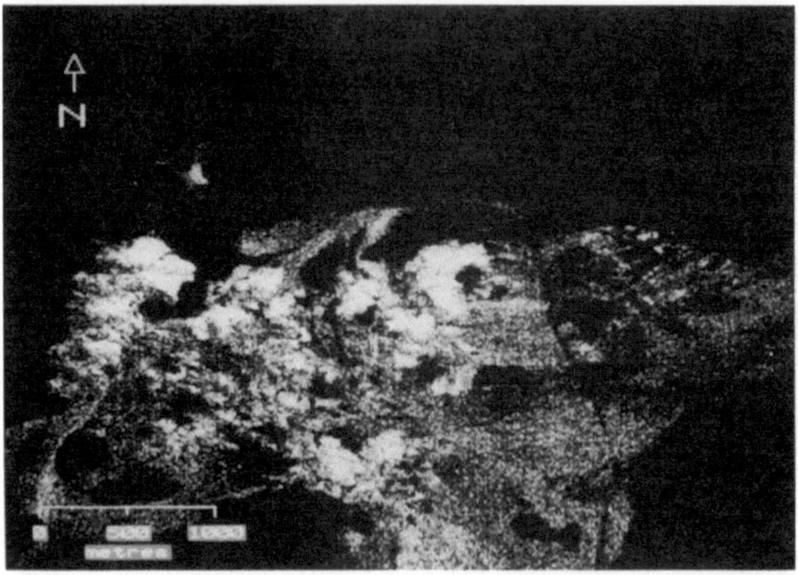

Figure 8. 11-14 band difference image

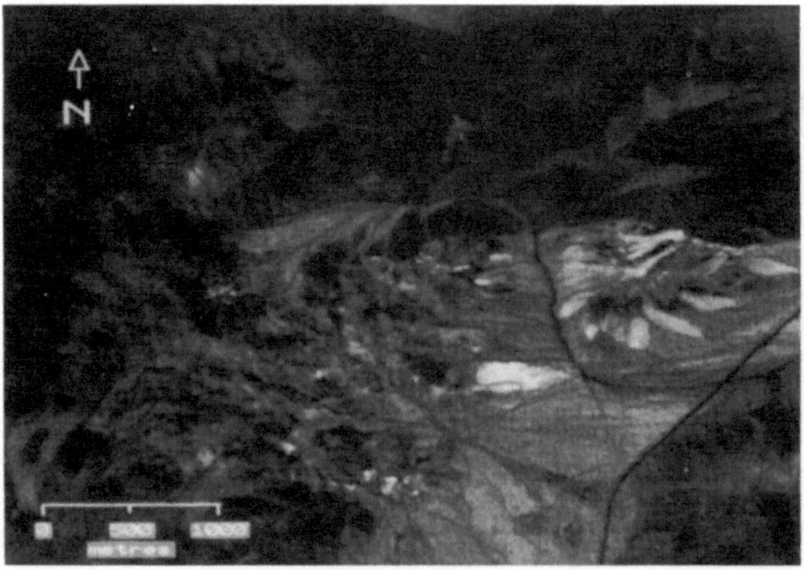

Figure 9. 22-20 band difference image

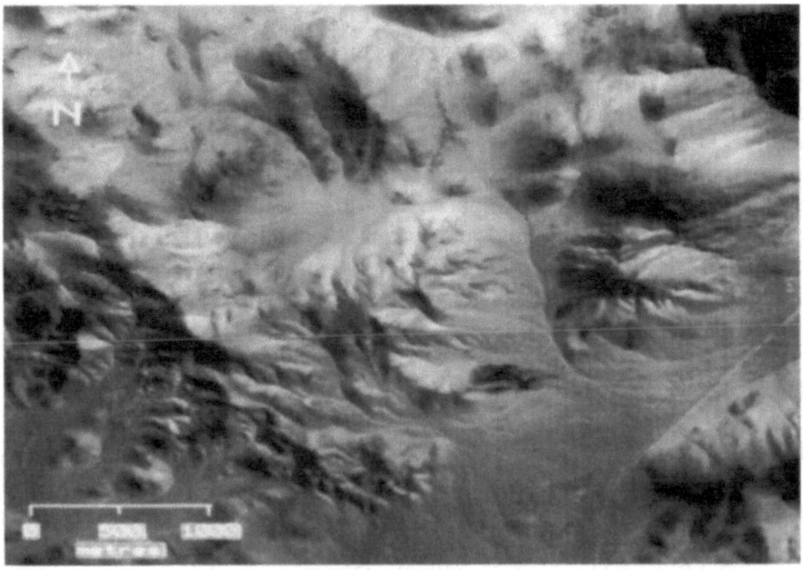

Figure 10. Bands 19 20 23 composite image

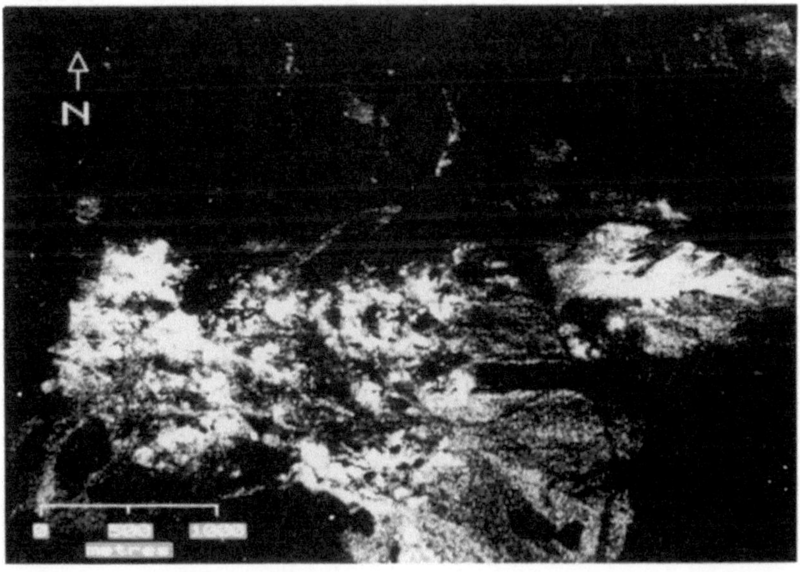

Figure 11. 6-8 band difference image

5 Conclusions

Processing of Geoscan AMSS imagery is capable of detecting hydrothermal alteration systems in arid areas where there is a low percentage of vegetation cover. Furthermore, the system is capable of distinguishing between aluminium bearing and magnesium bearing alteration minerals. Lithological discrimination and structural details may also be extracted from the data. The system therefore has significant potential to aid in exploration for mineral deposits such as epithermal gold and porphyry copper systems and is especially suited to covering large areas of arid terrain in a relatively short timeframe when compared with traditional field mapping techniques.

6 References

1. Bowen R. and Gunatilaka, A. (1977) **Copper: Its Geology and Economics.** Applied Science Publishers, Barking, England.
2. Edwards, R. and Atkinson, K. (1986) **Ore Deposit Geology.** University Press, Cambridge.
3. Lyon, R.J.P. and Honey, F.R. (1990) Direct mineral identification (DMI) with Geoscan MkII Advanced Multi-spectral Scanner (AMSS). **The International Society for Optical Engineering,** Vol 1298, Imaging Spectroscopy of the Terrestrial Environment.
4. Perez, G. (1993) Halcon prospect; geological review. Unpublished company report, Geoexploraciones S.A.
5. Vera, L.S. (1992) Halcon prospect. Unpublished company report, Sociedad Legal Minera Halcon.

Silver mineralization at San Bartolomé, Azuay, Ecuador

S. C. Mulshaw
Department of Geology, Imperial College of Science, Technology and Medicine, London, England
C. A. Puig
Armenonic del Ecuador, Quito, Ecuador

Abstract

This presentation focuses on mineralisation at San Bartolomé in Ecuador, a small silver deposit situated in one of the least studied metallogenic provinces of the Andes. It is an introductory study based on the results of fieldwork and initial laboratory findings.

The San Bartolomé mine, lies near the city of Cuenca in central Ecuador. A system of epithermal, Ag-bearing, polymetallic veins, having average grades of 20 oz/tonne Ag, 2.9% Zn and 1.15% Pb, with localised concentrations of Au up to 0.97 g/tonne, is exploited. The veins have formed in Tertiary volcanic rocks at the intersection of the N-S Peltetec suture and an E-W trending fault zone across which a marked change in the geology of the underlying basement occurs.

The mineral assemblage of the veins is made up mainly by the sulphides sphalerite, pyrite, pyrrhotite and galena, together with Ag-bearing sulphosalts, chiefly freibergite. The primary pyrrhotite in the assemblage is to a large extent replaced by pyrite/marcasite pseudomorphs. The gangue consists of quartz, rhodochrosite and ankeritic carbonates. The minerals form crustified bands and other open-space filling textures typical of epizonal veins in a Cordilleran setting.

Preliminary study of the fluid inclusions trapped in the vein quartz suggest that fluids of magmatic origin may have been diluted by meteoric waters. This is consistent with a model for mineralisation in which the flow of hydrothermal fluids supplied from an underlying intrusion has been focused along a tectonically fractured zone.
Keywords: Cañar-Azuay silver province, Ecuador, epithermal/epizonal mineralisation, fluid inclusions, rapid crystallisation, silver, vein textures.

1 Introduction

The chain of the Andes follows the western coast of South America, stretching from Venezuela in the north to Chile in the south. It was formed in response to plate collision and subduction and is therefore associated with magmatic upwelling, plutonism and volcanism. One consequence of this association is the presence of some of the world's largest mineral deposits within the Andean Cordillera, such as Chuquicamata in Chile. Ecuador, which is situated between Colombia and Peru coincides with a

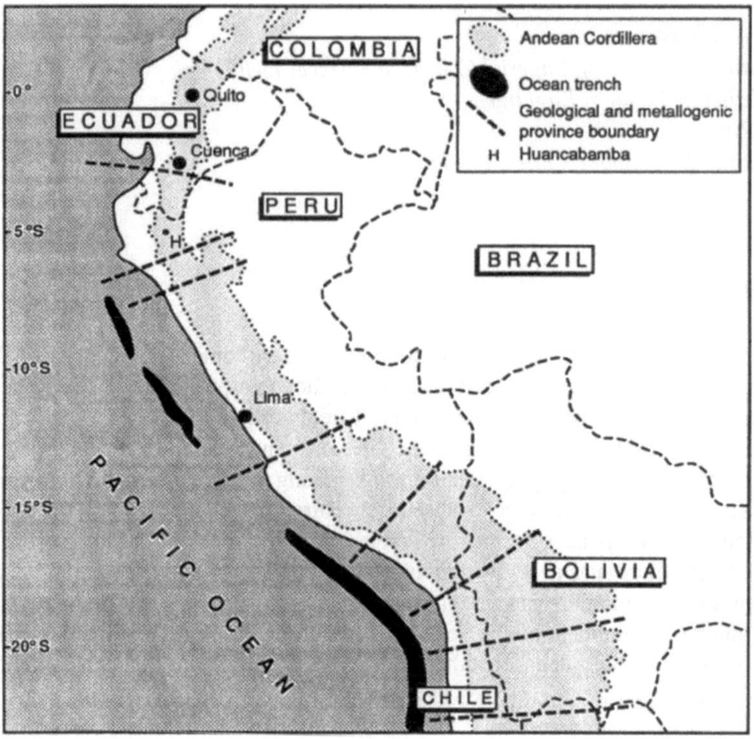

Fig. 1 Locality map of Ecuador and its geologic/structural setting within the Andes of South America.

relatively narrow part of the Andes (Figure 1). It is possibly for this reason that Ecuadorian mineral deposits seem to be both relatively small in scale and restricted in abundance when compared to those of its neighbours, such as Peru and Chile.

Reviews of the mineral resources of Ecuador at the turn of the century focused on the discovery of oil and the re-development of historical precious metal mining areas [1] [2]. Gold placer deposits were being extracted from the Santiago river in north-western Ecuador and the South American Development Company (SADCO) had begun commercial operations in the nationally famous gold district of Zaruma, El Oro Province [3]. Since then, a number of other prospective mining areas have been added to the growing list of mineral occurrences. Silver mineralisation seems to be concentrated in the province of Cañar [4]. Copper prospects, such as the stratabound Cu-deposit at La Plata, Pichincha Province [5], were reviewed by Stoll [6] and platinum group metals have recently been reported to be associated with the Santiago River placer deposits [7] (see Figure 2).

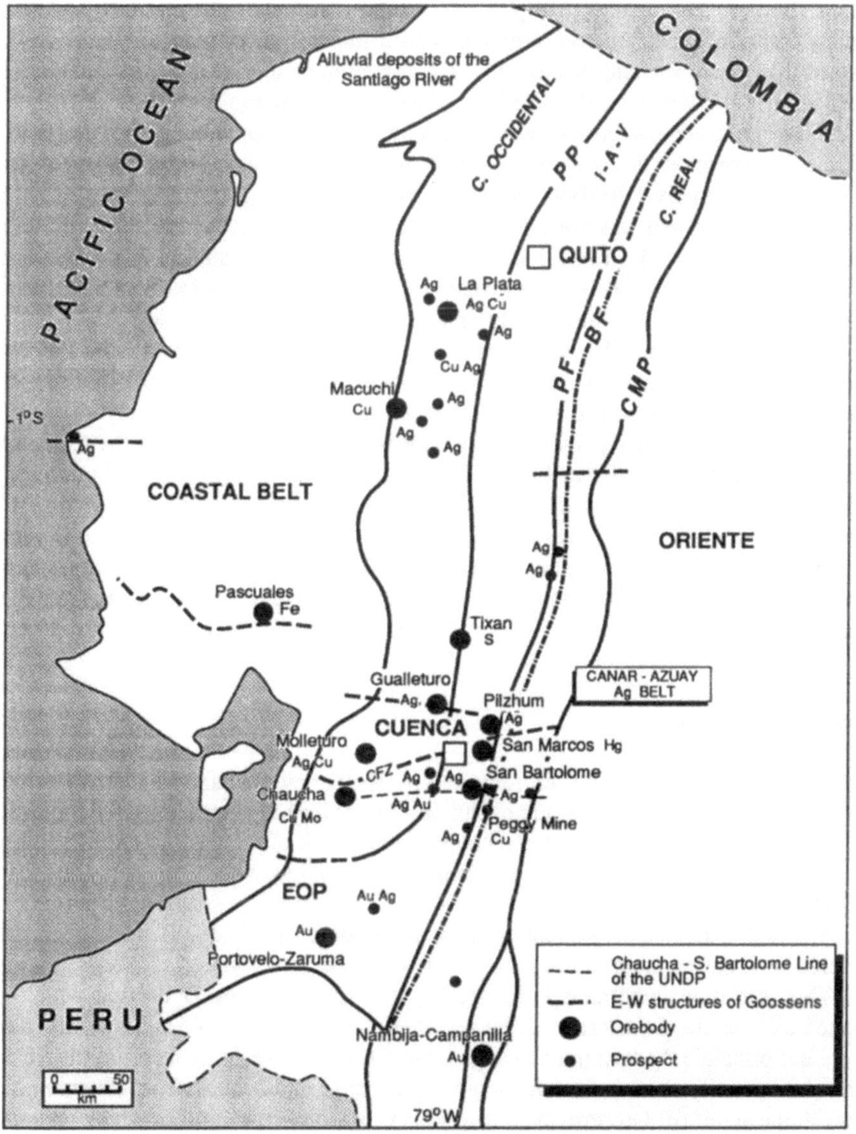

Fig 2 Generalised map of Ecuador showing major geologic provinces, N-S accretionary boundaries, E-W structures and the distribution of major orebodies and Ag-mineralisation. I-A-V - Inter-Andean Valley; EOP - El Oro Province; PP - Pallatanga-Pelenque Fault; PF - Peltetec Fault; BF - Banos Front; CMP - Cosanga-Mendez-Palanga Fault; CFZ - Chaucha Fault Zone.

In an attempt to promote further mineral exploration in Ecuador, the United Nations Development Programme (UNDP) initiated a number of exploration projects in the mid to late 1960's. The results of this work, together with those of earlier pioneering studies, contributed towards the production of a mineral index map for Ecuador [8] (revised by Paladines and Sanmartin [9]) and prompted a review of Ecuadorian metallogeny [4]. Recently, the Cordillera Real Geological Project (CRP), an Anglo-Ecuadorian joint venture involving members of the British Geological Survey (BGS) and the Corporacion de Desarollo e Investigacion Geologico-Minero Metalurgica (CODIGEM, formerly INEMIN) has made considerable progress in establishing the geologic evolution and mineral potential of Ecuador's eastern Andes, the Cordillera Real [10] [11] [12]. However, due to a combination of geographical, economic and socio-political problems, the growing catalogue of Ecuadorian precious and base metal deposits and prospects remains largely untapped on a commercial scale.

There are a few notable exceptions to this generalisation. The Portovelo-Zaruma gold district was successfully exploited until 1950 by SADCO (Figure 2). This deposit was reputedly the largest known concentration of veins and mines in any region of Ecuador [6] [13]. Similarly, the Macuchi copper mine was exploited by the Cotopaxi Mining Company from 1904 to 1947 [14]. In contrast, the Campanilla gold mine [15] which exploits part of the Au-skarn system which hosts the infamous "gold-rush" deposit at Nambija [16] [17], was forced into care and maintenance after only one year. A more recent mining venture, the extraction of silver from an epithermal vein system close to the town of San Bartolomé, in the province of Azuay, forms the basis for this paper.

2 Geologic setting

The geology of Ecuador has been reviewed by Kennerley [18] and Baldock [19]. In addition, the tectonic and magmatic evolution of Ecuador has been summarised by Hall and Calle [20]. More recently, the reconnaissance work of the CRP has considerably improved our knowledge and understanding of the geology of the Cordillera Real [10] [11] [12] [17] [21] [22].

Although encompassing a relatively small section of the Andes, Ecuador seems to coincide with an important transition area dividing the South American Andes (Figure 1). This transition is perhaps influenced by the Huancabamba Deflection which is situated close to the Ecuadorian border with Peru. This is an important hinge structure where the principle Andean trend swings around from the NE in Ecuador, to the SE in Peru. Additionally, there is a subduction-related change in basement, from metamorphic rocks in the south (El Oro Province) to younger, oceanic, crustal rocks which underlie the northern coastal belt [20] [23]. This fundamental change in basement geology is one of a series of major E-W structural and physiographic boundaries recognised as having metallogenic significance [24].

It is now generally accepted that Ecuador comprises a mosaic of accreted crustal slivers and wedges with associated ophiolites [21] [25] [26] brought together as a result of convergence between the South American craton and the Farallon and Nazca Plates of the Pacific [27]. This accreted terrane can be broadly divided into three geographic and geologic provinces (Figure 2): the relatively low-lying coastal belt or **Costa**; the sub-Andean belt and the Upper Amazon Basin collectively termed the **Oriente**; and the intervening Andean **Sierra**. A fourth, smaller province, **El Oro**, is situated in the south-west of the country. This metamorphic region is distinct because its orientation is oblique to the NE-trending Andes [28].

The Costa has a basement of Lower Cretaceous submarine basalts, pyroclastics and ultrabasic intrusives possibly representing arc-related oceanic crust [23] [29]. This is overlain by thick Upper Cretaceous to Pleistocene sediments deposited in several fault-bounded depressions within an extensive fore-arc basin [18] [19].

The mobile belt of the Sierra comprises two parallel, geologically distinct zones trending NNE which are separated by the Inter Andean Depression. The western Cordillera Occidental is dominantly composed of Cretaceous to Eocene volcanics and subsidiary sediments representing the marine Macuchi volcanic arc [18]. The Cordillera Real is made up of a variety of deformed and metamorphosed rocks. Geochronologic studies suggest that isotopic ages have been reset at least twice since the Jurassic. The original deposits were probably laid down sometime in the Palaeozoic [28]. Plio-Pleistocene volcanics cover parts of both the Occidental and Real Cordilleras and fill the Inter-Andean Depression. This block-faulted rift structure formed in response to a tensional regime caused by the arching of the Andean mountains in the late Miocene [18] [19].

The Oriente marks the western limit of the South American craton and forms part of the foreland platform or back-arc basin developed between the craton and the mobile belt. Continental and marine sediments were deposited in the Oriente Basin from Lower Cretaceous to the Quaternary but unmetamorphosed Palaeozoic rocks are also present in the sub-Andean zone. These probably overlie Precambrian basement (*ibid.*).

The principal mineral deposit types in Ecuador include disseminated mineralisation, polymetallic veins, fracture-fillings and stockworks, stratabound and alluvial deposits. They were formed at various times between the Palaeozoic and Pleistocene eras. Significantly, the distribution of many of these occurrences (Figure 2) shows a close spatial relationship with the major Andean crustal sutures (NNE). Some seem to be additionally controlled by intersecting E-W fault structures [4].

This association of mineralisation and E-W structures seems to be particularly significant in the Cañar-Azuay Ag-district [4] [14]. Here, a variety of vein, fracture-filling and porphyry deposits are spatially related to the intersection of the NNE Peltetec Suture/Baños Front systems [21] with the E-W trending Cañar and Chaucha fault zones [4]. The Ag-mine at San Bartolomé is also situated in this region (Figure 2).

3 The San Bartolomé mine

The San Bartolomé deposit was discovered as a result of detailed exploration by the UNDP [30]. The results of trench cutting, stream-sediment and soil sampling, geophysical surveying and diamond drilling delineated two areas of economic potential (Figure 3): Shunaste, immediately north of San Bartolomé town and Ocashuico,

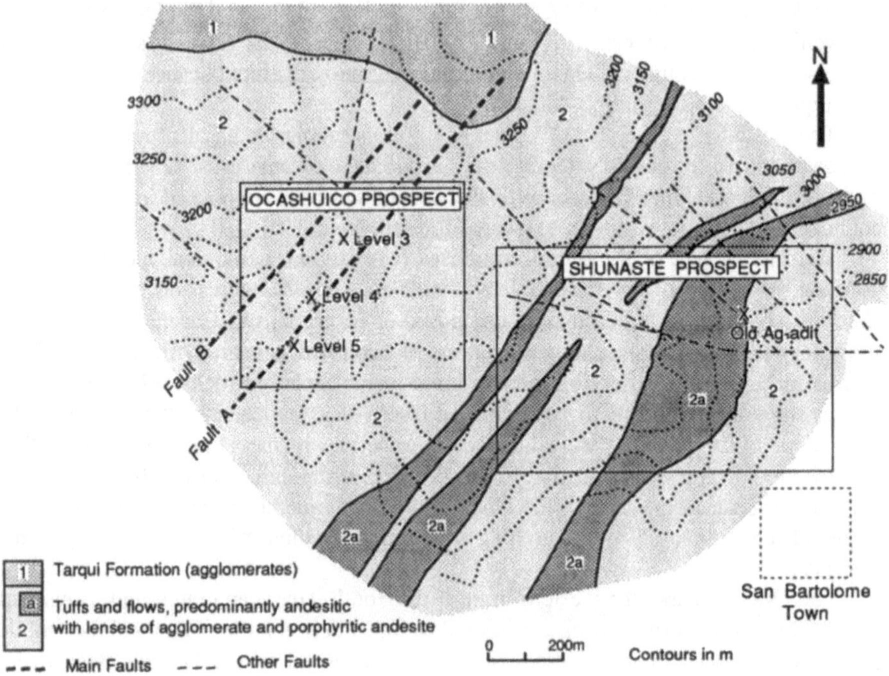

Fig 3 Regional geologic map of the San Bartolome area

approximately 2 km to the north-west. Evaluation of the results from the Ocashuico sector indicated the presence of a blind Ag-orebody. In 1975/6, more detailed studies of the San Bartolomé prospect were financed by the UNDP and subsequently by the UN Revolving Fund for Natural Resources Exploration (UNRFNRE). This period of investigation was completed by 1978 [31]. In the late 1980's, Armenonic del Ecuador entered a joint venture agreement with Nissho Iwai of Japan to further evaluate the economic potential of the San Bartolomé area. These investigations culminated in the development of the present mining operation in the Ocashuico valley.

The mine currently exploits a zone approximately 500 m by 100 m and has a vertical extent of about 90 m. It operates on four haulage levels with a number of sub-level stopes and a fifth level is currently being developed. Recent estimates state that the mineral reserves stand at 55,000 tonnes (proven) and 150,000 tonnes (probable), grading at 20 oz/tonne Ag, 2.9% Zn and 1.15% Pb. It also has the distinction of producing the highest ranked Pb/Ag content bulk concentrate in the world with a Ag concentration of more than 800 oz/tonne [32]. Additional reserves in the Shunaste sector, which is currently under consideration for further development, are estimated to stand at 500,000 tonnes of similar grade ore. San Bartolomé Mine is now wholly owned by Armeno Resources and is currently the **only** commercial venture of its type in Ecuador.

4 Regional geology

The San Bartolomé district can be divided into three geologic provinces [30]. In the east, the Cuenca-Azuay Basin is predominantly filled with Upper Cretaceous to Plio-Pleistocene sandstones, shales and conglomerates [33]. To the west is the NNE-striking metamorphic belt of the Cordillera Real. Between these provinces there is a major structural break which forms a relative high, probably representing the Peltetec Fault/Baños Front suture line. This is characterised by Tertiary intrusives and volcanics which perhaps have a modern representative in the deposits erupted by the active Sangay Volcano, situated in the same line, approximately 100km to the NE [30]. The transition between sedimentary basin and metamorphic belt is largely concealed by Pleistocene, rhyolitic to andesitic pyroclastics, tuffs and lavas of the Tarqui Formation [18] [19].

Mineralisation is hosted by the earlier Tertiary volcanics within the "suture-zone" in direct line with a second Ag-deposit (Pilzhum) situated a few km to the north. At San Bartolomé, the deposit has formed within an ENE fracture system which comprises mineralised fractures and faults, some displaying considerable thicknesses of stiff clay gouge. No geomorphologic manifestation of this fracture network is apparent at the surface.

Two major NE trending faults, which bound the Ocashuico valley, are also associated with the deposit and intersect the fracture zone at its western end (Figure 4). These have been named Faults A and B by the UNDP [30]. It is possible that dextral movements across these faults created an ENE tensional fracture zone which could subsequently have been exploited by mineralising fluids (*ibid.*). If this is the case and given the probable extension of the mineralisation to the east, similar structures may be present in the Shunaste sector.

The location of the deposit appears to be spatially related to the intersection of the Peltetec Fault/Baños Front suture line with the E-W oriented Chaucha Fault Zone [4]. Furthermore, it also lies on a sub-parallel lineament [30] a little to the south of the Chaucha Fault Zone which passes through the major Cu-porphyry mineralisation at Chaucha and polymetallic mineralisation at Angas, both situated in the Cordillera Occidental [14]. Therefore, overall, there seems to be a significant control of both E-W and N-S oriented structures on the localisation of mineralisation.

A small granodiorite intrusion is exposed at the surface in the mine area. Like all exposed rocks in the region it has been severely affected by weathering and is now represented by hard, exfoliated nodules, up to 2 m in diameter, set in a soft sandy matrix. The intrusion is not mineralised and has not been identified in the mine workings. However, it may be related to a larger, parent intrusion at depth which could have acted as a primary source for mineralising fluids.

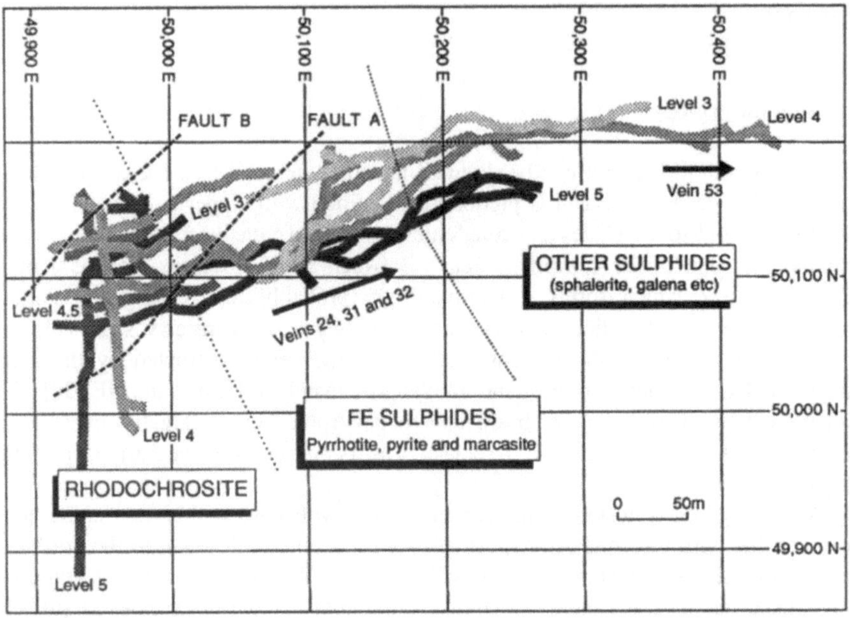

Fig. 4. Mine plans showing the distribution of the principal mineral veins and possible mineral zones.

5 Mineralisation

Currently, the area of the ore-zone being exploited measures 500 m by 100 m and extends over a vertical extent of 100 m. The four most important ore producing veins are the main focus for discussion here. They have been named by Armeno as Veins #24, #31, #32 (striking N080°) and #53 (strike N110°; see Figure 4). The common minerals found in all the veins are sphalerite, pyrite, marcasite, galena, boulangerite and pyrrhotite. Argentian tetrahedrite (freibergite) and a variety of other Ag-sulphosalts (owyheeite, freieslebenite, polybasite and pyrargyrite/pyrostilpnite) are the main source of Ag. Minor amounts of arsenopyrite, locally as coarse idiomorphic crystals, and more rarely, chalcopyrite are also present. The gangue is dominantly quartz, chalcedony and pink, spherulitic and idiomorphic rhodochrosite. Fe-Mn bearing dolomitic carbonates, calcite and minor siderite are also present. A late rhodochrosite stage is especially well developed in Veins #24 and #31.

Crustification of the veins (Figure 5) produces distinct banding which is the most obvious macroscopic feature of the veins. Three stages in the development of this banding are usually visible. A first stage composed mainly of sphalerite forms at the outer margins of the vein, although there is sometimes a narrow selvage of comb-

Fig 5 A - Banded vein sample showing all three principle mineral stages; black sphalerite, grey pyrite/marcasite and white quartz/rhodochrosite; B - Vein sample showing well developed quartz/rhodochrosite bands with black sphalerite rims.

structured quartz separating the sphalerite from the wall-rock. The framework of the second overlying stage consists of pyrite and marcasite which typically form an array of interpenetrating blades, pseudomorphing earlier pyrrhotite and extending towards the central area of the vein. Within this framework, aggregates of galena and Ag-bearing sulphosalts have been deposited. The third stage in the filling of the veins is marked by a central zone of quartz, locally with rhodochrosite.

Pyrrhotite was evidently an important component of the sulphide assemblage but is rarely visible in hand specimen. As described above, it has been almost completely replaced by marcasite and pyrite. Collomorphic "birds-eye" textures are common. In other places, granular masses of idiomorphic and anhedral pyrite form elongate pseudomorphs after blades of pyrrhotite (Figure 6). The texture of the pyrrhotite is reminiscent of interpenetrating and bladed calcite intergrowths observed in many epithermal systems. By analogy, the growth of the pyrrhotite in this way could be ascribed to rapid crystallisation under supersaturated conditions, perhaps caused by boiling or unmixing in response to changes in pressure.

Sphalerite crystallises in coarse-grained aggregates, forming bands on the vein margins and intergrowths with the pyrrhotite/pyrite laths towards the vein centres. It is marmatite with an average Fe content of 12% and locally contains microscopic inclusions of freibergite and chalcopyrite. Galena is present at the transition between the pyrite and quartz stages (Figure 6) and is commonly overgrown and/or replaced by Ag-sulphosalts. This suggests deposition relatively late in the paragenetic sequence.

The Ag-sulphosalt assemblage is made up of freibergite, owyheeite, freieslebenite, polybasite and pyrargyrite/pyrostilpnite. Freibergite was the earliest Ag-bearing mineral to be deposited and is found intergrown with sphalerite and galena. The other Ag-sulphosalts were precipitated with quartz and rhodochrosite of the last stage of mineralisation, often as acicular clusters along grain boundaries.

In general, late-stage, pink rhodochrosite, associated with quartz and other Fe-Mn carbonates, was deposited in the central portion of many veins. Although frequently occurring as zoned idiomorphic crystals, one late generation forms spherulitic masses (Figure 6B) which appear brown and cloudy in transmitted light. This spherulitic morphology is perhaps a further indication of rapid crystallisation taking place during mineral deposition. Quartz occurs at different times in the paragenetic sequence and is present as comb-structured selvages, granular mosaics, banded and massive chalcedony. In addition, veinlets of chalcedony commonly cut the earlier mineral stages and voids filled by spherular chalcedony are reminiscent of the recrystallised gel textures illustrated by Herrington and Wilkinson [34].

The occurrence of brecciated sulphide and gangue fragments set in a fine grained sulphide matrix shows that episodic movements within the fault-fracture zone accompanied the successive stages of mineralisation. In places, the pattern of microfractures in re-cemented sulphide clasts suggests that shearing was followed by break-up of the sulphide fabric due to cataclastic and hydraulic processes. Therefore, the banded textures in some veins have been modified to varying extents as a result of these dynamic processes.

Preliminary studies suggest that pyrrhotite-pyrite-marcasite assemblages are more abundant in the central and western parts of the deposit whereas sphalerite and galena

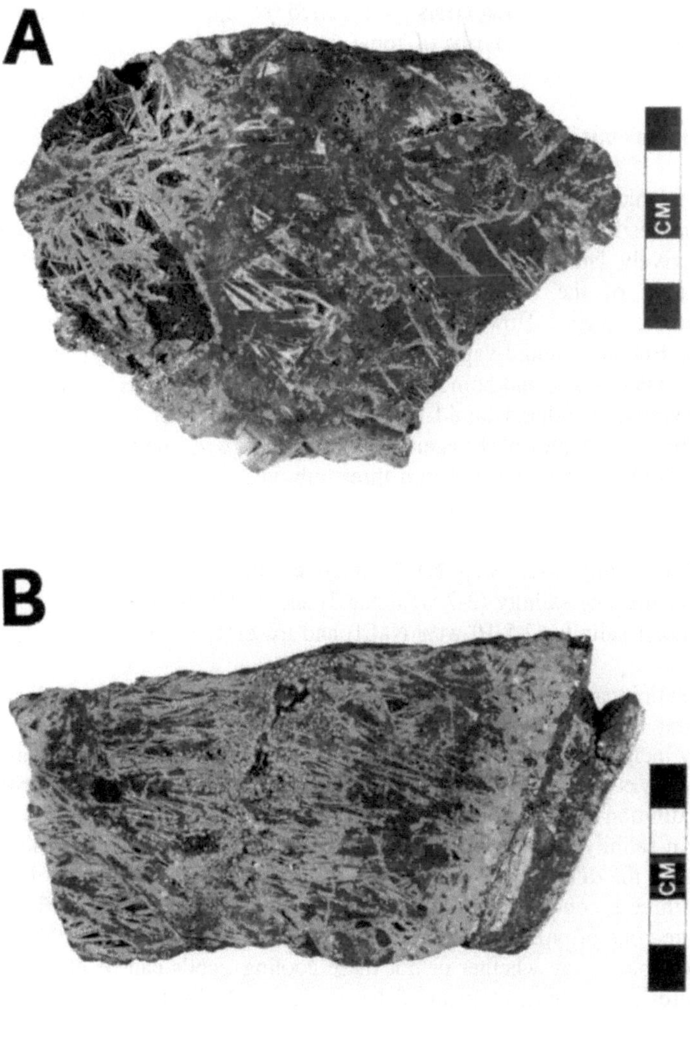

Fig 6 A - Interpenetrating bladed texture of pyrrhotite (slightly lighter grey cores) repleced by amorphous "birds-eye" pyrite/marcasite (darker grey). Light grey granular aggregates of pyrite and marcasite have replaced pyrrhotite most completely in the left hand section of the sample; B - Radiating bladed texture extending towards the vein centre showing the same pyrrhotite, pyrite/marcasite relationships as A. Spherulitic rhodochrosite is present in the central zone.

tend to increase towards the east (Figure 4). A well-developed stage of rhodochrosite is found in the western end of the deposit, between Faults A and B, though this is observed only on the deeper levels of the mine. A more comprehensive study of structure and mineral paragenesis is required so that these observations can be interpreted more clearly in terms of zonal theories.

6 Fluid inclusions

Preliminary microthermometry has been carried out on fluid inclusions in quartz and calcite samples from San Bartolomé. Since both quartz and carbonate phases generally occur relatively late in the paragenesis, the fluid inclusion data is not necessarily representative of the temperature/pressure conditions corresponding to the main mineralisation stage. However, two distinct types of fluid inclusion have been identified. **Primary**, liquid-vapour-solid inclusions characterised by high salinity fluids (32 equiv. wt% NaCl) and homogenisation temperatures (T_h) in the range 220-260°C. **Secondary**, fracture related, liquid-vapour inclusions characterised by low salinity fluids and variable T_h. A plot of T_h against salinity (Figure 7) suggests that the secondary inclusion fluids may be divided into three sub-types, all characterised by decreasing salinity with a decrease in temperature:

1. Low salinity (0-1.5 wt% NaCl), relatively high T_h (260-350°C)
2. Intermediate salinity (2-7 wt% NaCl) and T_h (190-310°C)
3. Higher salinity (2.5-10 wt% NaCl) and lower T_h (150-210°C)

The high salinity and relatively high temperatures associated with the primary inclusions is consistent with their formation from a magmatic-hydrothermal fluid. In contrast, secondary inclusions are most likely to have been formed from a low salinity, meteoric fluid which was subsequently introduced into the hydrothermal system, perhaps leading to some dilution of the fluid already present. In Figure 7, the overall trend of a decrease in salinity with decreasing temperature for the secondary inclusions could be explained by the dilution of a cooling fluid. Furthermore, the apparent division of the data into three secondary inclusion sub-types may represent the periodic introduction into the ore-zone of pulses of a progressively cooler and more dilute fluid. At present, it is impossible to say whether or not these cooling trends can be extrapolated to the temperatures and salinities typical of the primary inclusions.

7 Conclusions

The results of preliminary mineralogical and fluid inclusion studies, combined with field observations suggest that the San Bartolomé Ag-vein mineralisation is similar to other polymetallic, epizonal vein systems found in Cordilleran settings (eg, [37]). Some textures show that mineral precipitation took place rapidly, probably as a result of sudden changes in pressure leading to fluid supersaturation and/or boiling.

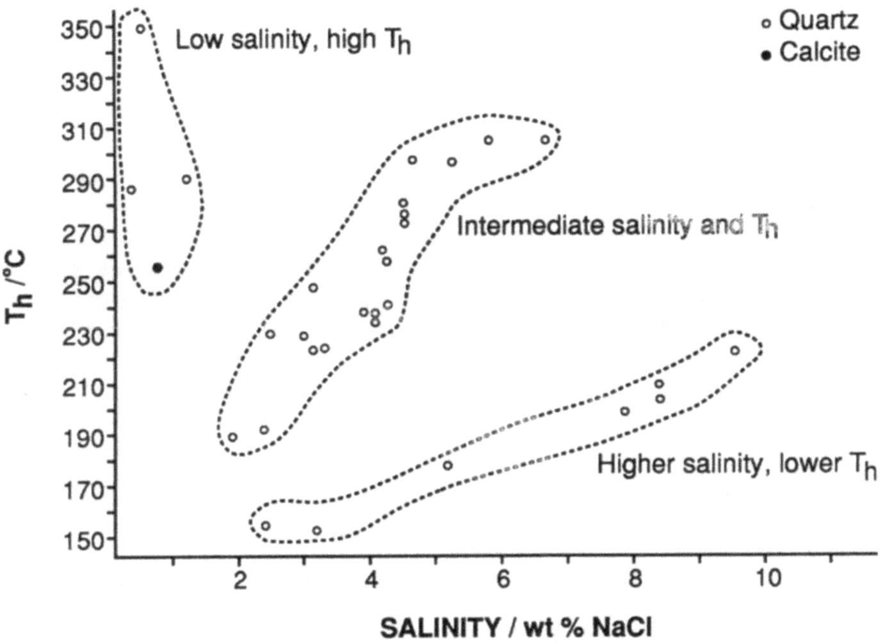

Fig 7 Plot of T_h vs salinity for secondary fluid inclusions in quartz and calcite samples from the San Bartolome mine.

The San Bartolomé veins could be part of a mid-high level epizonal system genetically linked to a sub-surface intrusion. Mapping and geophysical exploration has not yet proved the existence of a deep intrusion in the San Bartolomé area. However, the small granodiorite body exposed at surface in the western part of the mine area may be an indication of a larger underlying intrusive which could have provided energy and fluids for hydrothermal circulation and mineralisation. An additional consideration is the Peggy Cu deposit, near Sigsig. This polymetallic system is situated only 8km to the south-east of San Bartolomé and displays many characteristics often associated with porphyry-Cu type mineralisation [12]. Perhaps this system has a genetic relationship with the Ag-mineralisation at San Bartolomé.

Whatever the reasons for the formation of the deposit at San Bartolomé, the veins are situated within a distinct E-W structural zone which coincides with a marked change in basement geology and intersects the N-S, accretionary sutures of the Peltetec Fault and Baños Front. Since active subduction is taking place all along the western coast of South America, a weakness in the crust caused by the intersection of two major structural trends would be a favourable position for magmatic upwelling and migration of rising hydrothermal fluids. The localisation of precious and base metals, especially

Ag, in vein and porphyry deposits associated with both structural trends tends to support this hypothesis. It further suggests that a genetic link between deeper magmatism and epizonal mineralisation is possible at San Bartolomé.

Acknowledgements

This investigation was carried out while SCM was in receipt of an NERC-EIPS grant awarded to Prof. D. Buchanan at Imperial College, London. Fieldwork was supported by funds from the G. Vernon Hobson Bequest of the IMM. Additional funding towards the costs of attending the Mining Latin America conference were generously provided by the Hilary Bauerman Trust of Imperial College. The authors are extremely grateful for the assistance and encouragement received from John Aspden, Richard Jemielita and other personnel associated with the CRP in Quito. The helpful comments of Dennis Buchanan, Jamie Wilkinson, Chris Halls and an anonymous referee are also gratefully acknowledged.

References

1. Mercer, J.W. (1916) Mining in Ecuador. **Engineering and Mining Journal**, 101 (8), 343-346.
2. Miller, B.L. and Singewald, J.T. (1919) **The mineral deposits of South America** (Chapter 7 - Ecuador). McGraw-Hill, New York.
3. Billingsley, P. (1900) Geology of the Zaruma gold district of Ecuador. **American Institute of Mining and Metallurgical Engineering (AIME),** 74, 255-275.
4. Goossens, P.J. (1972) Metallogeny in the Ecuadorian Andes. **Economic Geology,** 67, 458-468.
5. Lehne, R.W. (1990) The Cu-Zn-Au-Ag deposit of La Plata (Toachi), Ecuador, in **Stratabound ore deposits in the Andes**, (eds. L. Fontboté, G.C. Amstutz, M. Cardozo, E. Cedillo, and J. Frutos), Springer-Verlag, pp. 389-393.
6. Stoll, W.C. (1962) Notes on the mineral resources of Ecuador. **Economic Geology**, 57, 799-808.
7. Wieser, T. and Schmidt-Thomé, M. (1993) Platinum group minerals from the Santiago River, Esmeraldas Province, Ecuador. **Canadian Mineralogist**, 31, 61-73.
8. Goossens, P.J. (1969) **Mineral index map of the Republic of Ecuador**, 1:1,000,000.
9. Paladines, A. and Sanmartin, H. (1980) **Mapa metallogenico de la Republica del Ecuador** (Explanation to the 1:1,000,000 metallogenic map of Ecuador).
10. Litherland, M., Aspden, J.A., Burmudez, A.R., Viteri, S.F. and Pozo, P.M. (1990) **The geology and mineral potential of the Cordillera Real, Ecuador.** Summary of the results of the Cordillera Real Geological Project, 1986-1990, INEMIN/Mision Britanica, Quito. BGS Open File Report.
11. Jemielita, R.A., Aspden, J.A. and Litherland, M. (1992) The metallogenesis and mineral potential of the Eastern Cordillera (Real) of Ecuador. **Programme with**

Abstracts, Mineral Deposits Studies Group AGM, Univ. Aberdeen, 6-7 January, 1992, pp. 11.
12. Jemielita, R.A. and Bolanos, J. (1993) **Mineralisation, mineral potential and metallogenesis of the Cordillera Real of Ecuador.** Final report of the results of the Economic Geology Survey of the Cordillera Real Geological Project, 1991-1993; CODIGEM-Mision Britanica, Quito. BGS Open File Report, 1993.
13. Herrera, J.Z and Mosquera, C.F. (1969) **Gold and base metal sulphides (operation No. 2 - Portovelo).** Technical Report #2, Survey of metallic and non-metallic minerals, United Nations Development Programme.
14. Mosquera, C.F. (1976) Mineral resources of Ecuador: Their development and prospects for new finds - Summary, in **Circum Pacific Energy and Mineral Resources,** (eds. M.T. Halbouty, J.C. Maher, and H.M. Lian), American Association of Petroleum Geologists Memoir #25, pp. 552-555.
15. Anon. (1991) Mineralization rediscovered after 200 years pays off. **Engineering and Mining Journal**, Dec. 1991, 6.
16. McKelvey, G.E. (1991) Interest shown in Nambija gold deposits, Zamora Province, Ecuador. **Mining Engineering**, 12, Dec 1991, 1412-1414.
17. Litherland, M., Fortey, N.J. and Beddoe-Stephens, B. (1992) Newly discovered skarnfields in the Ecuadorian Andes. **Journal of South American Earth Sciences,** 6 (1/2), 67-75.
18. Kennerley, J.B. (1980) Outline of the geology of Ecuador. **Overseas Geology and Mineral Resources**, 55, (with map).
19. Baldock, J.W. (1982) **Geology of Ecuador**. Explanatory bulletin of the National Geologic Map of the Republic of Ecuador, 1:1,000,000.
20. Hall, M.L. and Calle, J. (1982) Geochronological control for the main tectonic-magmatic events of Ecuador. **Earth Science Reviews**, 18, 215-239.
21. Aspden, J.A. and Litherland, M. (1992) The geology and Mesozoic collisional history of the Cordillera Real, Ecuador. **Tectonophysics**, 205, 187-204.
22. Aspden, J.A., Fortey, N., Litherland, M., Viteri, F. and Harrison, S.H. (1992) Regional S-type granites in the Ecuadorian Andes: Possible remnants of the break-up of Gondwana. **Journal of South American Earth Sciences**, 6 (3), 123-132.
23. Henderson, W.G. (1979) Cretaceous to Eocene volcanic arc activity in the Andes of northern Ecuador. **Journal of the Geological Society**, 136, 367-378.
24. Sillitoe, R.H. (1974) Tectonic segmentation in the Andes: Implications for magmatism and metallogeny. **Nature**, 250, 542-545.
25. Litherland, M. and Aspden, J.A. (1992) Terrane boundary reactivation: A control on the evolution of the northern Andes. **Journal of South American Earth Sciences**, 5 (1), 71-76.
26. Feininger, T. (1987) Allochthonous terranes in the Andes of Ecuador and northwestern Peru. **Canadian Journal of Earth Sciences**, 24 (2), 266-278.
27. Daly, M.C. (1989) Correlations between Nazca/Farallon plate kinematics and fore-arc basin evolution in Ecuador. **Tectonics**, 8 (4), 769-790.
28. Aspden, J.A., Harrison, S.H. and Rundle, C.C. (1992) New geochronological control for the tectono-magmatic evolution of the metamorphic basement of the Cordillera Real and El Oro Province of Ecuador. **Journal of South American**

Earth Sciences, 6, 77-96.

29. Goossens, P.J. and Rose, W.I. (1973) Chemical composition and age determinations of tholeitic rocks in the Basic Igneous complex, Ecuador. **Bulletin of the Geological Society of America**, 84, 1043-1052.
30. United Nations Development Programme (1972) **The San Bartolomé Silver prospect (operation #5, Azuay Province)**. Technical Report No. 17: Survey of metallic and non-metallic minerals (Phase III), Republic of Ecuador. New York.
31. Fozzard, P. (1985) San Bartolomé, Ecuador, in **Discoveries of epithermal precious metal deposits** (ed. V.F. Hollister), American Institution of Mining Engineering, 115-117.
32. Puig, C.A. (1993) Ecuador. **Mining Annual Review, 1993**, 128.
33. Sheppard, G. (1934) Geology of the interandine basin of Cuenca, Ecuador. **Geological Magazine**, 71, 356-370.
34. Herrington, R.J. and Wilkinson, J.J. (1993) Colloidal gold and silica in mesothermal vein systems. **Geology**, 21, 539-542.
35. Stanton, R.L. (1972) **Ore Petrology**. International Series in the Earth and Planetary Sciences, McGraw-Hill.
36. Guilbert, J.M. and Park, C.F. (1986) **The geology of ore deposits.** WH Freeman and Company, New York.
37. Gemmell, J.B., Simmons, S.F. and Zantop, H (1988) The Santa Niño silver-lead-zinc vein, Fresnillo district, Zacatecas, Mexico: structure, vein stratigraphy and mineralogy. **Economic Geology**, 83, 1597-1618.

Development of the Collahuasi copper deposits—a world-class copper project in the Andes of northern Chile

L. J. De Beer
L. A. Dick
Compañía Minera Doña Inés de Collahuasi S.A., Santiago, Chile

Abstract

Recent exploration successes at Collahuasi have placed the district as the world's largest, as yet undeveloped, copper resource. Now, in the final feasibility stage, two separate porphyry copper deposits, each with important secondary enrichment zones, contain a combined total tonnage of over 1.5 billion metric tons, with an average grade exceeding 1.0% copper. The first deposit to be mined, Ujina, contains over 168 million metric tons of strong secondary sulphide enrichment, grading 1.6% copper. The second deposit to be mined, Rosario, contains a minimum of one billion metric tons grading over 1.0% Cu. In addition to the porphyry copper deposits, high grade, copper-silver, copper-gold, and silver vein systems as well as exotic copper oxide deposits increase the resource potential of the district.

The location and geological characteristics of the deposits present a host of challenges for project development. Situated in the arid Andean altiplano, at a mean elevation of 4,500 m.a.s.l., Collahuasi is the highest major mining project in the world.

Designing for altitude must consider significant efficiency derating of both equipment and personnel. The relative remoteness of the project, combined with the altitude, has required creative solutions to the problems of material and product transport and to the work cycles, including adaptation time, for personnel. Water exploration, in one of the world's most arid deserts, has had to secure a long-term, sustainable yield to the project.

Infrastructure design and siting must consider the highly variable climatic conditions, as well as geologic hazards, including seismic activity and proximity to active volcanoes. Sequencing, sizing and process selection trade-off studies have had to consider that not just one, but several deposits are available to the project in close proximity to each other, and that both leaching and flotation could be applied to the processing of the higher grade secondary sulphide ores.
Keywords: Chile, Collahuasi, Ujina, Porphyry copper deposit.

1 Introduction

The Collahuasi Project represents one of the largest and most significant copper development projects in the world. Two major porphyry copper deposits, Rosario and Ujina, are currently in the final feasibility study stage. The Rosario deposit has been known since 1979, while the Ujina deposit, 7 km to the east of Rosario, was only discovered in 1991. Both deposits will be developed. However, Ujina will be mined first because of higher grades and superior early year economics.

At present, Ujina has been drill proven with 320 diamond drill holes on 70 meter centers. Over 110,000 meters of drilling was completed between 1991-1993. A shaft has been excavated to obtain bulk samples for metallurgical testing and provide access for underground, horizontal drilling. Pilot plant scale metallurgical testing on-site to provide data on leach kinetics for chalcocite and copper oxide ores is in progress. Rosario, a much larger deposit, has been delimited with approximately 65,000 meters of diamond drilling from surface and underground but still requires additional drilling to take the ore reserves to the same level of confidence as Ujina.

The project represents a challenge for development and production. It is the highest altitude major mining project in the world and, as such, planning and design must consider problems associated with the derating of machinery and decreased manpower productivity. The deposit is located in a seismically and volcanically active region within the Andean cordillera. Development of a sustainable, long term groundwater resource, without harm to the fragile high altitude environment, is a highest priority.

Two major deposits are present in close proximity to each other, and will share common infrastructure. However, the deposits differ greatly in ore grade, size, mineralogy, morphology of ore facies, ratio of potentially leachable versus non-leachable copper ores, stripping ratio, and geotechnical characteristics. These differences resulted in nearly 30 different development sequences being studied as part of the prefeasibility. These considered alternative process options of the secondary ores at variable recoveries, alternate process rates, initiating mining at Rosario, Ujina or both in parallel. The base case being developed in feasibility is a 60,000 tpd flotation operation with a separate oxide leach and SX-EW operation. Cathode production from the oxide ores is planned at approximately 50,000 tpy of fine copper. Should pilot plant testing prove sulphide leaching to be viable at Collahuasi, the merit of increasing the size of the leaching operation to include chalcocite-bearing ore will be revisited during the feasibility stage.

2 Location

The Collahuasi district is located 180 km SE of Iquique in the First Region of northern Chile (Figure 1). Mean elevation of the project is 4600 m.a.s.l. in a region of rolling to steep topography close to the border with Bolivia. At this altitude, Collahuasi will be one of the highest mining projects in the world.

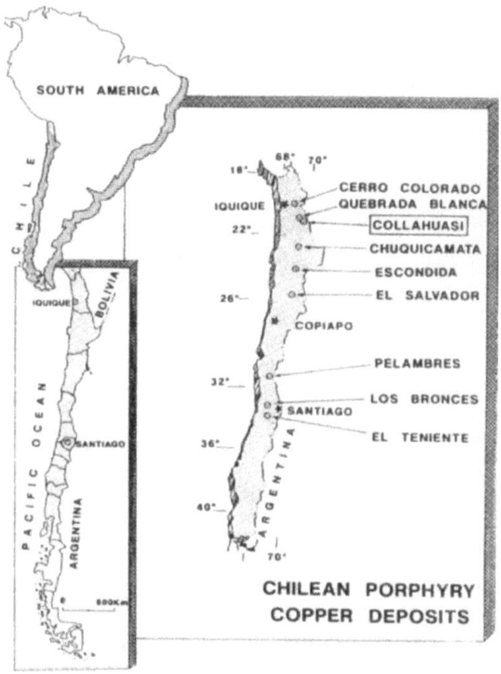

Fig.1.Location of Collahuasi and major Chilean copper deposits.

The Collahuasi deposits lie within a N-S striking belt of porphyry copper deposits which follows the trace of the Chilean Andes, a trend which contains such other major porphyry copper deposits as El Teniente, Disputada, El Salvador, Escondida, and Chuquicamata. The newly-developing Quebrada Blanca deposit is located just 7 km west of Collahuasi and forms part of the overall Collahuasi district.

Access is provided by three roads connecting the project with the major port cities of Antofagasta and Iquique via

Calama, Pica and the town of Huatacondo (Figure 2). In July of 1993 Quebrada Blanca completed construction of a new road linking the district with Iquique, via Pozo Almonte.

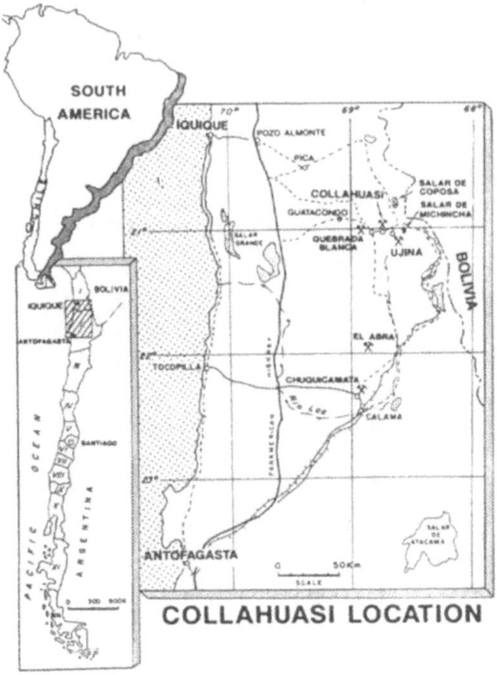

Fig.2. Location of Collahuasi and regional infrastructure.

A railroad spur line to Ujina was constructed by the Antofagasta-Bolivia Railroad Company in the 1920's to transport direct shipping ores from the historically mined copper vein deposits at Collahuasi to Antofagasta. This narrow gauge railroad is still in operation, periodically taking native sulphur ores, hand mined from the nearby volcanic craters on the Bolivian border, to the port of Antofagasta.

Collahuasi and Quebrada Blanca jointly operate an airstrip situated between both projects. At an elevation of 4200 m.a.s.l. it is one of the highest licensed airstrips in the world.

The high altitude has an important effect on the climate. Extreme temperature fluctuations from day to night characterize the district. Strong winds contribute to high wind chill factors. The mean average daily temperature at

Rosario is 4°C, varying from -25°C to +15°C. Temperature is the main concern for the ongoing sulphide leach testing, the bacteria at Collahuasi having been shown to become dormant below 8 degrees centigrade.

While bordering the Atacama desert, the climate is less arid, the main precipitation falling in the form of snow during the "Altiplano Winter", associated with severe electrical storms. Annual precipitation is 200 mm per year, although several years can pass with almost no precipitation. Evaporation rates and ultraviolet radiation levels are high, and humidity extremely low.

At the elevation of Collahuasi, the air has approximately 40 percent less oxygen than at sea level. Physiological and mechanical problems resulting from oxygen deficiency, high radiation and low humidity significantly affect project design. Design must also take cognizance of the high level of seismic activity in this region and the presence of three volcanoes which exhibit fumarolic activity within 15 km of the project.

3 History of the project

The multiple deposits and occurrences which comprise the Collahuasi district are shown in Figure 3. The district was first explored and exploited by the Incas, whose ancient copper tools and smelters are found near the Michincha Salar, adjacent to the Ujina orebody. Commercial development of the high grade copper-silver vein systems at Poderosa and La Grande began in the 1880's and continued until 1930. Recorded production is estimated at 330,000 tons of hand sorted ore grading 25% copper, 180 g/ton silver and 2 g/ton gold. No significant work was carried out from 1930 until 1976, when The Superior Oil Company and Falconbridge optioned Quebrada Blanca from ENAMI and acquired Collahuasi in 1978.

Exploration in 1978 and 1979 identified the key components of the Rosario system and the partially exposed alteration zone at Ujina.

In 1983 The Superior Oil Co. sold their mineral interests to Mobil, and in 1985, a new three-way Joint Venture was formed which included Chevron, Shell and Compañía Minera Doña Inés de Collahuasi (CMDIC), a wholly owned subsidiary of Falconbridge.

The new partners explored the high grade, bornite-rich Rosario vein system, an extension of the Poderosa vein system, for underground mining. Poor ground conditions and inadequate reserves led to the decision, in 1990, to focus on the larger porphyry copper resource, it having been determined earlier that the veins were hosted by a large disseminated copper deposit. By the end of 1991, a geological reserve of 1 billion tons of ore, grading approximately 1% copper and 0.025% molybdenum, at a 0.45% copper cut-off, had been defined at Rosario, not including the vein contribution.

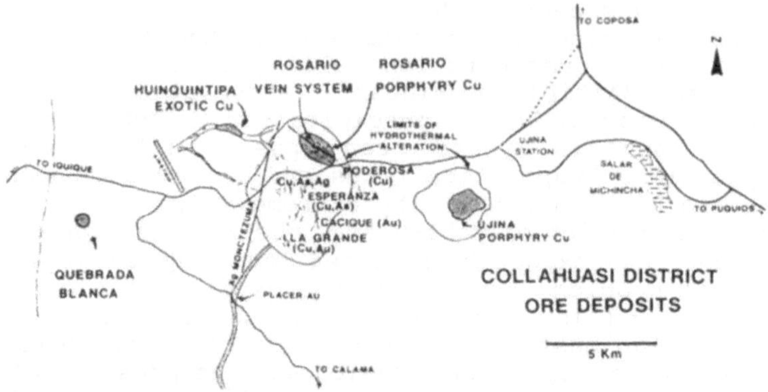

Fig.3. Ore deposits comprising Collahuasi District.

Concurrent with the definition of the Rosario resource, exploration was carried out to determine the potential of the entire joint-venture land holdings. In 1991, geophysical surveying, in conjunction with satellite image analyses were completed, and exploration was targeted on the Ujina area where an induced polarization geophysical anomaly, similar to that associated with the Rosario deposit, was identified beneath a young, unaltered volcanic cover. Encouraging drill results on the buried target led to a concentrated drill programme, which by year end 1992, had outlined a zone of strong secondary enrichment containing 168 million tons of chalcocite ore grading 1.57% copper, at a cut-off of 0.60% copper. The enriched chalcocite-bearing ore overlies a partially delineated primary (chalcopyrite-bornite) reserve of well over 400 million tons, grading 0.80% copper, at a 0.60% copper cut-off.

In December 1992, Chevron disposed of its 1/3 interest in the project to Minera Mantos Minorco Limitada, and in 1993, Shell announced its intention to sell its 1/3 interest in the project.

4 Characteristics of the Collahuasi orebodies.

4.1 General geological setting

Rosario and Ujina are porphyry copper deposits with significant tonnages of secondary enriched sulphide and oxide mineralization overlying much larger tonnage primary chalcopyrite-bearing ores. Spatially related to these large deposits are copper-silver, gold, and silver-bearing sulphide vein systems as well as an exotic oxide deposit called Huinquintipa (Figure 3).

Representative geological cross-sections through Rosario and Ujina are shown on Figures 4 and 5.

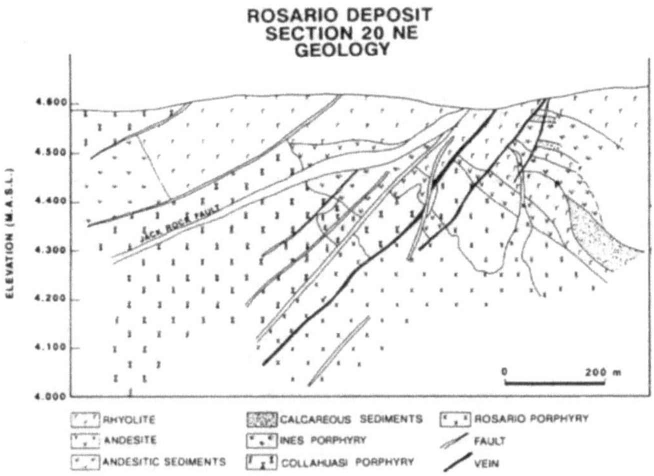

Fig.4. Geology, Section 20 NE, Rosario.

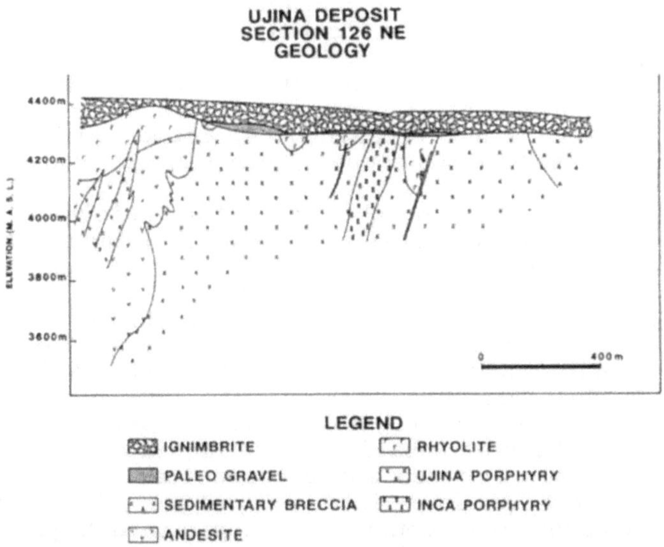

Fig.5. Geology, Section 126 NE, Ujina.

Mineralization at Rosario is hosted by Permo-Triassic age volcanic rocks, primarily andesites and rhyolites which have been intruded by three separate intrusive phases, the Inés, Collahuasi, and Rosario porphyries. All of these rock types host disseminated and fracture-controlled copper mineralization. Cutting the mineralization is a major set of bornite- and chalcopyrite-bearing veins, the Rosario vein system. Rosario is structurally complicated by extensive faulting. A major, post-mineralization normal fault, the Jackrock fault, has removed part of the deposit and juxtaposes weakly-altered and non-mineralized strata against the orebody.

The Ujina deposit is hosted by similar andesites and rhyolites, intruded by the Ujina porphyry. The Ujina and Rosario porphyries are petrographically similar and are likely apophyses off the same parent magma. Unlike Rosario, Ujina has not been subjected to the same post-mineralization structural complications.

Huinquintipa is an exotic copper oxide deposit, the copper having been dissolved from the Rosario deposit by groundwaters and redeposited in the form of chrysocolla and copper wad as cement in the interstices of a Quaternary gravel deposit, approximately 5 kilometers down the paleodrainage. Subsequent erosion has removed a large part of this deposit, which is exposed on surface and will require little stripping.

4.2 Mineralization

The porphyry orebodies are characterized by a variety of copper-bearing ore facies, subdivided on the basis of relative ratios of chalcocite (+ covellite) to the combined chalcopyrite and bornite content. Both deposits contain significant tonnages of copper oxide minerals, predominantly chrysocolla with minor brochanthite, copper pitch, copper wad and associated native copper. Huinquintipa contains only copper oxides.

The distribution and morphology of the various copper ore facies at Rosario and Ujina are shown in Figures 6 and 7. Geological resources of the individual ore types, for all three deposits, are shown in Table 1.

Detailed identification of the various facies of ore was given priority since chemical and biological leaching of both oxide and secondary enriched sulphide mineralization is potentially feasible.

For almost all drill holes at Ujina, and a high percentage of those at Rosario, heavy liquid concentrates of sulphide minerals for ten meter composites of mineralized drill core were pressed into briquettes, polished, and subjected to 2000 point count microscopic investigation. The relative ratios of pyrite, chalcopyrite, bornite, and chalcocite (+covellite) were then used to classify the ore facies based on the ternary classification scheme shown in Figure 8. This classification categorizes the ores into Strong Secondary Enrichment, Weak Secondary Enrichment and

Primary ore, based on the ratio of chalcocite (+ covellite) to chalcopyrite (+bornite).

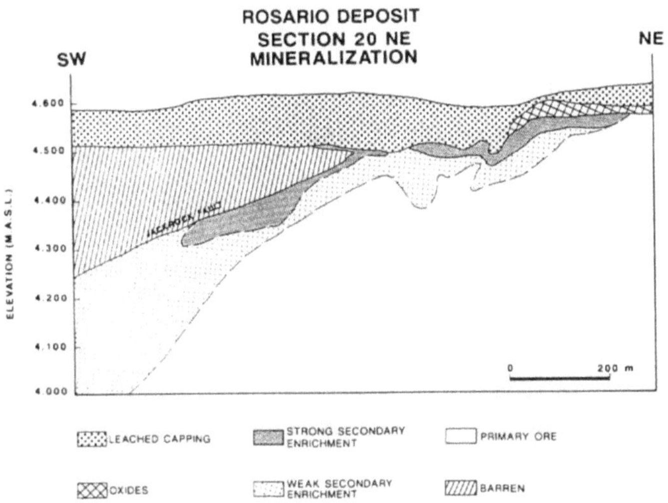

Fig.6. Distribution of ore facies, Section 20 NE, Rosario.

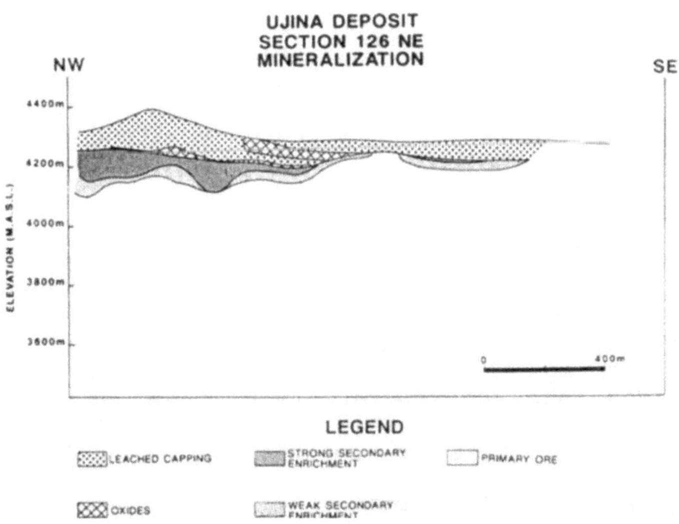

Fig.7. Distribution of ore facies, Section 126 NE, Ujina.

Table 1. Tonnage and grade of copper ore facies at Collahuasi at a 0.6% copper cut-off and potential process options.

Deposit	Ore Facies	Tons (millions)	Grade (% Cu)	Process
Ujina	Oxide	21	1.42	Chemical leach
	Strong Secondary Enrichment	168	1.57	Flotation/bacterial leach
	Weak Secondary Enrichment	51	0.86	Flotation
	Primary	> 400	0.80	Flotation
Rosario	Oxide	11	1.25	Chemical leach
	Strong Secondary Enrichment	23	1.3	Flotation/bacterial leach
	Weak Secondary Enrichment	136	0.96	Flotation
	Primary	> 600	0.97	Flotation
Huinquintipa	Oxide	7	1.9	Chemical leach

Oxide Facies

Oxide mineralization at Rosario and Ujina amounts to approximately 33 M tons combined with an average grade of 1.48% copper at Rosario and 1.25% copper at Ujina. A further 7 M tons grading just under 2% copper is available from Huinquintipa. Oxide mineralization at Rosario and Ujina overlies strong secondary enrichment and in both deposits tends to be displaced from the main secondary enrichment zones. Oxides represent zones where the secondary enrichment process was incomplete.

Secondary Enrichment

Secondary sulphide enrichment is defined as those parts of the orebodies where chalcopyrite (and, to a lesser extent, pyrite) has been replaced to varying degrees of intensity by chalcocite and covellite, therefore rendering them potentially amenable to bacterial leaching. Secondary enrichment in the Collahuasi ore deposits has been subdivided into two separate types based on the relative degree of replacement of the chalcopyrite (non-leachable) by chalcocite and covellite (potentially leachable)

Strong Secondary Enrichment

Strong Secondary Enrichment contains chalcocite as the principal copper-bearing mineral, on average the chalcocite accounting for over 95% of the copper minerals present. On Figure 8, a ratio of chalcocite/chalcopyrite greater than 75/25, regardless of pyrite content, classifies the ore as Strong Secondary Enrichment. A further subdivision based on pyrite content is also made (Very Weak, Weak, Strong and Very Strong Pyrite) to facilitate metallurgical testing and categorization. Strong Secondary Enrichment is a prime candidate for chemical and bacterial leaching. The morphology of the Strong Secondary Enrichment at Rosario is extremely irregular due to the channeling of supergene solutions and deposition of chalcocite along faults (Figure 6), in contrast to the more blanket-like morphology at Ujina (Figure 7) where fault control was not as strong an influence on secondary sulphide deposition. The Ujina enrichment ore would therefore be amenable to selective mining and would be the preferred source for application of leach SX/EW technology.

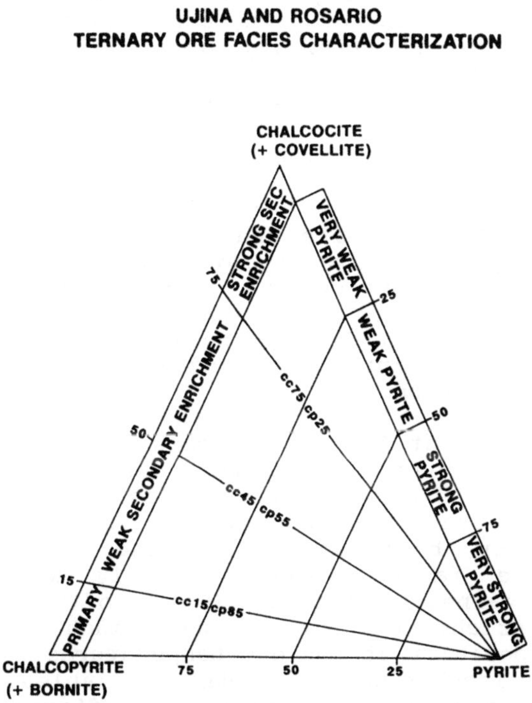

Fig.8. Ore facies classification, Rosario and Ujina.

Weak Secondary Enrichment

This ore facies contains a much lower chalcocite/ chalcopyrite ratio (less than 75/25) and lies beneath and peripheral to the Strong Secondary Enrichment. The chalcocite occurs as incomplete replacements in the form of rims on chalcopyrite and, more locally, pyrite grains. Copper grade is considerably lower and flotation is the only process option that could be applied. The contact between Strong and Weak Secondary Enrichment is sharp which will allow selective mining of the former. The contact, however, can only be determined with certainty microscopically, since the ore is visually identical regardless of the degree of replacement.

Primary ore

Primary ore at both deposits occurs beneath the secondary enrichment and accounts for the bulk of the tonnage of mineralization. Ore minerals are principally chalcopyrite with subordinate bornite and molybdenite. Primary grade at Rosario increases with depth, where a substantial tonnage (over 100 million tons) of Primary ore grading 1.4% copper occurs 300 meters below surface. Table 1 excludes the contribution to the Primary ore of the Rosario vein system, which will increase the grade by approximately 10 percent.

The Primary mineralization grade at Ujina averages 0.8% copper. Mining of Primary ore from Ujina is excluded from current development plans, the economic analyses indicating that mining should begin at Rosario once the Strong Secondary Enrichment at Ujina has been exhausted.

Metallurgy

The presence of chalcocite and copper oxide minerals as important ore components at Collahuasi complicates the selection of process option. The Strong Secondary Enrichment is a candidate for both flotation and chemical/biological leaching, the Oxide facies can be leached chemically, and the Weak Secondary Enrichment and Primary ores are only amenable to flotation.

To date, extensive testwork, both bench and pilot scale, has been carried out for both process options.

Extensive leach testing has been carried out in various sized columns and boxes in laboratories in Santiago, and at site. Samples were provided by PQ (3.5 inch) diameter drill core and, later, bulk samples from the Rosario and Ujina shafts. Oxide facies leaches readily with excellent recoveries obtained in less than 60 days of leaching. Strong Secondary Enrichment samples show much slower leach kinetics, especially at ambient temperatures beneath 10°C. The effect on leach kinetics of other variables, such as the quantity of acid in the cure step, different levels of ferric iron in the feed solution, solution flow rates, and crush size, have also been tested. Preliminary results confirm that the chalcocite leach kinetics are strongly dependent on temperature and the availability of ferric iron.

The potential of sulphide leaching at Collahuasi remains

the most critical outstanding metallurgical consideration. Only after significantly more pilot scale results are obtained during 1994, will the viability of a large scale leach operation be determined. Meanwhile, plans are to produce 50,000 tons per year of cathode copper from chemical leaching of the Oxide facies from all three deposits.

Flotation bench scale testing was carried out in Santiago and in the U.S.A. Samples were subjected to mineralogical characterization, MacPherson grindability testing and the determination of Bond Rod and Ball Mill Indices, Impact and Abrasion Indices, Flotation and settling and filtration rates. Flotation testing included continuous sand/slime rougher, rougher concentrate regrind, and locked cycle cleaner/recleaner testing.

The Ujina and Rosario ores are soft, especially the clay-rich secondary enrichment, and amenable to semi-autogenous grinding. Flotation testwork indicates that the average copper recovery in the Ujina Strong Secondary Enrichment will be approximately 92 percent. The final concentrate grade ranges from 30 to 60 percent copper, depending on mineralogy, for both the Rosario and Ujina deposits.

Mine Planning

The base case mine plan selected for the Collahuasi deposits at the prefeasibility study stage is to process 60,000 tpd in a flotation concentrator to produce in the order of 300,000 tons per year of fine copper and to leach Oxide ores to produce 50,000 tons per year of high grade cathodes. Selection of the base case was based on economic trade-off studies for approximately 30 different development options which were ranked in terms of NPV, IRR, Pre-production capital requirements, cash operating cost per pound of copper, and Payback period.

The total combined mineable sulphide reserves from Ujina and Rosario was considered to be 1.5 billion tons, sufficient for more than sixty years production at the 60,000 tons per day rate. Operational mining stages have been defined which have a combined reserve of approximately 672 million tons, representing 30 years of sulphide concentrate production.

Economic studies indicated that the optimum mining sequence is Ujina Oxide and Strong Secondary Enrichment, followed by Rosario Oxide, Strong Secondary Enrichment, Weak Secondary Enrichment and Primary ores. The extreme complexity of the morphology of individual sulphide ore facies at Rosario will make selective mining difficult and therefore flotation will likely be the process option for Strong Secondary Enrichment, even if sulphide leaching proves viable and is utilized for Ujina ores. Prestripping requirements at Ujina are much higher than at Rosario, but this is compensated for by the much higher, early year copper grades (the average ore grade for the first two years of production at Ujina will be 2.0% copper). After depletion of the Ujina Strong Secondary Enrichment in approximately eight years, a large tonnage of Weak Secondary Enrichment and Primary ore would be available without prestripping at the bottom of the pit, however the higher grade of the Rosario ores will then justify the abandonment of Ujina in favour of Rosario.

Mining of the Huinquintipa Oxide resource is scheduled to ensure a constant feed to the Oxide plant during the duration of cathode production.

Water Resources

For the base case mining scenario, approximately 600 liters per second of water will be required, assuming a conservative 30 percent recycling.

Definition of a high yielding, sustainable water resource for the project is, therefore, of highest priority. In the virtual absence of surface water, exploration is focused on two aquifers. Nearest Ujina, the Michincha high altitude salar is the surface expression of an aquifer that was drill tested by Collahuasi early in the 1980's. Water rights are shared by Collahuasi and Quebrada Blanca. Efforts are being directed towards maximizing this water resource given its close proximity to the operation.

Due to the magnitude of the planned Collahuasi production, a larger aquifer is required to ensure an uninterrupted supply of water without impacting on the high altitude environment, specifically the water level in the salars.

To define this resource, exploration drilling and pump testing has been carried out in the Coposa basin north of Ujina. The basin is a major graben feature bordered by the volcanoes on the east and the older rocks which host the copper deposits on the west. Up to 400 meters of young, semi-consolidated volcanic ash, debris flows and sediments infill the basin and comprise the aquifer. Aquifer modelling using the twenty six exploration and production wells drilled to date is ongoing and indicates that a large water resource is available to the project.

Effects of altitude

One of the main challenges towards a safe mining operation at Collahuasi is the effect of the extreme altitude. Operations will be carried out up to 5000 m.a.s.l., with mining beginning at 4200 m.a.s.l. at Ujina and later at 4600 m.a.s.l. at Rosario. Fuel-burning equipment must be derated due to less efficient combustion. Most important, however, is the effect of the altitude on human health and productivity. For most individuals, one full day is required for adaption or re-adaption after time off. As the majority of the staff are planned to be housed at sea level in Iquique, the change in altitude is severe. This has to be considered in designing working schedules, such that critical tasks as the operation of heavy equipment for example, must only be carried out after complete adaption.

In addition to adjustment to reduced oxygen, the effects of large daily temperature variations, very low humidity, high winds and high ultraviolet radiation need to be considered during operational design and towards the health and safety of personnel.

Collahuasi has developed a rigid pre-employment health screening and is studying how to limit the effects of abrupt altitude changes and other environmental factors on human health. As the most severe effect of the first days at high altitude is the difficulty in sleeping, the injection of oxygen into sleeping quarters is being considered. Cabins in heavy equipment will also be pressurized to improve ergonomics.

5 Conclusion

The Collahuasi project represents a world class copper project that will be an important copper producer by the end of the century. Pre-production stripping is due to begin in 1996 with production late in 1998 or early 1999. The distinctive geological and mineralogical characteristics of the multiple deposits which are present in the district, along with their location in an arid, high altitude environment, have presented a host of geological, metallurgical and engineering challenges and trade-offs. To develop and sustain a safe and efficient operation will require particular attention to the physiological effects of high altitude on the operational personnel, and an understanding of how best to minimize the effects of rapid changes in altitude.

6 Acknowledgements

The dedicated work of the entire Collahuasi staff is acknowledged, and in particular Guillermo Ossandón, Luis Catalán, Marcelo Villouta and their respective staff.

Solvent Extraction

Sulphuric acid plant installation—Ilo smelter, Peru

William Torres
Southern Peru Copper Corporation, Smelter Division, Ilo, Peru

Abstract

The Southern Peru Copper Corporation smelter at Ilo uses conventional Reverberatory-Converter process, smelting 1,050,000 short tons (950,000 MTPY) of concentrate per year. The three Reverberatory furnaces are equipped with waste heat boilers producing steam to generate electrical power. Off-gas of the Reverberatory furnaces are cleaned in electrostatic precipitators and discharged through two stacks. Off-gas of the seven converters are vented via hoods connected to a balloon flue plenum and cleaned in electrostatic precipitators before they are discharged through two stacks. Four holding vessels and two casting wheels are used to cast 300,000 tons of blister copper per year. Smelter facilities include a separation plant and a lime plant.

The installation of an acid plant is planned for 1995 and will produce 175,000 short tons per year (150,000 MTPY) of acid. The fresh water requirements for the new facilities will be met through the delivery of secondary treated effluent from a new sewage treatment plant at the Ilo townsite 15 Km. away.

The product acid will be used by the SX-EW projects in Toquepala and Cuajone Mines owned by SPCC. Excess acid will be exported by ship from the smelter via submarine pipeline.
Keywords: Acid system, consumption, gas purification, project description, storage.

1 Introduction

The Southern Peru Copper Corporation Smelter, in Ilo, operates a conventional plant with a nominal smelting capacity of 1,050,000 short tons per year (952,545 MTPY) of concentrate and 300,000 short tons (272,156 MT) of copper blister production per year.

The Reverberatory Department consists of one bath charge, silica brick suspended arch furnace with a nominal capacity of 750 DST per day (680 DMTPD) of concentrate, and two basic suspended arch, side charged furnaces, each with a capacity of 1,150 DST per day (1,043 DMTPD) of concentrate.

The Converter Department is equipped with seven Peirce Smith converters; the Casting Department is equipped with two 25 mould casting wheels and four vessels, having a casting capacity of 850 ST (771 MT) copper blister per day.

Offgas of reverberatory furnaces are cleaned in electrostatic precipitators and discharged through two stacks.

Offgas of the converters are vented via hoods into a common balloon flue plenum and then cleaned in electrostatic precipitators before being discharged through two stacks.

The smelter auxiliary facilities include a bed preparation plant thermal power plant, two sea water desalination plants, a lime plant, mechanical and electrical maintenance and fabrication shops and a pier at the Ilo port area.

Southern Peru Copper Corporation is currently modernizing the copper smelter.

This program will include a new Teniente Converter (CMT) with offgas cooling and cleaning equipment, an oxygen plant, and a single absorption sulfuric acid plant. Off gas from the CMT will be provided as feed gas to the new sulfuric acid plant. Present and future process flowcharts are presented in figures 1 and 2.

2 Acid Plant Project Description

2.1 General

A single contact sulfuric acid plant is planned at the Ilo Smelter facilities. This acid plant will produce 175,000 STPY (150,000 MTPY) of 98% sulfuric acid operating 330 days per year from the offgas of the new Teniente Converter (CMT). The CMT will have 90% "in stack" time. The acid product will be either 93% or 98% sulfuric acid, depending upon customer requirements.

The acid plant will be located north of the converter blower building and will be a metallurgical type contact plant. The acid plant will consist of gas purification system single contact acid section, and related installations.

The acid plant feed gases will be prepared by the installation of a spray cooling chamber (evaporative cooler) followed by an electrostatic precipitator.

Fresh water requirements of the new acid plant will be met through the delivery of secondary treated effluent from a new sewage treatment plant at the Ilo townsite, 15 Km away.

Seawater will be used for indirect cooling purposes, an additional vertical turbine pump (23,000 GPM = 5,220 m3/hr) will be installed at the existing pumping station.

2.2 Evaporative Cooling Chamber-Electrostatic Precipitator

The offgas from the CMT will be cooled in an evaporative cooling chamber with Sonic sprays in the upper part and dust collection system in the lower part. The gas stream will enter the cooling chamber at 1,260°F (682°C) and will leave at 630°F (332°C).

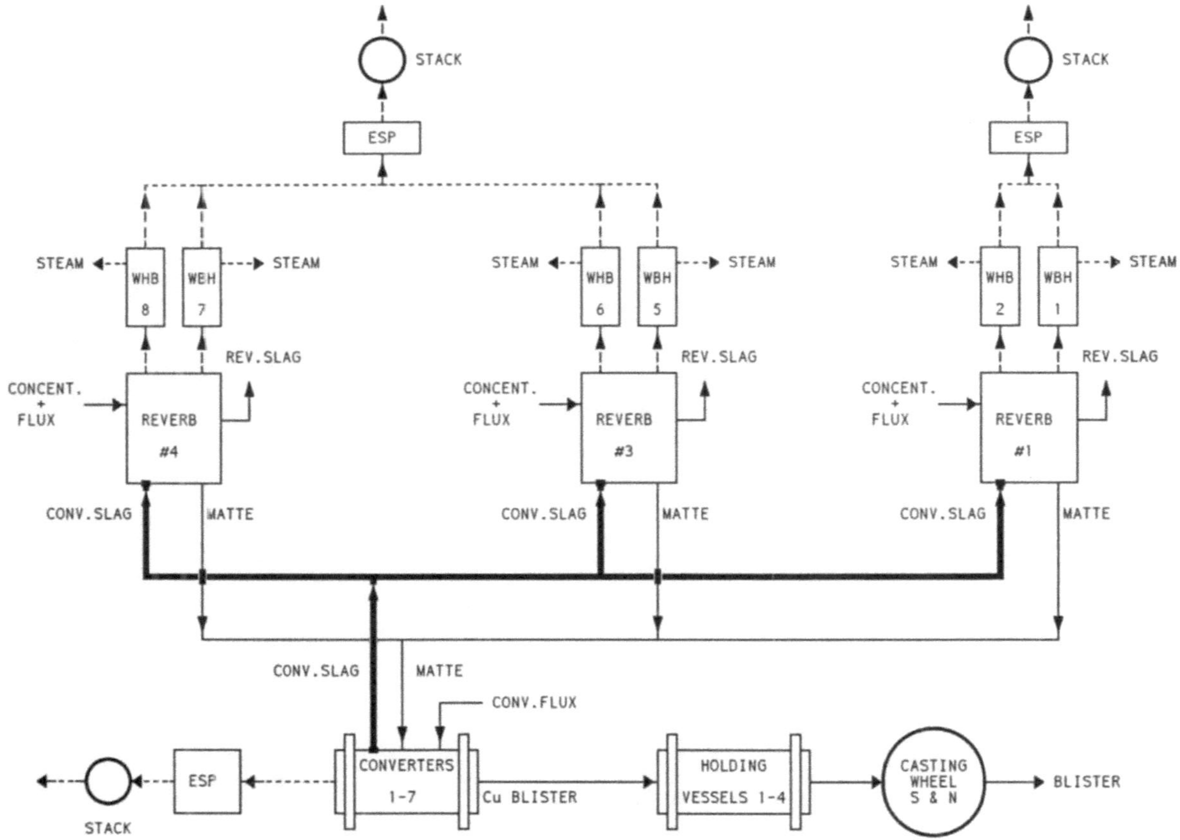

FIG.1 ILO SMELTER EXISTING PROCESS FLOWSHEET

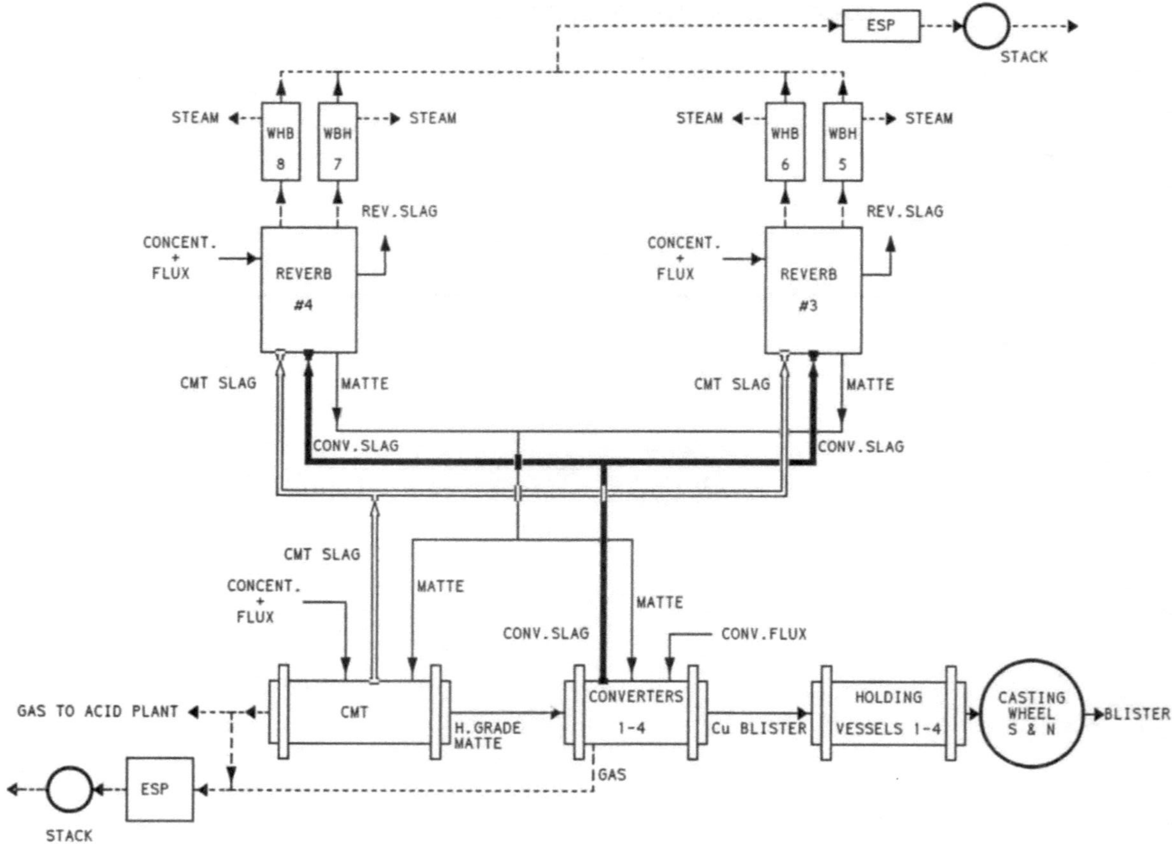

FIG.2 ILO SMELTER FUTURE PROCESS FLOWSHEET

The discharge gas from the cooling chamber will pass through an electrostatic precipitator (ESP) for solids removal before passing to the acid plant. The ESP will be a heavy duty hot dry electrostatic precipitator with three field in series each with its own separately adjustable electrical supply and rapping system and a collection efficiency of 98%.

The process offgas composition passing to the acid plant is presented in table 1.

2.3 Gas Purification System

The gas purification system will include an installation to clean the SO2 gas to allow the production of sulfuric acid in the contact section of the plant. The system will include a gas humidifying and cooling tower with associated weak acid systems and electrostatic mist precipitators to prepare the gas for the contact acid section (see figure 3). The system will remove dust, excess water, SO_3 or acid mist, and other materials deleterious to the equipment, its operation or product quality. The weak acid systems will include required heat exchangers, surge tanks, pumps, weak acid stripper, and all necessary accessory equipment.

The offgases from the CMT after primary cooling in an evaporative cooling chamber and cleaning in a hot electrostatic precipitator, are drawn into the inlet of the gas cleaning section, under slight negative pressure by the suction from the Main Blower.

Table 1. Gas Composition to Acid Plant

Wet Basis	Desing Conditions		Range
Gas Composition			
SO_2	Vol%	8.5	7.5 to 9.5
O_2	Vol%	13.0	11.0 to 14.0
N_2	Vol%	Balance	Balance
H_2O	Vol%	4.8	4.0 to 6.0
SO_3	Vol%	0.025	0.01 to 0.05
Particulate	Grains/SCF	0.5	0.1 to 1.0
	g/Nm^3	1.14	0.23 to 2.29
Particulate Size	Microns	100%<1	-
Pressure	H_2O	-0.5	0.0 to 1.0
	KPa	-0.125	0.0 to 0.249
Temperature	°F	630	600 to 700
	°C	332	316 to 371
Flowrate	scfm	38,800	42,700 to 35,300
	Nm^3/h	65,922	72,548 to 59,975

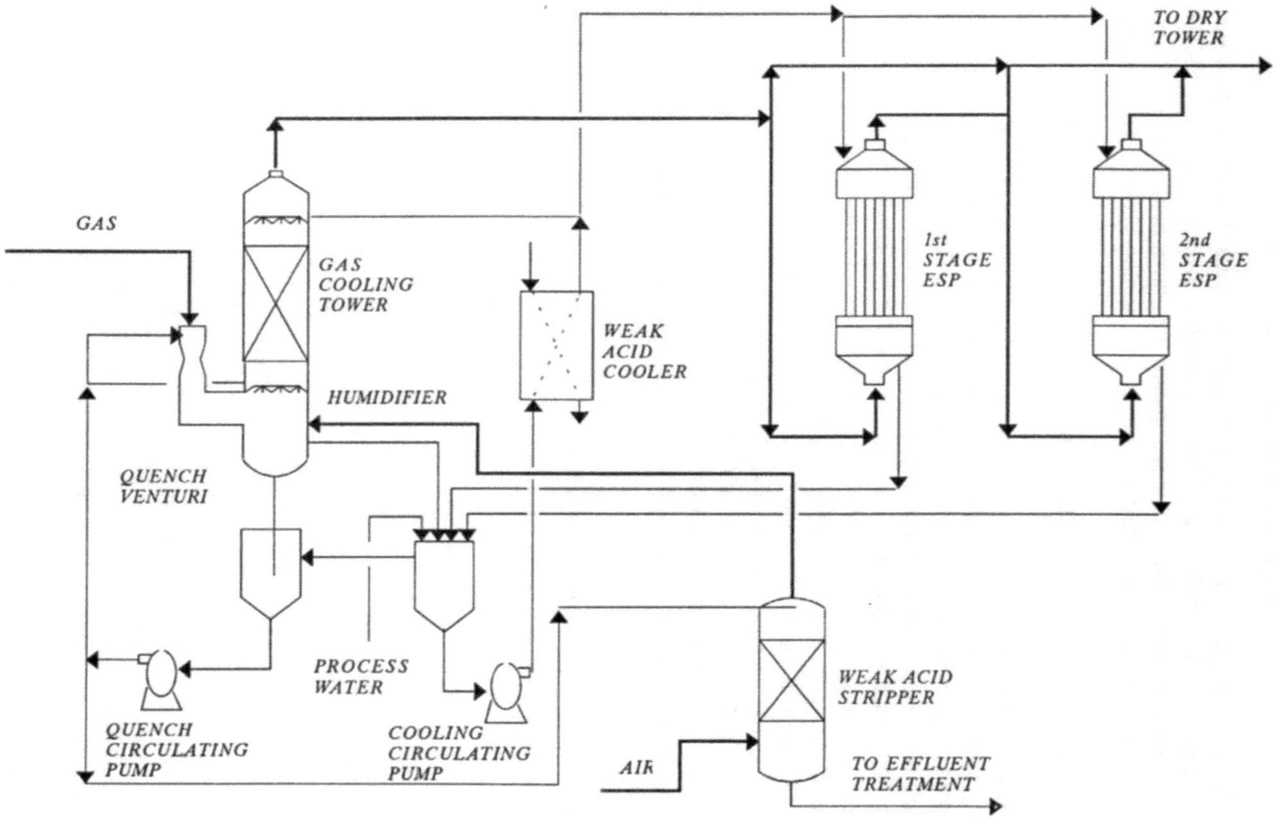

FIG.3 SULFURIC ACID PLANT,GAS CLEANING

The inlet gas temperature will be 600-700°F (315-370°C) and will contain normally 0.5 grains/SCF (1.14 grams/Nm3) of metallurgical, non volatile dust. The inlet pressure to the gas cleaning section will be regulated by controlling the guide vanes on the inlet to the main blower.

The hot offgas will first be drawn into the quench venturi (carbon steel shell, rubber lined, acid brick line and carbon brick-lined) where it will be quenched using sprays of circulating quench liquor to the adiabatic temperature of about 140°F (60°C). The ventury effect will promote quenching and separation of dust and liquid particulate.

The partially cleaned, quenched gas together with the spray liquid will pass to the humidifier. The humidifier comprises the bottom section of the gas cooling tower. Here the gas will be further contacted with sprays of quench liquor to ensure the gas is completely saturated and to remove additional particles. The liquor droplets will settle to the bottom of the humidifier and drain by gravity to the quench pump tank, the saturated gas will pass up into the gas cooling tower.

The gas will be cooled in the gas cooling tower by contact with a circulating stream of cooled liquor. The gas will leave the tower at 95°F (35°C) with a water content set by the water balance requirements to produce 98% acid at the autothermal (6.5% SO2) gas strength.

The cooled gas will pass through two stages of wet electrostatic precipitators (FRP shells and lead tubes) to complete the removal of the acid mist and dust particulate. As the gas flows up through the tubes, an electrostatic charge will be imparted to it by a high voltage electrode located in the center of each tube.

2.4 Contact Acid System

The contact acid section will include a drying tower with mist eliminator, main blower, gas-gas heat exchangers, converter with catalyst, absorbing tower with mist eliminator, stack, drying, absorbing and product acid coolers, preheater with combustion fan, product acid stripping tower, acid pump tanks, acid circulating, product and drain pumps, and all necessary accessory equipment and related facilities (see figures 4 and 5).

The gas from the gas cleaning system will be optically clear and saturated with water at 95°F (35°C). The gas will be drawn into the drying tower under suction from the main blower. Dilution air will be added as required at this point to maintain an oxygen to SO2 ratio greater than 1.1 to 1 in gas entering the Converter.

The gas will be dried in the dry tower (carbon steel shell, acid brick-lined, acid resistant ceramic packed tower) by counter current contact with a steam of 93% acid.

A blower located after the drying tower provides gas flow through the gas purification and contact section. The gas leaving the blower passes through the shell side of a series of heat exchangers, via ducts and by-pass ducts with valves as required for flow and temperature control. The heat exchangers will permit continuous operation of the acid plant with 6.5 percent SO2 without the use of the gas preheater. The heated gas then enters the converter catalyst beds.

The SO2 in the gas stream is oxidized to SO3 in the presence of a vanadium pentoxide catalyst and oxygen. The SO3 gas then flows through the tube side of the same heat exchangers providing heat for the incoming SO2 gas and cooling for the

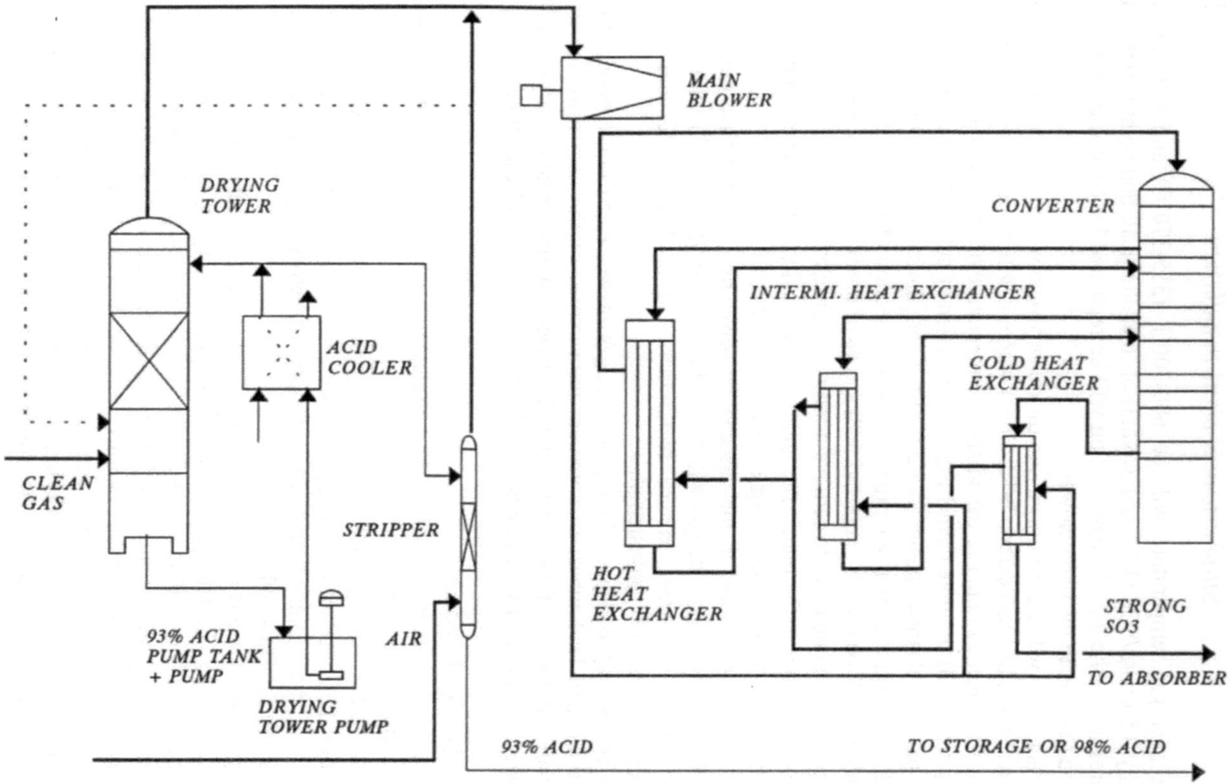

FIG.4 SULFURIC ACID PLANT,DRYING AND CONVERTING PROCESS

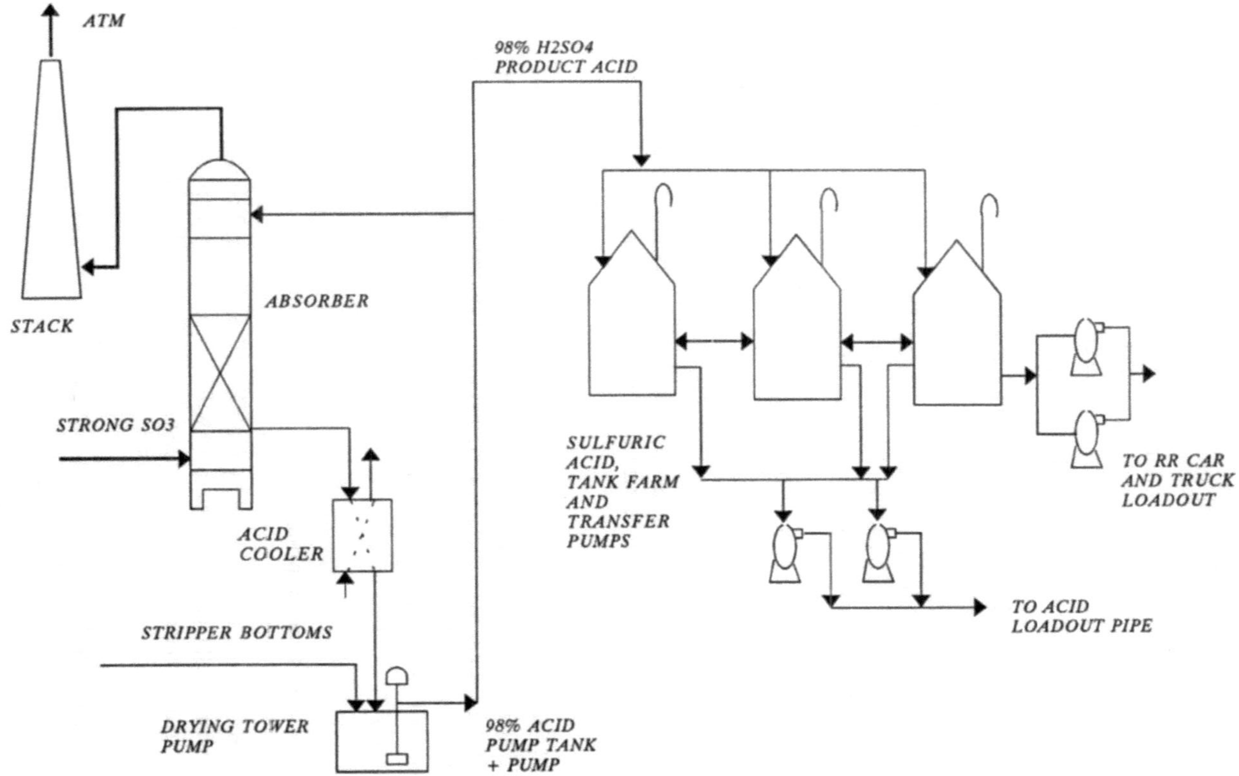

FIG.5 ACID PLANT, ABSORPTION AND STORAGE PROCESS

SO3 gas. Before entering the absorbing tower, the SO3 gas passes through a heat exchanger that lowers the temperature of the SO3 gas as required for absorption.

The cooled SO3 enters the absorbing tower which is circulating 98 percent sulfuric acid. The SO3 is absorbed by the 98 percent acid and remaining gas passes to the atmosphere through a mist eliminator and stack. The plant will be designed to convert a minimum of 96.5 % of the available SO2 to SO3.

A direct fire gas preheater system is provided to bring the converter up the catalyst reaction temperature at startup. This consists of a preheat exchanger, preheat furnace, and associated fans, controls and ducting. During the preheat sequence, air is drawn into the plant through the drying tower and passed through the cold heat exchanger. The preheater system is sized to heat up the plant from cold in approximately 24 hours.

The Contact Acid system will be designed to produce 175,000 STPY of 98% sulfuric acid operating 330 days per year.

2.5 Effluent Neutralization

Sulfuric acid will be formed in the gas cleaning and cooling system , the effluent will have a maximum of one percent of sulfuric acid. Burnt lime will be used for neutralization.

2.6 Acid Storage and Loadout

The acid storage facilities will consist of three carbon steel tanks having a capacity of 10,000 ST (9,000 MT) each. A common acid containment dike will be provided with sufficient capacity to contain the failure of any single tank.

Ships will be loaded through a new submarine pipeline (under evaluation). Additionally truck and rail loadout facilities will be installed.

3 Sulfuric Acid Consumption

The acid production will be used by the SX-EW project in Toquepala and Cuajone mines owned by SPCC. The excess acid will be exported by ship or sold locally.

After an acid market study, the following distribution of the acid production was estimated.

Table 2. Acid Distribution

	STPY	(MTPY)
Toquepala/Cuajone SX-EW	23,600	21,400
Local Sales	42,400	38,500
Export	109,000	98,900

4 Acknowledgment

The author wishes to express his gratitude to Southern Peru Copper Corporation for the opportunity to present this paper.

Reduction of leach–solvent extraction–electrowinning plant costs by attention to engineering

W. R. Hopkins
Davy International, Nonferrous Sector, San Francisco, California, U.S.A.

Keywords: Dump leaching, electrowinning, heap leaching, solvent extraction.

1 Introduction

In the past 25 years, the recovery of copper by leach/SX/EW techniques has matured as a technology. It has developed from an operation of secondary importance which treated old waste dump materials to produce a finished cathode of inferior quality to that of conventional electrorefineries into one of the bright stars of the future of the copper industry. Plants now process ore that is "mined for leach" in highly efficient plants to produce premium quality electrolytic cathodes at operating costs as low as \$0.35/lb.

During this time, there has been a continuing improvement in plant design as developed by both operations and engineering companies recognized in the field. New ideas and experience resulting from leach/SX/EW plant construction and operation are now being incorporated in the newest plant designs to lower first cost and so make marginal projects attractive. The leach/SX/EW plants for the 1990s show some significance variation from the plants of the last decade (Fig. 1). This paper discusses some of these aspects.

2 Dump and Heap Leaching

2.1 Crushing and Ore Placement

Most heap leaching operations (as distinct from dump leach) are based on careful analysis and pilot testing of the ore to be leached. They often involve a comminution step to achieve the desired size range for leaching and ore "agglomeration" or pre-curing with sulphuric acid and water prior to ore delivery and placement on the heap.

For several of its projects, Davy is employing the low-cost Hilfiker face wall construction system around crushing plants (Fig. 2). These produce significant savings over traditional concrete and are being used in seismic areas. Concrete is replaced by steel cages enclosing compacted fill or rock which extend progressively farther back into the face wall as depth increases. The resultant design can show savings of 50 percent over conventional reinforced concrete.

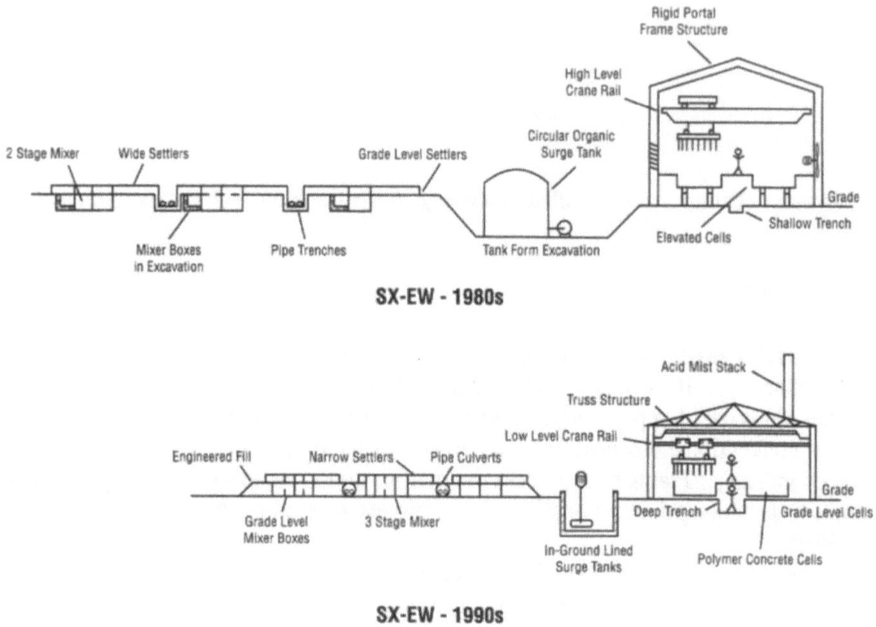

Figure 1 Changing SX/EW Plants

The coal industry, particularly in Britain and Germany, has been a rich source of new techniques for conveying and stacking. To reduce the cost of the long, conventional scissors conveyor systems between crushers and screens and also to reduce space requirements, innovative crusher layouts using "flexi-wall" conveyors or dual belt conveyors which allow material to be conveyed very steeply or vertically are now being considered (Fig. 3).

Another conveyor innovation with special potential for the movement of acid-agglomerated ores is the "pipe conveyor" where the loaded conveyor belting is formed by overlapping into a closed flexible tube (Fig. 4). This can be routed around curves and inclines so reducing the number of transfer points which can break up the agglomerated particles. Such units do not require rigourous alignment and the absence of spillage minimizes acid attack with acidulated materials.

Figure 2 Typical Hilfiker Type Wall Installation

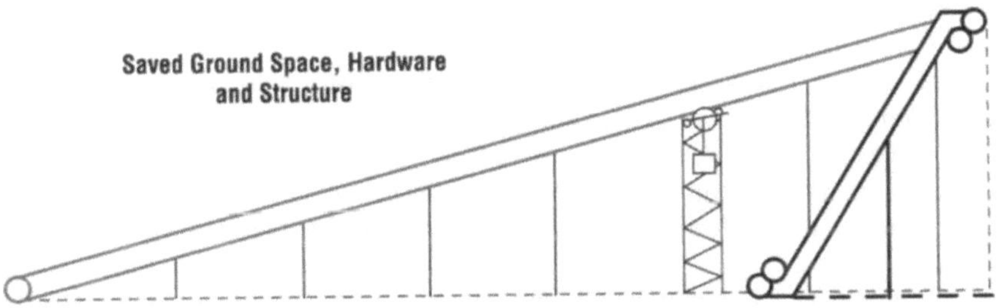

Figure 3 Space and Equipment Savings with Vertical Life Conveyors

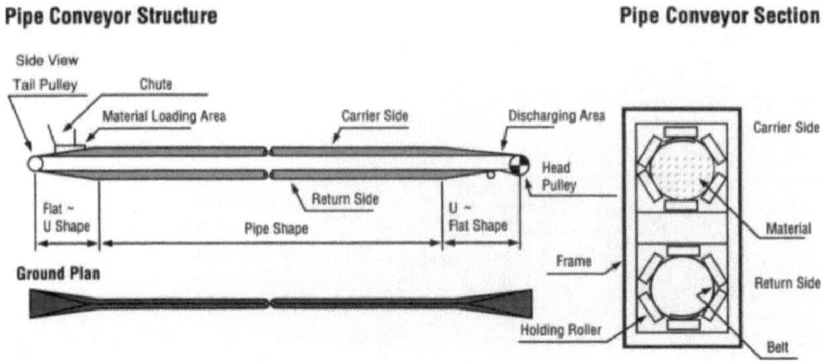

Figure 4 Pipe Conveyor Dischage (Courtesy Japan Pipe Conveyor)

Heap placement by conveyor stacking has been favoured for crushed and agglomerated ores because of its gentle handling. Trucks dump onto a primary conveyor that feeds a relocatable "zig-zag" set of intermediate conveyors which deliver ore to a radial stacker. The use of acidified pre-treated copper ores may preclude the use of usual conveyor equipment, particularly in a windy location. As an alternative to the radial stacker which usually works in retreat, track mounted spreader conveyors with a jockey conveyor feeding a stinger conveyor can be used for heap placement. The stinger places the heap edge and the whole tracked unit moves forward as the heap edge advances. These units allow high heaps to be placed. Long throw stackers based on coal industry design permit leach heap placement by a long boom stacker that may be truck dumped or conveyor fed (Fig. 5). The independent hydraulically powered device is mounted on its own tyred transporter. The discharge point from the boom conveyor can be raised and lowered to ground level so allowing friable agglomerates to be gently stacked to form the heap.

Additionally, the boom length permits the heap membrane to be rolled out between the stacker and the advancing heap toe. As a result, the stacker never travels over the membrane and with the reduction of impacts from the boom lowering capability, the thickness and cost of the membrane can be significantly reduced.

Systems of ore "agglomeration" or preconditioning systems are often included in the process. This can be done at the transfer points on the conveyor system, in a rotating drum agglomerator, or at the final stacker discharge using concentrated acid and water in spray chambers. The last system allows crushed ore to be successfully agglomerated just prior to heap building by passing the discharge from the final conveyor into a chamber in which it falls through horizontal curtains of acid and water spray (Fig. 6). The material then agglomerates effectively as it rolls down the stockpile heap. This technique allows the more usual agglomeration drum to be avoided and is particularly suited to low-cost installations. Agglomeration normally is more effective than initial wetting of placed material with high acid strength lixiviant as it produces a more homogeneous heap.

Figure 5 - Long Boom Stacker (Courtesy Macmin)

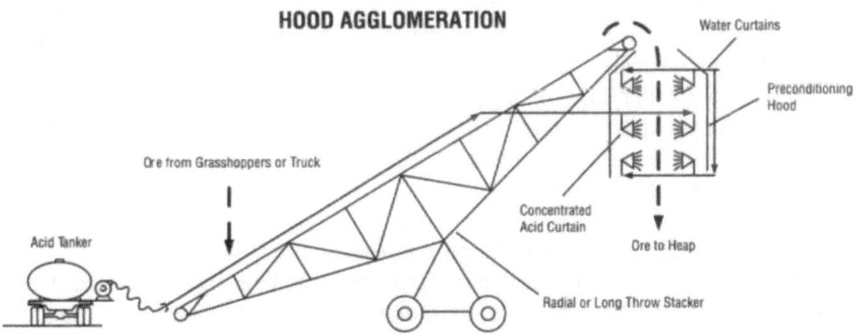

Figure 6 Hood Agglomeration

If the heap leaching operation is not in an arid area, a significant cost can be the provision of an emergency and rainwater run-off collection pond, especially for large areas of heap construction which will exist well into the life of the plant. This cost can often be deferred to later in the plant life when the heap areas are larger. It may be worthwhile to check the site contours to see if the heap can be constructed in a "valley" type arrangement. Here, the reclaim pump would be located over the low point of the valley to pump out PLS during normal operation. During a shutdown or high rainfall period, the valley can be used to store drain down liquor, thus saving costs on emergency storage ponds.

2.2 Solution Collection and Pumping

The leaching solution (recycled and perhaps acidified raffinate from solvent extraction) is returned to the top of the leaching dumps and fed to a network of irrigation headers. Large pipelines in stainless steel or high-density polyethylene are used. High-pressure sections of the lixiviant (and leachate) lines which may exceed the allowable pressures for HDPE pipe have normally been constructed in stainless steel. This can be expensive, especially if the presence of chlorides requires higher grade stainless such as 317L or 904L. Several plants are now employing HDPE pulled through carbon steel pipe for the high-pressure sections. This can provide significant cost benefits and is suitable when placed away from the acidic solution spraying areas.

The alternative of using only HDPE lines delivering to intermediate booster stations which ensures the maximum pressure for HDPE is never exceeded and is usually more expensive due to the additional pumps and installation required.

2.3 Solution Stacking

Dump and heap leaching requires careful coordination in the design of mining and the downstream SX/EW recovery plant. The most direct approach is single leaching combined with a water rinse to produce a single PLS stream. This results in a large flow of low-grade solution which requires a rather large SX plant. An improvement to this process is two-stage leaching or "stacking" of solution. With this option, an intermediate solution is produced by rinsing previously leached dump areas with fresh water or raffinate. This intermediate solution is then used to leach new areas to produce a smaller flow of higher grade PLS with a consequently smaller and cheaper solvent extraction plant and solution pumping/piping system. This benefit must, of course, be offset against the additional ponding and pumping required for the "stacking technique."

2.4 Pond and Pump Station Design Approach

Experience with lined ponds is that perforation of the liner by pipe entries and exits should be avoided where possible, as should the impact of entering streams on the liner material. As a result, cost saving techniques have been developed which allow entering solution flows to join the pond contents in a quiescent manner and to avoid impacts on the liner by allowing the feed HDPE pipeline to simply float out on the pond surface. Also, removal of solution can be performed by on-shore, simple, horizontal centrifugal pumps without liner penetration. This latter technique usually requires the use of an on-shore sealed prime tank which is fed by HDPE lines weighted to rest on the pond floor or below the varying solution surface level (Fig. 7). For smaller capacity plants, low-cost all polyethylene floating pumps have been developed in Chile which are suitable for the low heads encountered in many small heap leach situations when the leach heaps are nearby.

Figure 7 On Shore Pumps with Prime Tanks

3 SOLVENT EXTRACTION

3.1 Reagent Development and Plant Configuration

New salicyladoxime-based extractant reagents, which can be customized to suit particular extraction conditions, have undoubtedly been the main factor in reducing solvent extraction plant cost. They offer very fast extraction kinetics and very strong copper extraction over other ions with rapid phase disengagement. Their steeply rising and sharply bowed distribution isotherms for extraction allow very low copper raffinates and high percentage recoveries to be achieved with fewer stages in the low aqueous copper range.

Thus, two stages of extraction can replace the three or four extraction stages of earlier designs to recover at least 90 percent of the copper in the PLS; mixer residence times are now 2-3 minutes rather than the 3-4 minutes of earlier plants and copper selectivity over iron is now often better than 2000:1. Copper is removed from the organic phase more efficiently by two-stage stripping than by a single-strip stage because of the higher average acid content of the strip solution with two stages. To overcome this, additional extractant is needed in the single stage strip. This cost of additional reagent is much less than the cost of an additional strip mixer-settler, but when evaluated against organic losses over a period of years the second strip stage can often be justified.

The improved new reagents have made various alternative configurations, such as cross flow circuits, more attractive. Of special note is the series-parallel circuit. With this concept, the leach solution flow is divided into two halves. One stream is mixed with stripped organic in a single mixer settler and produces a raffinate. The partially loaded organic passes forward to counter currently contact the remainder of the leach solution in two other mixer settlers to produce a second raffinate stream and loaded organic. The two raffinates are similar in copper content, but copper recovery is approximately 6 - 7 percent lower than for a conventional 2 + 1 plant for the same organic extractant concentration. However, as the mixer-settler throughput is only half that of the conventional plant, settler areas may be 30-35 percent smaller, so saving inventory of organic and plant area.

By increasing the reagent concentration, overall copper recovery can be brought up to that of the conventional plant. Series-parallel is most effective for processing leach solutions of approximately three grams per litre of copper or less. For higher copper concentrations, the relations of desired copper recovery, reagent concentration, effective entrainment recovery and use of after settler devices must be studied.

Changing a conventional series circuit to a series parallel system allows inexpensive plant expansion. Minimum capital investment in one additional mixer settler will double PLS flowrate with recovery probably maintained. This has been done effectively at several plants in the Southwest U.S.

A new development has been the "double concentration" dual organic circuit. In this scheme, incoming PLS passes in series across two mixer-settlers each having a dedicated organic loop connecting it to a stripping stage. The organic loops use different reagent concentrations with that in the E_1 loop being higher than in the E_2 loop.

Electrolyte circulates in series across the two stripping mixer-settlers from the E_1 loop to the E_2 loop. This system produces the following effects:

- The higher organic concentration of the E_1 loop, plus an organic recycle lowers the overall organic flow compared to the E_2 loop. Thus the stripping mixer-settler for E_1 is smaller than that for E_2, and this can more than compensate for the higher organic reagent concentration.
- High concentration extractant organic entrainment in aqueous from E_1 is caught in E_2 and is substituted by entrainment of the lower concentration E_2 loop organic in the leaving raffinate.

Earlier SX plant designs often used a stripped or barren organic surge tank. Here loaded organic carrying leach liquor impurities as aqueous entrainment passes directly into the stripping mixer settlers where these impurities will contaminate the electrolyte. If chloride ion is present in the leachate, as is common from atacamite in Chilean ores, the use of stainless steel permanent cathodes may require significant chloride bleed to maintain the low levels required.

An organic wash stage prior to stripping overcomes the contamination problem, but an additional mixer-settler is needed. However, for light contamination, changing the stripped organic tank to a correctly designed loaded organic tank can reduce aqueous entrainment loss to the electrolyte to 10-50 ppm. These tanks have a format and feed system which produce flow down the tank length and use de-entrainment baffles to provide aqueous entrainment coalescence and removal.

Much higher organic layer space velocities than the 3.5 cm/s used normally are now being considered. Used in conjunction with new designs of aqueous take-off to replace the traditional aqueous weir which is difficult to adjust, this can reduce organic depths with significant savings in organic inventory. Also the settler configuration is more elongated which reduces the cost of roofing and allows easier distribution of flow into the settler. Conversely, there is now some evidence that increased organic layer thickness reduces entrainment significantly with the increased inventory soon paying for itself by reduced operating cost. In addition, the settler is elongated over the conventional design.

3.2 Multi Stage Mixers

Multiple mixing tanks in series improve contacting and approach to equilibrium, plus any short-circuiting is reduced, with a more efficient mixing action overall. The first tank in conventional systems contains the pump mixer-impeller with low-power hydrofoil turbines in the following tanks.

Using this concept, extraction stage efficiencies in modern mixer-settlers have improved from approximately 80% per stage to approximately 95%. Together with the improved isotherm shape for modern extraction reagents, this has eliminated extraction stages in some cases or results in generally improved recoveries.

Two and three mixer boxes per stage are used in most modern designs. Two discrete mixing stages give such high stage efficiencies (93-95% ate in extraction and 94-100% ate in strip) that a third stage offers little benefit from the mass transer standpoint.

However, the use of three boxes for the same total residence time can offer savings in civil costs in some forms of mixer settler construction. The depth of pipe trenches and excavations for mixers around the mixer settlers (or height of fill for settlers), impeller shaft length, impeller pumping requirements, impeller diameters, etc. are all reduced.

With the newer reagents, stripping is so rapid that only one or two mixer boxes with a shorter total residence time are sufficient in the stripping stages.

With large plants, there still remains a significant height differential between the settler and mixer box floors even with the lowest aspect ratio of mixer box advisable. Rather than place the mixer boxes in a well with the settler floor at grade, recent plans have preferred to place the settlers on compacted fill with the mixer boxes at grade. This simplifies maintenance, leak detection and plant housekeeping, and provides sufficient head for the gravity flows to reach the tank area without having to undertake large-scale excavation to lower the plant tankage.

3.3 Reverse Flow Mixer-Settlers

The layout of mixer settlers should minimize the length of large gravity flow lines and locate the pump mixer units as close as possible to the points in the adjacent settler from which they draw solution. The traditional form of mixer settler with its mixing boxes at alternate ends going from stage to stage achieved this apart from the recycle lines needed in strip stages to maintain the desired mixer box phase ratio of close to 1:1.

Although it is possible to remove the recycle volume from some intermediate point in the settler by a perforated cross settler pipe, the potential for flow and dispersion band disruption as well as blocking by crud has made designers and operators wary of this approach. A more elegant solution is to carry out the final weir discharge adjacent to the mixer boxes. This is part of the concept of the Krebs mixer settler, where dispersion is pumped by the pump mixer-impeller up to a down-settler launder, fed into the settler proper at the end opposite the mixer boxes and the separated phases are taken off at weirboxes at the mixer box end.

A similar concept lay behind the design developed originally by Falconbridge in their nickel/cobalt SX plant at Kristiansand. Here dispersion is fed into one side of the settler which acts as a launder, allowing the flow to turn at the settler end and return down the other half of the settler to discharge adjacent to the mixer box. This concept has now been developed by Davy in conjunction with Mexicana de Cobre for use in the copper industry. By using the side settler launder concept rather than the elevated launder conventional pump mixer, impellers with low head requirements and consequent low dispersion shear can be used. The new SX/EW facility for Mexicana de Cobre at Nacozari will employ these new "reverse flow" mixer settlers.

3.4 Mixer Settler Construction

The traditional design for most large mixer settlers has used 316L stainless steel lined concrete or freestanding 316L stainless steel construction. If chlorides are present in the leach solution, much more sophisticated and expensive stainless steels will be needed. To overcome this constraint, new designs have been developed for chloride-containing liquors for large units in HDPE-lined concrete, FRP, and recently, polymer concrete. In most countries, these are more economical than conventional stainless steel designs.

Mixer tanks are usually freestanding stainless steel, FRP or lined concrete. For the last, the stainless "liner" can be shop-fabricated as a tank and leak-tested. It can then be used as the inner form to pour the concrete walls to form the tank. The construction schedule can so be improved and leak repair in the field for these complicated tanks is avoided. Primary mixing tank internals such as false floors, division baffles and draught tubes are usually fabricated from stainless steel. The simpler internals of the secondary mixers allow them to be made with lower cost FRP or HDPE lining.

The settlers are usually roofed using translucent paneling. This minimizes organic evaporation losses, helps to lower solution heat loss, reduces contamination by dust, and prevents wind turbulence. To reduce initial capital investment, roofs can be added later as part of an operating maintenance budget. They are usually justified by two to three years' higher kerosene evaporation losses without them.

3.5 Tank Area Design

Typical copper SX/EW facilities contain 10-12 surge and holding tanks and equipment items grouped together into a tankfarm.

The tank area can be modified away from the traditional form, especially with short-life plants. Several innovative features in tank requirements, location and elevations, together with the use of low-cost agricultural and domestic tanks have achieved savings, especially for clients who desire a facility at minimum cost.

Some reagent and diluent storage can be eliminated by delivering make-up directly into the process tanks, and several new operations are using organic float columns in series to avoid costly electrolytic filters.

With newer reagents and construction methods, modern SX plants use fewer mixer settlers which seldom need to be drained or repaired over the plant life. It is possible to avoid provision of a holding tank by utilizing freeboard storage in the mixer settler units, but the procedure is cumbersome and operators will often elect to have a holding tank for operational ease. This tank offers possibilities for the use of low-cost lined agricultural tanks.

By careful selection of relative levels of grade, mixer box floor, settler floor and cell overflow elevation, the large excavation usually needed for these tanks can be avoided with all gravity fed tanks simply mounted partially below ground. The tanks are usually more economically constructed as HDPE lined rectangular concrete tanks rather than as stainless steel or FRP. By locating tanks, receiving pumped flows at grade, the size of the excavated tank farm is reduced. Davy is also now offering lead electrowinning techniques and new pumping systems, which avoid the common gravity-fed tanks entirely.

3.6 Organic Recovery Equipment

To recover organic entrainment from electrolyte, SX/EW facilities now tend to use column flotation units and dual media sand filters in various arrangements. With columns alone used in the present parallel configurations, some organic may slip by and contaminate the tankhouse electrolyte. As an alternative to dual media sand-type filters, Codelco-Division Chuquicamata has now developed a new lower cost organic entrainment coalescer. These are atmospheric pressure units which use a packed bed of organphilic materials. These "Chuqui filters" have been successfully trialed and are now supplementing and replacing sand filter installations in the Chuquicamata solvent extraction facilities.

To avoid the cost of sand filters for low-cost plants, two flotation columns in series with no dual media sand filters are effective. Used in conjunction with de-entrainment baffles in the settler, or an after settler and scavenger cells in the electrowinning plant, organic slip into the tankhouse can be almost eliminated. Organic removal by columns arranged in series can be expected to exceed 85%, down to 5 ppm organic in strong electrolyte.

Until recently, organic removal flotation columns have been fabricated from 316L stainless steel, but high-density polyethylene spiral-wound pipe or FRP units are now being proposed.

4 Electrowinning

4.1 Tankhouse Cranes

Technology developed for automated handling in zinc cell rooms is now appearing in copper tankhouses. The stainless steel blades of permanent cathodes are a consistent part for handling, and this facilitates gang loading of whole bale loads of cathodes into the cells.

Conventional dual-box girder electric overhead-type cranes are usually employed, but their control systems have been improved to allow very smooth acceleration and braking. Larger cell sizes have necessitated wider crane spans. For these, crane cabs move with the load by mounting them on the crane trolley. Operator vision is improved and "crane chaser" personnel requirements are reduced, especially for crane bales fitted with rotating hooks. Many new designs have dispensed with cabs completely and employ radio control by a floor-located operator. Drip trays which are remotely controlled allow dripping loads to cross freshly loaded cells without dripping electrolyte on contacts.

These improved crane designs and the use of permanent cathodes have allowed greater levels of automation of copper tankhouses. Addressable cranes with PLC units are used in some new installations and designs are available for fully automated systems. Two cells in different half sections can be served by two trolleys at the same bridge position, or a single trolley is available that can lift from two adjacent cells and interleave the load to feed the cathode machinery. Remote-controlled cranes can be operated in a "pitch and catch" mode by two part-time operators, one at cathode handling and the other at cell loading and dedicated crane operators are not needed.

Low-cost cranes using monorail hoists on a single traveling crane girder have been developed for the lower loads and less intense duty cycles of small tankhouses. These use radio control and are very cost-effective, being of the order of 50 percent of the cost of a conventional double-box girder design.

Virtually leak-proof polymer concrete cells, which are proposed for modern plants, can be placed at or near floor level. This permits the use of gantry crane systems running on rails at grade. Significant building cost savings can be realized in extreme seismic areas, as the scheme allows for low-cost agricultural or pre-engineered buildings.

Larger tankhouses may use a two-parallel runway configuration with multiple cranes. A crane transfer car in this situation can be very beneficial to move a disabled crane into a maintenance bay, allowing unimpeded operation of the remaining units. It can also permit crane transfers which allow concentration of cranes in one aisle if required and reduce the number of cranes needed.

Suspended transfer car designs can relocate a crane between aisles in less than one minute and take up no floor space.

4.2 Cell Construction

Since the late 1960s, the normal cell construction has been concrete lined with a flexible PVC liner with "buffer" sheets to protect the lining from electrode impact.

The past two to three years have seen increasing acceptance of individual cell boxes cast from vinyl ester polymer concrete, which do not require a liner. These are well-proven in Africa, the U.S.A. and Chile. Although of similar cost to the plastic-lined concrete cell, they are economically attractive because there are cost savings in areas apart from the cells alone which result from their leak-free character. For example, they can be placed directly on grade so lowering tankhouse buildings. The locations of electrode machine conveyors and take-off points for cathode stacks may determine the building height in these modern electrowinning plants. Correctly fabricated, the polymer concrete is several times stronger than conventional reinforced concrete and is fully corrosion-resistant to the electrolyte and any organic entrainment throughout its thickness. Several North American and African copper and zinc tankhouses have used these cells for some years successfully and most modern new plants will use this type of cell. The principal manufacturing facilities for the cells are in the U.S.A., Australia, Poland, Austria and Chile. There are variations which utilize FRP tanks within the polymer concrete and polymer concrete panels making up a tank which is supported in a steel tank skeleton.

The "jumbo" cell design places all the cells in a half section in one large jumbo cell tank with no dividing walls. A carbon steel tank is used with a glued-on liner of plasticized PVC. In the Russian Republic, a similar tank concept is fitted with internal walls of acid-proof brick, with flow slots alternately at top and bottom. Such multiple cell tanks have not been actively considered in electrowinning plants due mainly to the convenience of the newer cell systems and apprehensions regarding stray current losses.

4.3 Tankhouse Buildings

Low-cost building designs can now be considered for tankhouses. Such designs use low-cost "agricultural" type structures suitably strengthened and protectively coated. They can be pre-engineered, standard designs and employ innovative methods of fixing roofing and sliding.

By employing shorter cells, low-cost plants can be designed to allow standard off-the-shelf building widths to be chosen, thereby reducing the cost of the tankhouse building. For a given tonnage more of the smaller cells are required, but this cost is partly offset by increasing savings in the current power component of the rectifier and busbar units as compared to additional transformer voltage cost.

Prepackaged, relocatable buildings can offer savings for control rooms, offices, laboratories and administration buildings. Many countries' electrical standards allow motor control centres to be locked off set of MCC cabinets into which a low cost PLC-based control computer and VDU are incorporated. The combined "control room" and MCC assembly can be housed in a low-cost prefabricated structure.

4.4 Tankhouse Ventilation

For large tankhouse installations, control of the acidic mist created by evolving anodic oxygen is becoming of crucial importance. Sulphuric acid has been listed in some locations as carcinogenic and this will make the present strict regulations on acid-mist levels even more stringent. The acid-mist source control and building ventilation system of the tankhouse should be considered from early in the project.

Single-aisle tankhouses can use horizontal air curtain schemes fairly effectively. These have been shown to produce good removal of mist with primarily only a movement of air about one metre above the cells. They are a great improvement over the large building air change concept used in both natural and forced ventilation schemes in earlier designs. However, wide multiple crane aisle installations should always be hydraulically modelled to develop optimum fan placement and air flow type. Source control for the mist is receiving wide attention, with various types of gas bubble coalescers and built-in cross cell suction systems under review. These usually supplement the floating ball or bead blankets commonly used.

Effective control of mist level and easy means of collection of mist-laden air for scrubbing and discharge can be achieved by controlled and collected air flow in a new scheme known as "Close Capture." This concept is based on steel and plating industry technology for localized fume collection. Special jets at the cell feed end provide individual horizontal air curtains to a suction plenum formed by the space between two rows of cells. It is therefore largely independent of the building configuration. By controlling the air around the cells, the rest of the building is contaminant-free with consequent savings on structure, paint systems, etc.

The above mechanical draught air flow control schemes allow the capture and handling of misted air by ducts and cleaning in scrubbers before discharge.

5 Conclusion

The foregoing examples illustrate many potential cost-saving areas for new leach/SX/EW projects. The discussion has not been specifically quantitative as many of the ideas suggested will depend on local conditions and costs such as steel, concrete, plastic materials, etc. It is seen, however, that there are many facets of a new design which can be refined to cut investment costs and/or improve operating changes.

Southern Peru Copper Corporation's Toquepala and Cuajone leach projects, Peru

Igor Gonzales
Southern Peru Copper Corporation, Toquepala, Tacna, Peru

Abstract

Southern Peru Copper Corporation (SPCC) is to install copper leaching facilities to process 432.6 million mt low grade sulphides (0.176% Cu) at its Toquepala mine and 13.6 million mt high grade oxides (1.0% Cu) at its Cuajone mine. Cuajone pregnant leach solution (PLS) will be pumped to Toquepala and treated with Toquepala PLS to produce 97.4 mt/day (35,064 mt/yr) of high quality cathode copper in a single SX-EW plant.

The integrated project, which will cost $103 million, should start operation in the middle of 1995. Operating costs are expected to fluctuate between 23 and 30 c/lb copper over the 15 year project life at Toquepala, and 20 year project life at Cuajone.

Introduction

Southern Peru Copper Corporation (SPCC) is planning the installation of an integrated copper leaching project consisting of a dump leaching and solvent extraction-electrowinning facility at its Toquepala mine, and crushing and heap leaching at its Cuajone mine. Pregnant leach solution from Cuajone will be pumped to a single SX-EW plant at Toquepala.

Sampling of the sulphide dumps at Toquepala, together with agitated bottle and column leach tests, began in 1985. In 1986, dump leach pilot testwork began at Dump 3250 to evaluate the main process variables such as irrigation rates, rest periods, and curing acid concentrations. Dump effluents were treated in a small scrap iron precipitation plant to produce cement copper. The first feasibility study on leaching the Toquepala dumps was carried out by Bechtel International of San Francisco, California, during 1986.

Testing of the Cuajone oxide material was first carried out in 1968 during mine development investigations. Testing was resumed in 1988 and Engineering & Construction International of Tucson performed a feasibility study in 1990.

The metallurgical results showed that the Toquepala and Cuajone ores were amenable to dump and heap leaching respectively. Both feasibility studies concluded that leach-SX-EW operations would be technically and economically viable.

TABLE 1
Toquepala Dump Material: Tonnages and Grades (September 1993)

Material Type	Unit	Dump No.1	Dump No.2	Dump No.3	Dump No.4	Dump 3250	Dump T-11	Total
Leach	'000 Tons	20,231	42,270	91,725	24,671	5,550	10,102	194,549
Grade	% Cu	0.332	0.340	0.352	0.340	0.356	0.350	0.346
Waste	'000 Tons	21,390	74,828	21,256	64,082	32,722	68,581	282,859
Grade	% Cu	0.077	0.065	0.082	0.066	0.057	0.048	0.062
Total	'000 Tons	41,622	117,098	112,981	88,753	38,272	78,683	477,409
Grade	% Cu	0.201	0.164	0.301	0.142	0.100	0.087	0.178

TABLE 2
Mineralogical Composition of Dump Material (January 1992)

Mineralogy (Mineral)	Unit	Dump No.1	Dump No.2	Dump No.3	Dump No.4	Dump 3250	Dump T-11	Total	% Total Tonnage	% Total Contained Cu
Enriched	'000 Tons	3,177	8,039	10,973	4,524	1,283	1,753	29,749	6.73	12.72
(Chalcocite)	% Cu	0.315	0.324	0.319	0.313	0.322	0.288	0.317		
Transitional	'000 Tons	1,021	3,720	4,336	2,132	884	neg.	12,093	2.74	5.12
(Bornite)	% Cu	0.292	0.321	0.314	0.303	0.335		0.314		
Primary	'000 Tons	28,064	94,005	86,879	74,834	35,601	30,367	349,750	79.16	79.74
(Chalcopyrite)	% Cu	0.181	0.131	0.301	0.114	0.085	0.134	0.169		
Leach Capping	'000 Tons	2,995	240	287	136		46,563	50,221	11.37	2.42
(Oxides)	% Cu	0.040	0.085	0.067	0.068		0.035	0.036		
Total	'000 Tons	35,257	106,004	102,475	81,627	37,767	78,683	441,813	100.00	100.00
	% Cu	0.184	0.152	0.303	0.130	0.099	0.079	0.168		

In April 1992, SPCC awarded the basic engineering studies for both projects to Bechtel International and by July the same year the completed studies concluded that a 13.6 mt/day cathode copper facility would be suitable to process Cuajone oxides over 10 years, and a 83.8 mt/day plant in Toquepala to treat sulphides for 15 years.

However, in 1993 this scope evolved to incorporate additional ore reserves at Toquepala and Cuajone, and the proposal to transfer pregnant leach solution from Cuajone to Toquepala for treatment in a single SX-EW plant. This new approach resulted in a longer project life for Cuajone, estimated at 20 years, with significant reductions in capital and operating costs. Detailed engineering is scheduled for early 1994 and it is expected that a 97.4 mt/day facility will be operating at Toquepala by the third quarter of 1995.

Toquepala

The Toquepala open pit copper mine began operations in 1959. Conventional crushing, milling, and flotation techniques are employed for copper and molybdenum recovery.

At various periods, cut-off grades of 0.35%, 0.45%, and 0.55% Cu (current cut-off) have been used. Material below the cut-off grade is classified as either waste (less than 0.1% Cu) or leach material. The six dumps to be leached, both waste and leach material (See Table 1), contain 432.6 million mt at an average grade of 0.176% Cu. The copper is present mainly as chalcopyrite and chalcocite. Iron averages 4% and is present mostly as pyrite. Gangue minerals average 65% silica and 15% alumina.

The historical composition of the Toquepala dumps indicates that 80% of the copper is present as chalcopyrite, 13% as chalcocite, 5% as bornite, and 2% as oxide copper (See Table 2). Approximately 142 million mt of material grading an average 0.31% Cu will be added to these dumps over the next 15 years. 95% of the contained copper in this material will be chalcopyrite and the reminder chalcocite and bornite.

Successful leaching of primary sulphides such as chalcopyrite requires the alternation of leaching periods with long oxidation (rest) periods. During the oxidation stages the copper solubilizes with the aid of bacteria, and ferric sulphate and sulphuric acid are liberated during the leaching stage. In general, it can be said that the longer the rest period the more available the copper salts become. At Toquepala, rest periods of 26 weeks are planned.

A program of dump conditioning was initiated in 1987; water with a low acid concentration (0.8 g/l H_2SO_4) being sprayed on the dumps at rates of 7 to 15 l/hr/m^2 to promote bacterial activity for increased copper recoveries (See Table 3). It is estimated that a moisture content of 6-8% and a pH close to 2 will provide suitable conditions for bacterial growth. Dump conditioning will be complete by the middle of 1995, just before plant start-up.

TABLE 3
Toquepala Dump Conditioning Program
Water and Acid Consumption (April 1993)

Year	Dump No.1		Dump No.2		Dump No.3		Dump No.4		Dump T-11		Yearly Total	
	Water ('000 gal)	Acid (mt)	Water ('000 gal)	Acid (mt)	Water ('000 gal)	Acid (mt)	Water '000 gal)	Acid (mt)	Water ('000 gal)	Acid (mt)	Water ('000 gal)	Acid (mt)
1986			44,723								44,723	
1987			20,586		6,219	18.84					26,805	18.84
1988					61,411	185.70	46,345	111.76			107,756	297.46
1989					47,041	152.09	35,817	113.83			82,858	265.92
1990	677				22,166	108.63	17,043	60.33	4,346	32.32	44,232	201.28
1991	5,134		24,181	133.84	9,080	44.95	27,288	92.74	14,603	64.81	80,286	336.34
1992	36,999	138.50	81,619	299.69	43,375	131.00	76,330	293.91	14,685	66.82	253,008	929.92
1993*	10,863		20,884	60.90	45,123	140.66	58,679	182.73			135,549	384.29
Total	53,673	138.50	191,993	494.43	234,415	781.87	261,502	855.30	33,634	163.95	775,217	2434.05

(*) Consumption calculated until April 1993

The dump surface will be ripped to a depth of approximately two metres prior to the start of leaching and rest cycles. Ripping permits better oxygen and solution penetration, reduces solution ponding, and promotes better aeration during rest cycles. The feasibility study is based on sprinkler irrigation but due to probable future water shortages drip irrigation will be tested before plant start-up. Irrigation tests on the dump slopes has demonstrated that the slope contribution to the PLS copper grade is significant.

Dump areas will be divided into leaching blocks, which will in turn be subdivided into eight strips. Each strip will be leached for eight weeks. As leaching progresses, a new strip will be incorporated every week while the strip that has completed the eight week cycle will be put to rest. Table 4 shows this distribution by dumps.

The eight week cycle is based on field data and the experience of other operations. It is the estimated time required for the PLS grade of the strip to fall below 0.45 g/l Cu. Thus, the simultaneous leaching of strips at different levels of copper contribution should yield the average 1.43 g/l required for solvent extraction.

During operations, controlled pH SX raffinate will be distributed on the dumps at a rate of 53,400 l/min (14,100 gal/min). Considering the 15% loss due to evaporation and seepage established during pilot tests, this will provide a 45,200 l/min (12,000 gal/min) flow of pregnant solution grading 1.43 g/l Cu to the SX plant. Approximately 20% of the total dump area will be under leach at any one time.

PLS will be collected in two HDPE membrane lined dams. Leak detection systems, emergency dams, and monitor wells will be installed to ensure that no PLS spills are lost to the ground. Copper production is estimated at 83.8 mt/day for the first five years and 66.8 mt/day thereafter for a total recovery of 36% over 15 years. Testwork on Dump 3250 has corroborated this calculation as 30% extraction has been obtained in six years of pilot leaching.

Cuajone

The Cuajone copper complex, located 45 minutes from Toquepala, came on stream in 1976. Like Toquepala, Cuajone is a conventional open pit porphyry copper mine using milling and flotation to produce copper and molybdenum concentrates.

A total of 13.7 million mt of copper oxides grading 1.0% Cu have been stockpiled on six dumps. One of these, Dump 3430 NF, is a higher grade stockpile with 2.3 million mt grading 2.1% Cu (See Table 5). The Cuajone oxide copper minerals are mainly chrysocolla and malachite, and the sulphide mineral chalcocite.

Material will be excavated using an existing 9 yd^3 P&H 1800 shovel and hauled by 36 mt trucks to the 180 mt/hr crushing plant. The amount of material to be crushed will vary depending on the grade of the various stockpiles being exploited at the time. Ore will be crushed to 100% minus 13 mm, cured, agglomerated, and transported to the permanent pad area where it will be placed in 2-m lifts. Curing and agglomerating will be carried out by reagent sprays during belt-to-belt transfers before the ore reaches the stacking conveyor.

TABLE 4
Available Leaching Area Distribution (January 1992)

Dump	Leachable Material at Start mill.ft^2	Surface Area at Start '000 ft^2	Number of Leach Strips: Active	Under Oxidation	Total	Total Flow to Active Dump Area (gal/min)
No.1	31.85	1,486.3	8	31	39	2,100
No.2	87.05	1,764.6	8	30	38	2,600
No.3	88.13	1,842.7	8	26	34	3,000
No.4	92.98	2,005.0	8	35	43	2,600
3250	36.71	1,057.6	8	29	37	2,200
T-11	59.50	2,321.9	8	51	59	1,600
Total	396.22	10,478.1	48	202	250	14,100

TABLE 5
Cuajone Oxide Material: Tonnages and Grades

Stockpile	'000 Tons	% Cu	Tons Cu
3430 NF & Fines	2,474	2.086	51,608
3460 SL	1,261	0.795	10,026
3470 NL	4,215	0.649	27,359
3490 SL	1,919	0.619	11,878
Panel Platform	916	0.892	8,173
RR#1L	4,324	0.994	42,976
Total	15,109	1.006	152,020

The permanent pad will be built in stages as the need arises. The first stage will cover approximately 50,000 m^2. Agglomerated ore will be cured for 14 days and leached for 14 days before the next lift is placed atop. Copper recoveries of around 80% per leach cycle are expected. The pad will be constructed with leak detection systems to make it environmentally safe.

Leach solution application rates will be approximately 1,100 l/min or 300 gal/min (at 20 l/hr/m^2). PLS recovery will average 1,060 l/min (280 gal/min) at 10 g/l Cu and 4-5 g/l H_2SO_4 and a pH of 1.5-2.0. Acid consumption is estimated at 2.5 mt/mt Cu.

Cuajone PLS will be transferred to the Toquepala SX-EW plant via a 35 km pipeline which will follow the existing railroad through a series of five tunnels with a total length of 30 km. The pipe will be mounted above the existing Cuajone tailings launder which also follows the railroad. Leak detection and control systems will be included. Two 1,060 l/min capacity pumps, one operating and one standby, will pump the PLS to the start of the railroad from where it will flow by gravity to Toquepala, and thence be pumped a further 5 km to the SX-EW plant.

Acid Balance

Since there will be no return of acidified raffinate from Toquepala to Cuajone, net acid consumption at Cuajone is estimated at 2.7 mt/mt Cu or (13.6/0.90) x 2.7 = 40.8 mt/day.

The estimated acid requirement at Toquepala is 2.0 mt/mt Cu due to biogeneration of acid in the dumps. However, as the SX raffinate will contain 1.54 mt regenerated acid per mt copper extracted, the net acid consumption will be 0.46 mt/mt Cu or 38.6 mt/day.

The 13.6 mt/day copper recovered from the Cuajone PLS at Toquepala will regenerate 13.6 x 1.54 = 20.9 mt/day H_2SO_4. Additionally, depending on the co-extraction of iron, it will be necessary to bleed lean electrolyte in order to reduce the iron concentration to a maximum 1.5 g/l Fe. For an estimated average Cu/Fe ratio of 1,000:1, the 46 l/min bleed volume at 172 g/l H_2SO_4 that reports to the raffinate would contain 11.4 mt/day H_2SO_4.

Thus to exactly balance the requirement of 38.6 mt/day, 38.6 - (20.9 + 11.4) = 6.3 mt/day H_2SO_4 would be required. To avoid installing large acid storage facilities at Toquepala this acid will be added at Cuajone. This means that the Cuajone PLS will need to have an acid concentration of 4-5 g/l H_2SO_4.

The main acid consumption will therefore be 40.8 + 6.3 = 47.1 mt/day (17,000 mt/yr) on a 100% acid basis.

TABLE 6
Toquepala / Cuajone SX-EW Project Capital Costs

TOQUEPALA	Cost US$'000
Leaching	1,353
Solvent Extraction	6,135
Tank Farm	3,187
Electrowinning	13,126
Site Development/Utilities	4,677
Main Power Substation	1,590
Raffinate Pump Stations	2,349
PLS Pump Stations	6,976
Acid Tank and Loading	117
Acid Storage	(1,144)
Pipelines / Power Lines	10,184
Direct Costs	48,550
Indirect Costs	37,490
New PLS Pumping System	(1,924)
Total Toquepala	84,116
CUAJONE	
Crushing	3,760
Leaching	490
Site Development/Utilities	294
PLS Pump Stations	231
PLS Pipeline to Toquepala	1,658
Direct Costs	6,433
Indirect Costs	6,247
Used Crusher (deleted)	(300)
New Crusher (added)	550
Jaw Discharge Conveyor (added)	259
Crushing (cost reduction)	(1,264)
Dynamic Pad (deleted)	(720)
Permanent Pad (added)	2,323
SX-EW Modifications (at Toquepala)	815
Total Cuajone	14,343
Toquepala + Cuajone (July 1992 $)	98,459
Toquepala + Cuajone (January 1994 $)	102,934

Solvent Extraction

The solvent extraction plant will consist of three parallel mixer-settler trains each with two extraction stages and one stripping stage. Nominal capacity will be 15,520 l/min/train (4,100 gal/min/train). The mixer-settlers will be of the low profile type. These are large units capable of operating under a wide range of PLS flows and grades and whose installation costs are low. Additionally, aqueous recycle capability is provided on each stage for operating flexibility.

The projected PLS grade, including the Cuajone PLS, is 1.62 g/l Cu. The organic to aqueous ratio in the extraction stage will be held at 1:1 and retention time will be 3 minutes. 90% extraction of copper is expected. Raffinate will be returned to the dumps and loaded organic stripped with strong acid (172 g/l H_2SO_4 and 30 g/l Cu) recirculated from the electrowinning tankhouse.

A tank area will be provided for the storage of diluent, barren organic and electrolyte solutions. This area will also house the elctrolyte heat exchangers and the column cells and filters.

Electrowinning

The copper-enriched electrolyte, at 45 g/l Cu and 149 g/l H_2SO_4, will be transferred to the electrowinning tankhouse. The tankhouse will have two parallel lines of polymer concrete electrowinning cells each supplied by its own rectifier.

Stainless steel cathode blank technology will be used for cathode production. This technology permits current densities greater than 215 A/m^2 and gives improved current efficiency (92% design), superior cathode quality, and reduces labour requirements. The finished cathodes will be harvested every 7 days and separated from the blanks on a semi-automatic stripping machine.

Combined production from Cuajone and Toquepala during the first five years of operation will be 97.4 mt/day (35,064 mt/yr). The cathodes will comply with LME Grade A and Comex Grade 1 quality standards.

Capital and Operating Costs

The initial capital investment for the Toquepala and Cuajone facilities is $103 million. Additional capital expenditures of $7.5 million over the following years are also expected, mainly due to the progressive construction of the permanent pad at Cuajone (See Table 6).

Operating costs will be approximately 23 c/lb Cu during the first years and a maximum of 30 c/lb in the later years (See Table 7).

TABLE 7
Toquepala / Cuajone SX-EW Project
Operating Costs US c/lb Cu

YEAR	1	2	3	4	5	6	7	8	9	10	11	12	13	14	15
COMMODITY / AREA															
LABOR															
Toquepala	1.65	1.65	1.65	1.65	1.65	2.02	2.02	2.02	2.02	2.02	2.26	2.26	2.26	2.26	2.26
Cuajone	4.03	4.03	4.03	4.03	4.03	4.03	4.03	5.42	5.42	5.42	5.42	5.42	5.42	5.42	5.42
OPERATING SUPPLIES															
Toquepala	6.34	6.34	6.34	6.34	6.34	7.37	7.37	7.37	7.37	7.37	8.01	8.01	8.01	8.01	8.01
Cuajone	7.37	7.37	7.37	7.37	7.37	7.37	7.37	8.30	8.38	8.54	7.94	7.75	7.74	7.74	7.74
MAINTENANCE SUPPLIES															
Toquepala	1.27	1.27	1.27	1.27	1.27	1.28	1.28	1.28	1.28	1.28	1.29	1.29	1.29	1.29	1.29
Cuajone	1.07	1.07	1.07	1.07	1.07	1.07	1.07	1.36	1.37	1.40	1.30	1.28	1.27	1.27	1.27
POWER															
Toquepala	12.83	12.83	12.83	12.83	12.83	14.28	14.28	14.28	14.28	14.28	15.19	15.19	15.19	15.19	15.19
Cuajone	0.96	0.96	0.96	0.96	0.96	0.96	0.96	1.90	1.99	2.10	1.61	1.51	1.50	1.50	1.50
ADDITIONAL															
Shovel Loading (C)	0.40	0.40	0.40	0.40	0.40	0.40	0.40	1.28	1.36	1.13	1.01	0.91	0.90	0.90	0.90
SX-EW Increase (T)	8.73	8.73	8.73	8.73	8.73	8.73	8.73	8.73	8.73	8.73	8.73	8.73	8.73	8.73	8.73
TOTAL															
Toquepala	22.09	22.09	22.09	22.09	22.09	24.95	24.95	24.95	24.95	24.95	26.75	26.75	26.75	26.75	26.75
Cuajone	22.56	22.56	22.56	22.56	22.56	22.56	22.56	26.99	27.25	27.32	26.01	25.60	25.56	25.56	25.56
COMBINED (July 1992 $)	22.16	22.16	22.16	22.16	22.16	24.62	24.62	25.23	25.27	25.28	26.65	26.59	26.58	26.58	26.58
(Jan 1994 $)	23.16	23.16	23.16	23.16	23.16	25.73	25.73	26.38	26.42	26.43	27.86	27.80	27.79	27.79	27.79

Conclusion

SPCC will build a 97.4 mt/day (35,064 mt/yr) SX-EW plant to process pregnant leach solution from the dump leaching of 432.6 million mt of low grade sulphides at Toquepala, and the heap leaching of 13.6 million mt of high grade oxides at Cuajone. The Cuajone PLS will be pumped 35 km to the plant which will be built at Toquepala. Integrating the two operations will allow better resource exploitation with lower capital and operating costs than separate plants. Operating costs are expected to be less than 30 c/lb Cu.

The zero emission plant will fully comply with environmental standards.

Technological innovations are present mainly in the design of the mixer-settlers, electrowinning cells, and in the use of stainless steel cathode blanks.

Bibliography

1. Toquepala Leach-SX-EW Basic Engineering Report, Bechtel International Inc., San Francisco, July 1992.

2. Cuajone Leach-SX-EW Basic Engineering Report, Bechtel International Inc., San Francisco, July 1992.

3. Cuajone/Toquepala Leach-SX-EW PLS Pipeline Study, Bechtel International Inc., San Francisco, May 1993.

4. Toquepala and Cuajone Leach-SX-EW Alternative Studies, SPCC, Toquepala, March 1993.

5. SPCC New Leaching Scope, SPCC, Toquepala, August 1993.

Advanced methods of coalescing and filtration

K. W. Severing
SpinTek Systems, Huntington Beach, California, U.S.A.

Abstract

This paper outlines the methods of coalescing and filtering of electrolyte presently being used and discusses new methods that have recently been developed to improve the performance and enhance the recovery of organic. Different methods of floatation have been tried in the last few years as a roughing coalescer to a media filter and this paper will present a new concept in treating electrolyte. We will also present new methods of coalescing raffinate and loaded organic.
Keywords: Coalescing, CoMatrix, Filters, Electrolyte, Organic Recovery.

1. Introduction

Most Copper solvent extraction plants today operate with a filter in the electrolyte stream to act as a coalescer and filter for the small amounts of organic left entrained after the settlers. This practice has been adequate but the loss of organic through the backwash step and the large quantities of waste generated during this backwash process have been the topic of studies to reduce the operating costs. The smaller SX-EW plants do not have the funds to install these filters and are attempting to reduce the organic content by means of a flotation column alone with the capability to retrofit a filter at a later date should it be necessary.

We developed a two stage coalescing filter, the CoMatrix Tower™, to address these issues. This unit consists of a matrix coalescing plate section in the upper zone of a pressure tank followed by a deep dual media bed. The combination allows operation at flow rates of 25 GPM/ft^2 (59 M^3/hr/M^2)which is 5 times the flow rate of conventional filters and of a cost usually half that of conventional filters. The backwash volume is also about half that of conventional filters with improved performance.. The coalesced organic is recovered in the upper zone as a clean stream that isn't tied up in the backwash waste tank requiring treatment for recovery.

The development of the CoMatrix Tower led to other applications for the matrix plate coalescer acting without the media bed called the Matrix Tower. These units perform coalescing prior to an existing electrolyte filter, for the raffinate, and for loaded organic. All these areas will be discussed in this paper.

2. CoMatrix Tower ™ Design

The conventional electrolyte filter, SX filter, is designed to operate at 5 GPM/ft^2 (12 M^3/hr/M^2) in a pressure vessel containing a media bed. This bed consists of two active layers, a 12" (305 mm) layer of fine anthracite above a 24" (610 mm) bed of fine garnet. The anthracite is crushed coal that has irregular surfaces to perform the coalescing portion of the filtration. The fine garnet is more uniform and performs the conventional filtration similar to a sand filter and removes suspended solids like fine clays. A cross section of the tank is shown in Figure 1. The organic is coated on the surfaces of the anthracite and some of the organic is coalesced in large enough droplets to rise to the upper zone of the tank. This bed becomes saturated in about 12 hours of operation when exposed to normal levels of organic entrainment requiring a backwash cleaning cycle to remove the organic and solids in preparation for the next cycle. This is a difficult process and a few steps are necessary to adequately clean the media. These are to lower the liquid level in the tank to a level just above the media and introduce air at 3 CFM/ft^2 (55 M^3/hr/M^2) for 10 minutes to completely scour and loosen the particles from the media. The entire bed is then backwashed at 10-12 GPM/ft^2 (24-29 M^3/hr/M^2) for 5-7 minutes to flush the particles from the bed and reclassify the media back into the two zones of heavier garnet on the bottom and lighter anthracite on top. There are many variations to this basic operation depending upon each plants operation preferences. Some of these consist of displacing the entire volume of the tank to recover the electrolyte and use water for backwash. The entire sequence usually takes about 30 minutes.

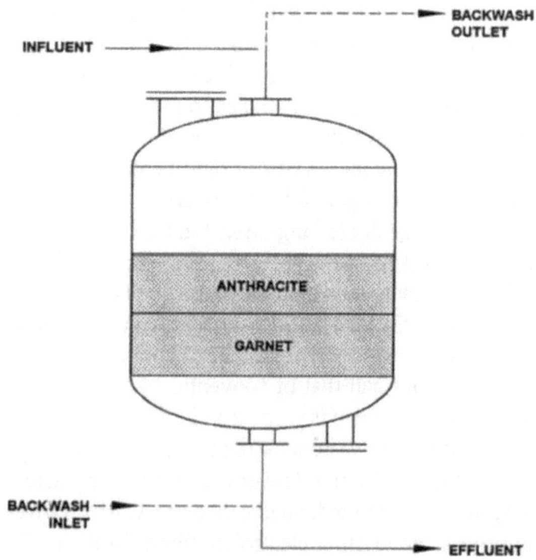

Figure 1 - SX Filter design

The CoMatrix Tower is quite different from the SX filters having a packed section of matrix plates 7' (2.1 M) deep in the upper zone to perform the coalescing of usually 70% of the inlet organic. A cross section is shown in Figure 2. This organic is allowed to rise in the tank to a quiet zone in the upper head of the tank and periodically removed by means of a level switch and an automatic valve. The matrix plates are made of PVC which is hydrophobic and acts as an excellent media for organic coalescing. The design of the plates is illustrated in Figure 3 and consists of corrugated plates at opposing angles which allow liquid to easily flow from one channel to the next and do not limit the effective area of the tank as would be the case with conventional parallel plates. The arrows show that the direction of flow for the liquid being coalesced is opposite the flow of the main influent stream. These matrix plates are welded together allowing for easy installation and removal and are strong enough for a person to stand on during maintenance.

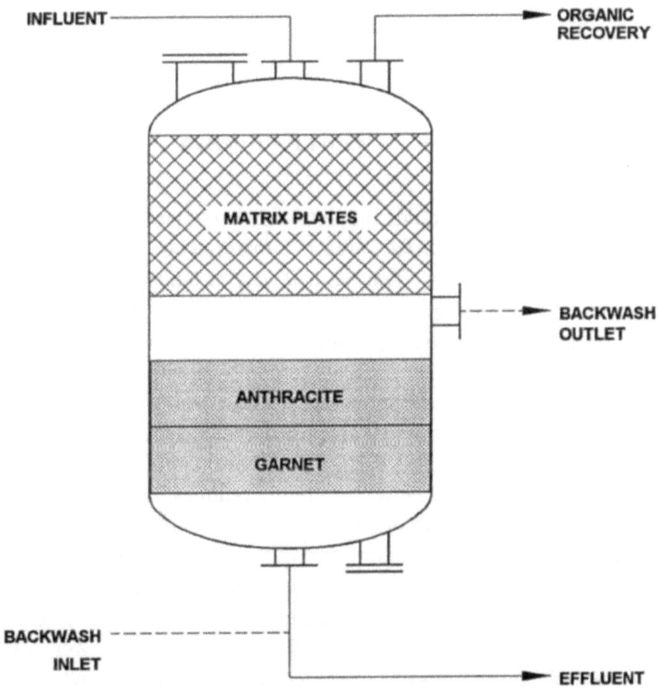

Figure 2 - CoMatrix Tower design

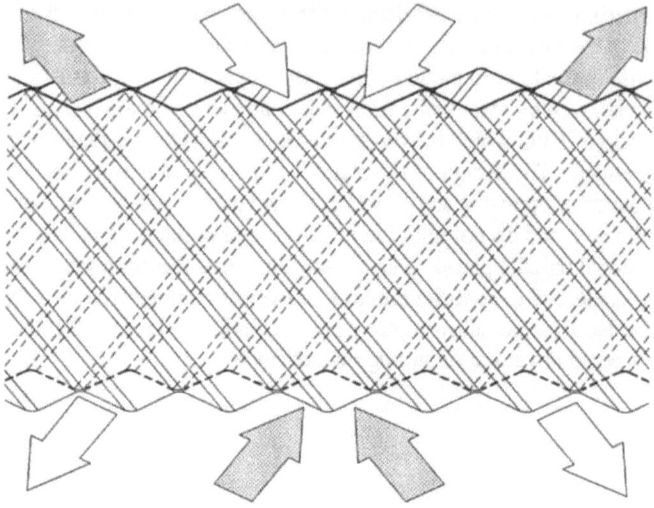

Figure 3 - Matrix Plate design

The practice of coalescing liquids on plates has been used for some time and relies on providing an increased surface area of a material compatible with the liquid being coalesced with corrugations. These create low turbulence areas in the upper area of the corrugation for the lighter liquid to collect. These droplets have only a short vertical distance to rise to reach this area rather than the entire tank if no plates were used and this area allows other droplets to combine with each other to form larger droplets that then rise to the upper zone without worry of being carried downward with the flow of electrolyte. This phenomena is illustrated in Figure 4. The added benefit with any organic that is carried down to the media bed is that it will already have been partially coalesced which greatly reduces the load on the anthracite. This organic can then combine with other organic droplets on the top of the anthracite bed and will rise back up to the plate zone and through the plates to the upper section of the tank. The result is a large reduction in the loading on the anthracite allowing for the increased operating velocity. (We should point out that velocity is very important to proper operation of the CoMatrix Tower and that operation at lower flow rates does not improve performance. In our initial testing, we found that when we increased the flow rate from 10 GPM/ft^2 (24 M^3/hr/M^2) to 25 GPM/ft^2 (59 M^3/hr/M^2) we actually recovered more organic and had little effect on the loading of the media bed below).

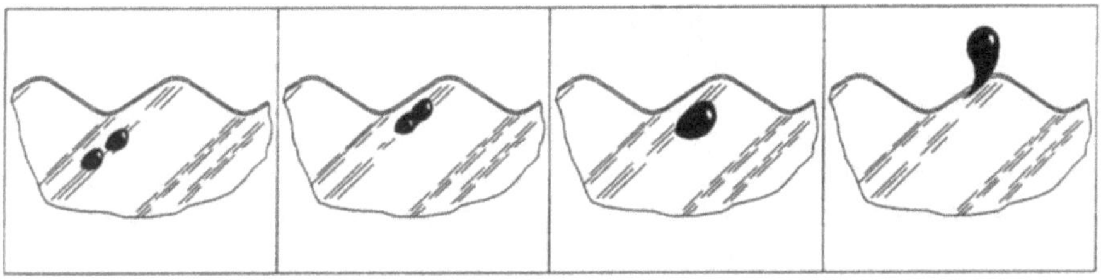

Figure 4 - Coalescing droplets on Matrix plate surface

An important factor in coalescing is the micron size of the droplet. When using plates or air flotation the size of the coalescing liquid droplet is more critical than when using an anthracite bed since the contact through a media bed offers far more surface area and more potential liquid contact. Generally a droplet size ranging from 10 micron and larger coalesces well in a matrix plate tower with 100% efficiency obtained at micron sizes 30 and larger. When the size decreases to below 10 micron, the efficiency is reduced and the media bed performs more of the removal.

The media bed consists of anthracite as in the SX filters except that the level is increased to 24" (610 mm) from 12" (305 mm). Since the anthracite performs the fine coalescing and protects the garnet from being coated with organic and losing solids removal efficiency, we have increased the depth to allow for longer run times and protection from breakthrough on organic. The garnet remains the same since the amount of suspended solids is usually not the controlling factor during an operating cycle. The area between the anthracite bed and the plates is high enough to allow for the cleaning cycle to take place below the plates. A backwash distributor is installed below the plate section with laterals for proper collection of the backwash waste and allows the bed to expand sufficiently to remove the loosened particles. A cross section of the CoMatrix is shown in Figure 4. As in the case of the SX filter, the media portion of the tower is air scoured and backwashed in a similar manner.

The plates are not disturbed during the backwash step as they operate better with the organic coating left in place. The performance of the plates improves during the first cycle while this coating takes place since the organic is attracted to other organic surfaces or droplets and removal of this surface during a cleaning step would require the coating process to reoccur before good performance is reestablished.

3. CoMatrix Tower Testing

Testing was performed at a few plants in Arizona that operate existing SX filters and provide an excellent model of a well run plant with organic levels normally around 30 PPM (MG/L). Our test unit was 12" (305 mm) diameter and was installed parallel to the existing SX filters. The test unit has an area of 0.8 ft^2 (0.07 M^2) and operates at 20 GPM (4.5 M^3/hr). The feed was taken from the same pumps that supplied the SX filters and the effluent organic was non detectable during the cycle as was the SX filters except that we were able to recover a pressurized stream of clean organic from the top of the CoMatrix Tower as a result of the plate performance. We consistently obtained non-detectable organic from the lower portion of the media bed and we monitored the organic level at points within the tank to verify the recovery amounts in the matrix plate zone.

The CoMatrix's operating cycle was 8 hours compared to 12 hours for the SX filter but the CoMatrix processed 333 % more electrolyte per cycle. Figure 5 is a graph of the performance cycle for a typical electrolyte stream with 30 PPM (MG/L) organic. The upper matrix plate section recovered 20 PPM (MG/L) or 66% of the inlet organic and the anthracite/garnet bed removed the remaining 10 PPM (MG/L). Subsequent cycles obtained the same performance. An important factor was the reduction in backwash volume when using the CoMatrix

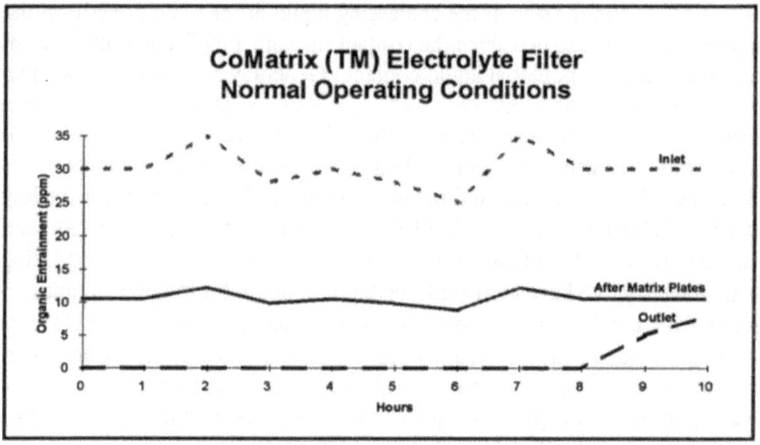

Figure 5 - CoMatrix performance

In an attempt to simulate an upset condition, we operated the CoMatrix from the effluent weir of the E1 settler and fed the unit with an average of 800 PPM (MG/L) organic. The graph in Figure 6 illustrates the performance when operated at these high levels. We were able to achieve the same non-detectable level of organic in the effluent of the unit and recovered about 75% of the organic in the upper matrix plate zone. The service cycle in this case was 6 hours compared to the 8 hours at low levels. Our sample point located below the plates was actually being contaminated with organic that was rising from the anthracite bed due to the high loading and the matrix plates were processing this organic as well.

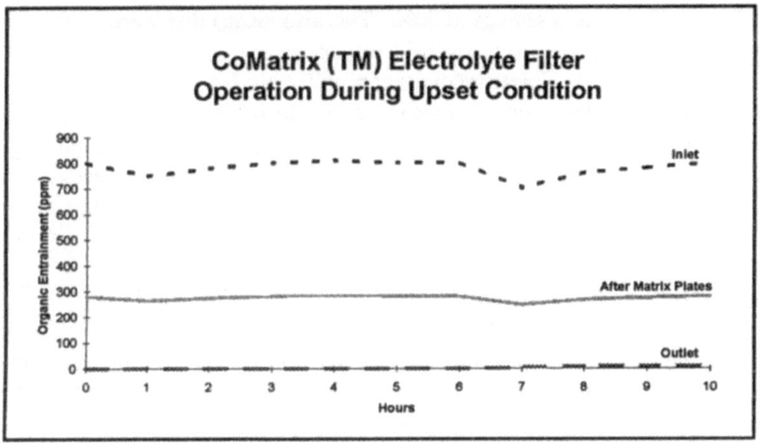

Figure 6 - CoMatrix Tower performance

4. Backwash Comparison

Using the service cycle referenced in Figure 5 we can see that the CoMatrix Tower will process 333% more electrolyte than a conventional SX filter per cycle. We must also compare the amount of backwash required to clean media beds. For our comparison we will consider that water is used for backwash of the beds and the SX filter and CoMatrix Tower are both 12' diameter. The service flow rate for a single CoMatrix is 2,800 GPM (636M^3/hr) and the flow rate for each SX filter is 560 GPM (127 M^3/hr) requiring five SX filters for this rate. The backwash rate and time is the same for the CoMatrix and SX filter beds therefore the following calculations apply.

4.1 CoMatrix Tower Backwash Cycles

Using 8 hour service cycles, the unit would backwash 3 times each day

(1) CoMatrix Tower • (3) backwash cycles/day = 3 backwashes/day

3 /day • 10 minutes • 1130 GPM (257 M^3/hr) = 33,900 GPD (128 M^3/D)

4.2 SX Filter Backwash cycles

Using 12 hour service cycles, each filter would require 2 backwashes each day

(5) SX filters • (2) backwashes/day = 10 backwashes/day

10 /day • 10 minutes • 1130 GPM (257 M^3/hr) = 113,000 GPD (428 M^3/D)

Using this comparison, in a one-year period of time the difference is 41.2 million gallons (3.8 million M^3) for the SX filter and 12.4 million gallons (1.1 million M^3) for the CoMatrix Tower or a savings of 70%. This also means that there is 70% less backwash waste to process to recover the organic.

Since the CoMatrix Tower recovers approximately 2/3 of the organic in a concentrated stream, less recovery time is lost for the retrieval of the organic. On a 2800 GPM electrolyte stream with 30 PPM (MG/L) of organic present, this represents 30,000 gallons (2,800 M^3) each year of recovered organic that is not recovered as part of the backwash water. The remaining 10,000 gallons (929 M^3)of organic recovered is part of the backwash water similar to conventional SX filters.

5. CoMatrix Cost Comparison

Using the example of 2,800 GPM above, the cost for each of the SX filters is approximately \$200,000 USD resulting in a total equipment cost of \$1,000,000 USD. The cost of construction must be added to the equipment cost for a real cost comparison. The cost for the single CoMatrix Tower is approximately \$350,000 USD and of course the construction costs would also be reduced by a similar amount. This means that the CoMatrix cost is 35% of the cost for SX filters.

We have assumed a normal design for both units with 316L SS vessels, piping, internal distributors and all 316 SS high performance butterfly valves. The controls included a Allen Bradley PLC and a high pressure blower is included.

6. Matrix Tower™ Design

This tower is the upper portion of the CoMatrix Tower described above and operates at 25 GPM/ft^2 (59 M^3/hr/M^2). Figure 7 illustrates the cross section design of the tower. The performance when processing electrolyte is the same as for the CoMatrix Tower, recovering 60-75% of the organic at the top of the tank. This unit is excellent at coalescing ahead of existing SX filters where a flotation column might be considered. The obvious advantage is that it uses plate technology rather than air flotation and doesn't need large volumes of air or an ejector that uses a considerable pressure drop to create air for it's operation.

An improvement can be expected in the SX filter performance since a major portion of the organic will be recovered prior to the SX filter. We have found that some of the organic that has been coalesced in the matrix plate section is still sufficiently small that it passes through to the anthracite media bed for removal. Some of this organic due to the pre-coalescing will further coalesce and be able to rise to the plate zone in the CoMatrix design. When operated in two separate tanks, this organic will either remain in the media bed or rise to the top of the SX filter tank and can potentially be recovered if the SX filter is modified to accommodate this recovery technique.

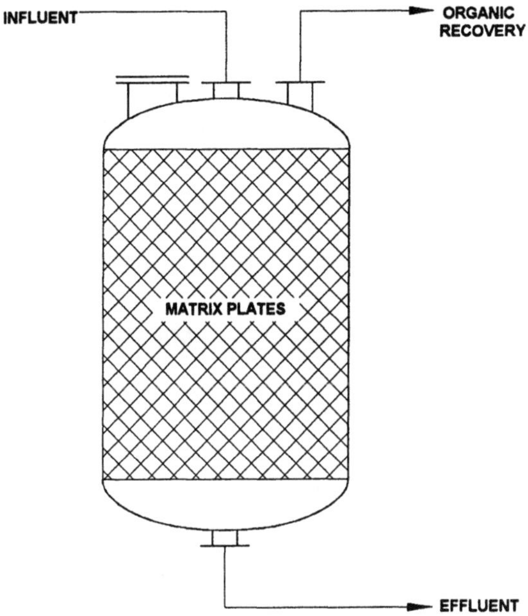

Figure 7 - Matrix Tower design

The same concept is applied to raffinate for recovery of organic now being carried to the raffinate pond. Using a Matrix Tower reduces the amount of organic exposed on the pond surface that evaporates and is more difficult to recover or lose. The increased EPA regulations will limit the amount of solvents that can evaporate in the future.

The Matrix Tower can also be used to coalesce the aqueous phase in the loaded organic stream when the materials are changed to 316 SS rather than plastic. We expect similar performance with this concept as with the electrolyte, providing the micron size of the aqueous droplet is within the 20-40 micron range. In this case, we use an atmospheric tank operated in the down flow mode to drop the aqueous to the bottom of the tank and any aqueous that is carried over to the loaded organic tank will be trapped in the baffled aqueous zone that is presently being used.

As with the CoMatrix, the flow rate is 25 GPM/ft^2 (59 $M^3/hr/M^2$) and this velocity, which is 3 ft/minute, is high enough to keep the plate opening clear of crud loading. The opening in the plates is 1/2" (13 mm) and with this velocity we have not yet had any cases where we have plugged the plates. If plugging ever became a problem, many methods could be used to clear them such as backwash, high pressure hosing, recirculating with diluent, or removal and use of steam or water.

7. Conclusion

We conclude from our testing and operating performance to date that the use of matrix plates enhance the performance of coalescing and filtration in the following ways.

a) Coalesces up to 70% of the entrained organic prior to the media bed.
b) Allows for clean recovery of the coalesced organic.
c) Improves the performance of the media bed.
d) Allows for increased flow rates of 25 GPM/ft^2 (59 M^3/hr/M^2) through the media bed.
e) Reduces the backwash flow rate to 20% of that of SX filters.
f) Reduces the backwash volume by 70% compared to SX filters.
g) Reduces the cost of equipment by 65%.
h) Operation of Matrix plates alone are comparable to flotation columns but do not require the use of air or a pressurized feed.
i) Materials of construction can vary for use with aqueous or organic.

Mining – Transportation, Conveyor Design and Practice

From mine to customer—the marine connection

P. A. White
Sandwell, Inc., Swan Wooster Division, Vancouver, Canada

Abstract

The port and deep sea shipping are key elements in the transportation of products from the mine to international markets. The paper describes aspects of loading and transport of bulk materials by ship, and other essential factors which must be considered in the design and development of ports.
Keywords: Environment, shiploading systems, simulation modelling, terminals, transportation

1 Introduction

The mining industry faces increasing challenges in the development of transportation and port facilities for movement of minerals to market. New developments are exposed to vigorous price competition against a background of tight capital markets and ever changing more stringent regulations.

The capital costs of transportation systems can be low relative to total project costs, whereas the unit cost of transportation is high relative to the total delivered commodity price. Pre-planning and simulation of throughput and variables in the transportation systems and bulk commodity terminals are tools used to meet the challenge of designing a system with the lowest overall cost.

This paper presents an overview of planning, environmental and operational issues faced in the development of new transportation and terminal operations for bulk commodities, illustrated with example projects.

2 The transportation system

Generally, the total transportation system for a bulk commodity consists of one or more mines or production plants, mine site storage, train or truck load-out facilities, a unit train railway or trucking fleet, an export terminal, a fleet of bulk carriers, an import terminal and the local distribution system to the end user. Annual throughput of mineral concentrates including copper, lead, zinc, iron, potash, sulphur or coal can range from 10,000 tonnes per annum to over 40 million tonnes per annum through a single system. The terminals may be dedicated to a single commodity or to multiple commodities.

The capital costs of transportation systems often represent less than 10% of the total project capital cost, whereas the transportation system component of the over-all delivered cost of a commodity can reach 40% to 60%. These ratios illustrate the potential cost savings over the life of a project and the importance of paying extra attention to the transportation system.

New mines are generally in frontier locations where existing transportation systems, port terminals and infrastructures are inadequate or unavailable. This expands the responsibility of the mine developers to become active participants in the development of transportation systems and to not become complacent about delivery costs perceived to be outside of the scope of a project.

The importance of the early planning and preliminary design stages cannot be over-emphasized. Only in the early stages of a project, when the overall design concepts are established, can significant cost savings (or reductions in environmental impacts) be made.

2.1 Terminals - the vital link in the transportation chain

Terminals act as transportation links and as buffers between two transportation modes at ports, or at inland distribution centres. Although terminals are a small link in the overall transportation system, they play a vital role in the overall system efficiency. They must provide sufficient storage capacity to avoid ship delay due to late cargo delivery to the port and similarly, trains delayed waiting for ships arrivals.

Terminal charges usually represent a minor percentage of the total delivered cost of a bulk product, however customer end costs are very sensitive to the terminal planning and investment. Terminals have a strong influence that extends throughout the whole length of the transportation system. An under-designed terminal which processes ships or trains too slowly, is limited to small ships, or has inadequate storage capacity causes chain reactions of delays and interruptions in the system which may far outweigh the intended savings. In addition to high demurrage costs, the delays or interruptions in the system can reach right back to the producers, resulting in a necessity to stop mine or plant production. An over-designed terminal, while able to handle the throughput, represents excess investment in the facilities thus resulting in a higher cost per tonne.

The terminal should be designed as part of the transportation system with the minimum total delivered cost from mine to customer. The planner must also consider less tangible criteria such as system reliability, redundancy, environmental concerns and operators preference. A list of principal planning parameters for the transportation system is given in Appendix A.

3 External factors influencing port design

The selection of the most economic transportation route from the mine and a site for a shipping terminal depends on many factors. Local conditions define many of the options but other factors beyond normal mining industry concerns must be considered. Some of these factors affect only copper and other heavy metal concentrates.

3.1 Ship sizes

Handysize and Panamax ships ranging in size from 35,000 DWT to 65,000 DWT are the workhorses of the bulk trades. For the very large tonnages of coal, Capesize ships of up to 150,000 DWT are also used. Copper concentrate which is of comparatively high value and low tonnage does not fit this pattern.

For the last twenty five years the typical ships in copper concentrate trades in South America have ranged in size from 12,000 DWT to 30,000 DWT but this is changing. Most of these ships are old since for many years the majority of new buildings have been Handysize ships of 35,000 DWT and larger. Smaller ships require the same size of crew and cannot compete in many markets. In the same manner that the Liberty type tween-deck ship disappeared so too will the 25,000 DWT bulk carriers. Terminal must be designed for larger ships which will replace the small bulkers, probably during the next decade and certainly during the economic life of a mine.

3.2 Shipping costs

The lower freight costs of large ships can be lost if the ship spends many days loading in port. During a recent review of copper concentrate terminals in Chile it was learned that the ships loaded at a terminal with 2,000 tonnes/ shift capacity were 25,000 DWT or less in size whereas at a terminal with 10,000 tonnes/ shift capacity the ships were all in the 40,000 DWT class. At both terminals the parcels were about 10,000 tonnes, High loading rates allow the shipper to take advantage of cost savings offered through the use of larger ships. An added advantage of a larger ship is that the parcel may be loaded into only one hold thereby minimising the delay of moving the ship if the shiploader is fixed

With a throughput of 400,000 tonnes/year it should be possible to benefit from the economies of long term charters and known freight rates to remove the uncertainty of going into the market on a shipment by shipment basis. Whether the concentrates are sold at point of loading or under the hook at port of destination affects ship chartering arrangements, it also has a bearing on the method of measuring quantities. It is wise to base the terminal design criteria on long term trends rather than the freight rates in effect at the time the terminal is built.

3.3 Trimming

The dangers of cargoes shifting at sea make copper concentrate a difficult cargo to handle. The holds of self trimming bulk carriers are shaped so that a low density material such as coal which fills the hold to the top is confined by the wing tanks and cannot move at sea. Because of its high density copper concentrate fills only the bottom of the hold and gains little advantage from the self trimming feature. In addition the material becomes unstable at the flow moisture point of about 9% and will slump below its normal angle of repose.

Many ore ships have been lost because the cargo has moved as the ship rolled in heavy seas. In 1987 alone seven were lost. To prevent this, IMO regulations call for copper concentrate "to be trimmed reasonably level to the boundaries of the cargo space".

In Chilean ports the ship's captain decides what is acceptable. Some may accept a stow which can be achieved by a shiploader reaching all parts of the hatch opening. Others will demand trimming the cargo flat which requires putting bulldozers or men on board. As a result provision for trimming must be made in dock design. If gearless ships are to be handled either the shiploader or a crane will be needed to put the bulldozer on board. The delays can be appreciable if the shiploader is used as a crane, particularly with a fixed shiploader where the cargo is loaded into several holds as the ship is moved along the dock.

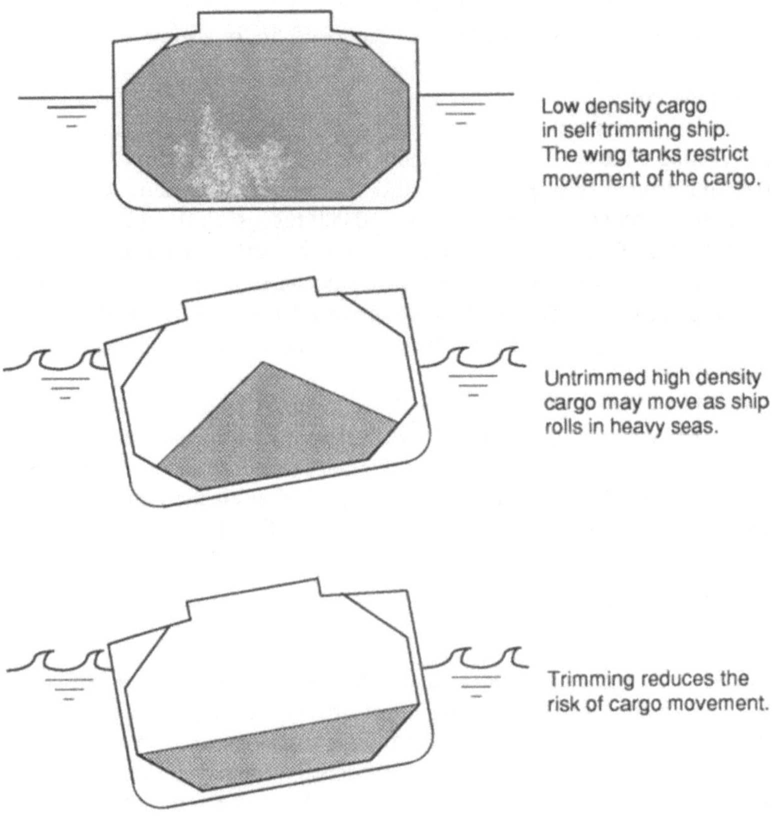

Fig.1 Cargo trimming in bulk carriers

3.4 Cargo measurement

There is an ongoing debate in the industry on two competing methods of cargo measurement, ships draft survey and conveyor scale measurement. The former is not very accurate and the latter can be unreliable, neither very comforting when the results have a direct effect on the mine's revenue.

A ship's displacement can be calculated from load lines marked at the bow and stern. For a Handysize, 35,000 tonne ship a 1.0 centimeter change in draft is equivalent to about 35 tonnes change in displacement. In calm seas during daylight, a surveyor can make a reasonable estimate of the draft by reading the water level at the load lines. But on a dark night with a choppy sea the accuracy of a surveyor leaning over the handrail with a flashlight is uncertain. A survey error of 3.0 centimeter in the reading of the water level at the load line for a 10,000 tonne loading represents a 1.0% error in tonnage. For a fully loaded 65,000 Panamax ship the 3.0 centimeter error is equivalent to 0.25%.

The survey procedure in reverse is carried out at the discharge port and the results compared. It is rare for there to be a significant difference. Advocates of scale measurement claim that these differences are far less than would be predicted by statistical probability and point out that the surveyor at the discharge port has access to the results of the survey at the loading terminal.

The accuracy of a high quality belt conveyor scale in good calibration is about 0.25%. However a minor defect can cause the scale accuracy to drift by 2.0% or higher. If the ship loading is based on the measurements of a single scale the uncertainty of a serious error exists, and there is little that can be done to rectify a scale defect during a loading even if is noticed. Two or three scales on different conveyors in series can be used to overcome this problem and legally approved belt weighing systems of this type are used in many countries. If this method is to be employed it must be incorporated at the beginning of the terminal design.

There are other methods of measuring ship load quantities. Bin scales have better accuracy and inherently are more reliable than belt scales. The disadvantage is cost and space required for the incline conveyor to carry the concentrate to the top of the bin scale.

At one terminal in Chile the concentrate is shipped from the mine by trucks which are weighed on a scale at the terminal gate. The cargo is accumulated for each shiploading in separated divisions of the storage building and the ship load quantity is calculated from the truck weights.

The terminal must be specially designed to accommodate any method of cargo measurement other than draft survey. In the absence of a specific decision on weighing, the mine must accept the uncertainties of draft survey, or worse, rely on the weight measured at the discharge port.

3.5 World Bank environmental requirements

The Abstract, reproduced below, from the World Bank Technical Paper Number 126 "Environmental Considerations for Port and Harbour Developments" published in 1990 speaks for itself.

"This paper, prepared in association with the International Maritime Organization (IMO), is aimed at both World Bank staff associated with lending operations involving port development as well as the staff of the Bank's borrowers - port authorities, port managers and engineering staff and also government officials responsible for port and maritime activities. The introduction highlights typical environmental problems that are likely to be present in developing country ports but the paper's primary purpose is to act as an aide memoire for those responsible for port and harbour development and who need to know the complete range of topics to be considered, which ones are applicable in their particular case and where to find more information on the subject. It is basically a checklist and the paper is produced so that it can be used as such. The checklist is followed by more detailed information on how a certain issue is normally resolved and resources of information and assistance."

It is apparent that the World Bank expects projects in developing countries to meet the highest environmental standards. Appendix B is the check list given in the paper for common environmental problems in port and harbour development.

3.6 Support services

Responsibility for the port infrastructure varies from country to country. In Chile the Navy carries out the function of general administration and regulation of ports on behalf of the Director General of Maritime Territory. They are responsible for safety, pollution control,pilots, maritime communications, security of facilities and operating permits. The Navy also provides hydrographic, oceanographic, wind and tide data.

Recent experience indicates that it takes about one year to get permission to build a dock and three to four months for modification of an existing facility.

The services required by a ship at a bulk terminal must be available on demand and for periods of a few days probably no more than once a month. Services can be shared by ports close together but if the terminal is a long way from other ports it must bear the cost of some services alone.

Pilots frequently serve several ports. All they require is a boat to take them to the ship. Ship's agents work from a central community and arrange documentation and communication with the shipping companies.

Tugs are needed for arrival and departure. If the shiploader is a fixed type, tugs may be held on standby for moving the ship during loading. The cost of bringing tugs from a neighbouring port is the governing factor in deciding whether to provide tugs at a new terminal. One day's sailing time between ports is not considered excessive on the coast of Chile.

4 Shiploading system

4.1 The fixed or moving shiploader

The main criteria for design of shiploading systems are the capacity in tonnes per hour and the sizes of ship to be loaded. However secondary requirements; trimming, dust control and spillage collection, can be of similar importance and dictate the selected design.

For a terminal with large throughput commodity such as coal, a moving shiploader is essential to achieve a high loading rate and adequate berth availability. Apart from cost, the delays which occur in moving the ship reduce berth availability. Sandwell has designed many terminals for coal and the considerations for optimum design are well established. For many installations the quadrant type of shiploader has proved to be the most economic for a piled structure where the dock is some distance from the shore line. In some situations the linear travelling shiploader is preferred.

Fig. 2 Ridley Terminals Inc. Coal Export Terminal

4.2 The fixed loader

Even though slackening ropes and warping a ship along a dock is a slow process and tedious procedure many terminals have fixed shiploaders. With a low annual tonnage the cost of delays moving the ship under a fixed loader may not justify the extra cost of a moving loader. A moving shiploader becomes economic at between about 0.5 million and 2.0 million tons per year depending on the loading rate required. Tonnages at a terminal handing the output from a single copper concentrate mine do not usually justify a moving shiploader.

However for a small increase in cost a fixed shiploader can be given limited slewing and shuttling motions to cover the area of a single hatch. The ship need only be moved to change hatches. The handling of bulldozers used for trimming concentrates adds another dimension to the shiploading and it is advisable to design the terminal to handle both gearless and geared ships. When loading gearless ships either the shiploader or a crane will be required to put the trimming bulldozers on board. When a fixed shiploader is used as a crane and the cargo is loaded into several holds the delays moving the ship to put bulldozers on board can be appreciable. The logistics of the shiploading operation need careful planning at the design stage.

Apart from moving between holds, the cargo must be distributed throughput individual holds. It is possible to load into a single point in the hold and use a bulldozer to spread the cargo. The reduction in shiploader capital cost must be weighed against the cost and extra handling of bulldozers. A disadvantage is that the shiploading rate is dependent on the performance of the bulldozer operators.

Trimming concentrates with belt throwers mounted on the bottom of the shiploader chute is not recommended. If the concentrate is dry enough not to plug the system it will be too dusty to meet environmental requirements.

For a typical concentrate terminal a fixed shiploader with slewing capability to cover one hold is the most economic alternative. The slewing feature eliminates the need to move the ship except to move from hold to hold. Depending on the circumstances the shiploader or a crane can be used for loading bulldozers on board. The shiploader designed by Sandwell for Escondida is an example of this type of loader configuration.

4.3 The Escondida shiploader

The shiploader consists of a truck-mounted, carriage structure which is capable of shuttling on rails aligned at right angles to the berth face. The transfer conveyor, within the carriage structure, feeds a slewing-luffing, boom-mounted belt conveyor. The combined shuttling-slewing-luffing motions give the shiploader three degrees of freedom. This feature results in a good coverage of hatch coamings for the full range of design ships - reducing lading time associated with manual trimming. It eliminates moving the ship except to move from hold to hold and is an effective compromise between shiploader capital cost and operational delays. The unique slewing-luffing system of the shiploader is of particular interest. By utilizing a gimbals-type suspension of the boom, the boom is allowed to slew and luff about a single point in space. This layout results in a compact material transfer area - eliminating the need for elaborate chute work and reducing the overall size of the structure. The state-of-the-art gimbals suspension uses standard bearings which makes maintenance uncomplicated.

The nature of the material being handled requires special design considerations since copper concentrate, when moist, is a material which tends to stick to most surfaces. A discharge chute was designed such that the head of the boom is protected, but no contact is made with the material. Belt cleaning devices ensure that the belt is wiped clean on the return run to the extent possible short of washdown.

The conveyor system itself is enclosed to minimize dust emissions during operation. The transfer conveyor is covered by a zipper-like device which seals the conveyor as the carriage shuttles. The boom conveyor is enclosed by arch-shaped hoods. Overall, the shiploader operates very cleanly with a minimum of impact on the environment.

The Escondida dock illustrates the use of mooring buoys and shows the hatch coverage obtained with the slewing shiploader.

Fig. 3 Escondida Project Coloso Port Facility, Chile
Copper Concentrate - Engineering by Sandwell

4.5 Dock design

Most bulk shiploading systems are built with a piled structure leading out from the shoreline to deep water. It is rare to carve the dock into a rock face or build an apron dock the length of the ship. The costs are very high and there is no need for an apron dock to bring containers or trucks along side the ship.

Wharf structures without berthing dolphins have been proposed for some terminals. From Sandwell's experience with berthing and loading operations a layout with a single berthing dolphin-wharfhead is often preferred. Such a layout has two main advantages: easier berthing operations and provision of a safer berth. This dolphin-wharfhead has more resistance to ship berthing impact than an isolated dolphin in a row of fenders along the length of the ship. Furthermore, the dolphin-wharfhead provides better protection of the shiploader structure from possible damage.

Handling the bulldozers used for trimming is a major consideration in dock design whether a crane, shiploader or ships gear is used to put the bulldozers on board. An area is needed for storage on the dock and a method must be provided for taking the bulldozers ashore. If the dock is far from the shoreline, a small barge may be preferable to bridging the distance with a structure capable of carrying a bulldozer.

5 Environmental considerations

While environmental protection may not cost more than about 10% of the total capital cost of the terminal, it can be the factor which governs approval of the project. The main considerations are spillage of material and contaminated water into the sea, and dust control.

5.1 Contaminated water handling

Although not as toxic as lead or mercury, soluble copper compounds are harmful to benthic organisms on the sea bed. Environmental requirements for discharge of water contaminated with copper vary widely. Many Canadian and European regulations specify a range of 0.1 to 1.0 ppm for soluble and 20 to 100 ppm for insoluble copper compounds. US regulations often specify that the effluent contamination must not exceed the background level of the receiving waters which are normally below the levels which can be achieved with conventional technology.

Coal is not normally considered toxic but spillage can smother the marine life on the sea bed and a discharge of "black water" into a harbour is not acceptable whether or not it causes any environmental damage

"Best Available Technology" calls for the prevention of all spillage from the shiploader and conveyors over the water which in turn requires collection trays beneath the conveyors and transfer points. Although some material can be removed by hand it is very difficult to carry out an adequate dry cleanup of spilled material collected in the trays under the conveyors. Washing is more effective provided the contaminated water is collected and pumped ashore for treatment. Pipe galleries for the conveyors ensure containment of water which can splash over the sides of open troughs during washing. They have the additional advantage of enclosing conveyors for dust control.

The water from a wet cleaning system must be treated to remove insoluble spilled material and recycled. For a material such as coal with a small environmental impact, an open stockpile may be practical. The rain water contaminated with coal would be treated to a level suitable for discharge to the receiving waters. Treating rain water contaminated with copper concentrate to match levels in the receiving waters is probably impossible and certainly uneconomic. The entire concentrate handling system including the stockpile must be covered otherwise a rainfall will flush material spilled and eroded from the stockpile into the sewers. A closed system with no discharge has the advantage that the recycled water need only be treated to a level suitable for wash water use.

5.2 Dust control

Conventional dust control methods are effective at a terminal except for the final discharge of material into the ship's hold. A chute or spout suspended from the shiploader boom to contain the material dropping into the hold acts as a pump with the falling material entraining air and dust which is discharged at high velocity at the bottom of the chute. At this point the only effective method for control of a dusty material is a device called a "choke feeder" to arrests the flow of material at the bottom of the shiploader chute and form a bed of material which prevents discharge of air and dust. A "sticky" material will not flow freely through a choke feeder; on the other hand sticky materials are not usually dusty.

Control of dust emissions from coal, copper and many other bulk products is effected by the moisture content and particle size. It is a factor to be considered in the design of the processing plant at the mine since the control measures necessary to satisfy environmental requirements during handling of the material from mine to customer can affect the cost of transport.

Fig. 4 Shiploading of Bulk Material Treated to Control Dust

5.3 Site selection

A rocky headland makes a poor place to site a port. It is generally too exposed and building piled dock structures on a rock seabed is expensive. River estuaries or sheltered bays are more attractive.

Until recently, economic, geological and technical factors were the main considerations in selection of a port site. This has changed and environmental factors are now equally or even more important. Most river estuaries have tidal flats and areas of flat land created by settlement of sand and mud carried down by the river. Apart from possible foundation problems caused by lenses of silt, sandy shores and tidal flats are excellent for the pile driving, dredging and filling of dock construction. However tidal flats are also the wetlands which are the source of prolific plant and animal life of all types; land and marine animals, fish and birds flourish and reproduce in this environment. The need for their preservation can rule out attractive sites at river estuaries. A sheltered bay with a sandy sea bed may be a better proposition.

It is said that "beauty is in the eye of the beholder". Except to the engineer and the odd economist, bulk shipping terminals are not visibly attractive. A terminal across the bay from a beach resort will be an irritant which attracts criticism even if it has a flawless environmental record.

Distance from the mine, dock development, dredging costs and suitable topography, particularly for rail transport, are essential considerations for selection of a port site. The environmental and sociological concerns must be weighed together with the normal technical and economic factors.

6 Simulation modelling of port facilities

6.1 Simulation goals

Simulation modelling attempts to mimic reality (i.e. sub-systems of the real world) by incorporating the major rules, procedures and operational properties of the real system in the model. No model can ever duplicate the real world because the infinite complexity of real systems precludes such creations unless the model is the real world. However, idealized representations of the real system can be made to perform similarly such that meaningful projections of the simulated system can be applied to the real system. The level and complexity of detail (i.e. the degree of realism) for the modeller is limited by the power of the computer, the usual budgetary restrictions and the overriding question of the necessity for greater refinement. If the scope of the model has been designed to include the expected extensions of the real system and can be shown to duplicate the response of the real system within acceptable bounds, then the model can be used to make significant investigations of that real system.

6.2 Scope

Whenever possible Sandwell uses simulation to model the design and operation of bulk commodity handling facilities, port developments and transportation systems. Simulation is a valuable tool to explore for the "optimal" solutions of such systems.

The models are used to assess equipment layouts and capacities, to identify bottlenecks and explore means to remove such bottlenecks, to determine efficient stockpile levels and storage volumes, to assess facility capacities, prepare future operational plans or operator assignments.

Examples of bulk handling systems examined by Sandwell are:

- The optimal berth extension length for an expanding facility
- Various loading options subject to tidal restrictions and channel dredging options.
- The number and size of proposed unit trains to transport coal from a mine to the port.
- Expansion for the required number of berths and ship unloaders (and their capacities) for a coal unloading facility.
- Entire coal export handling systems consisting of three major interactive sub-systems of mine production, railway transportation and port operations.

Simulation models are extremely valuable whenever major cost decisions are required. The saving in capital and operational costs are enormous if the model can show that a planned extra stacker/reclaimer, shiploader or ship unloader is not required. For a major project the cost for development and use of the simulation model is small in comparison to the possible savings.

6.3 Simulation technology

The technology for simulation modelling for the design and operation of bulk handling systems has evolved rapidly in the last few years. The trend is to an environment almost completely dedicated to personal computers or to more powerful work station computers, and away from mainframe computers, except for very complex detailed models. This is largely the result of more powerful and faster personal computers.

The other major advance has been the development of animation packages for existing simulation languages and to computer interactive modelling with "object oriented animation" languages such as AUDITION which is more powerful than earlier languages such as GPSS/PC, SLAM etc. The strength of these newer object oriented languages is in the capability of the modeller to create much more realistic, animated models with as much detail as required and with considerable ease in expansion and duplication than were previously possible with conventional simulation languages. Results can be evaluated on the screen during the model run without the need to analyze pages of computer printout.

The objects of one model can be copied or transferred to another model easily so that a "tide" object in one can become the "tide" object in another model. The copying of original objects, such as the duplication of a "ship unloader" object or the replication of a "berth" object permits the model to become multi-berth facility with additional unloaders.

The modeller, using a mouse, can re-arrange the "stage" by moving "berth objects", "queuing objects" and "unloader objects" around. The "ship objects" merely asks for the current location of the queuing object and the berth object. There is no need for re-programming.

By testing each object one at a time, bugs are quickly detected and corrected, allowing the construction of vastly complex systems in reduced time.

6.4 Example of a coal unloading facility

The first major application of the AUDITION software by Sandwell was in the construction of the Power Station Simulation Model (PSSM) for China Light and Power in Hong Kong. The model comprised a power plant, stockyards, berths and ship unloaders for the unloading, storing and supplying blended coal to the power plant. The power plant demand and ship arrivals were seasonal. The unloading of the ships was subject to delays because of night time restrictions, tides, storms and typhoons, and by equipment breakdowns. The tidal component for the passage through the Ma Wan Channel was modelled to simulate current ship transit rules depending upon the size of the vessel, the direction of ship travel, the tidal flow and the availability of the limited capacity channel. Similar tidal logic applied to the mooring of ships at the berths, whether berthed Port-Side-To or Starboard-Side-To. The PSSM model has graphical display objects of the sun and moon denoting day or night, a digital clock, and a tidal gauge. The Ma Wan channel is colour coded as are the conveyors and unloaders: red when unavailable, green when available, and the equipment: grey when idle. The state of ship unloading is displayed pictorially as a decreasing horizontal bar.

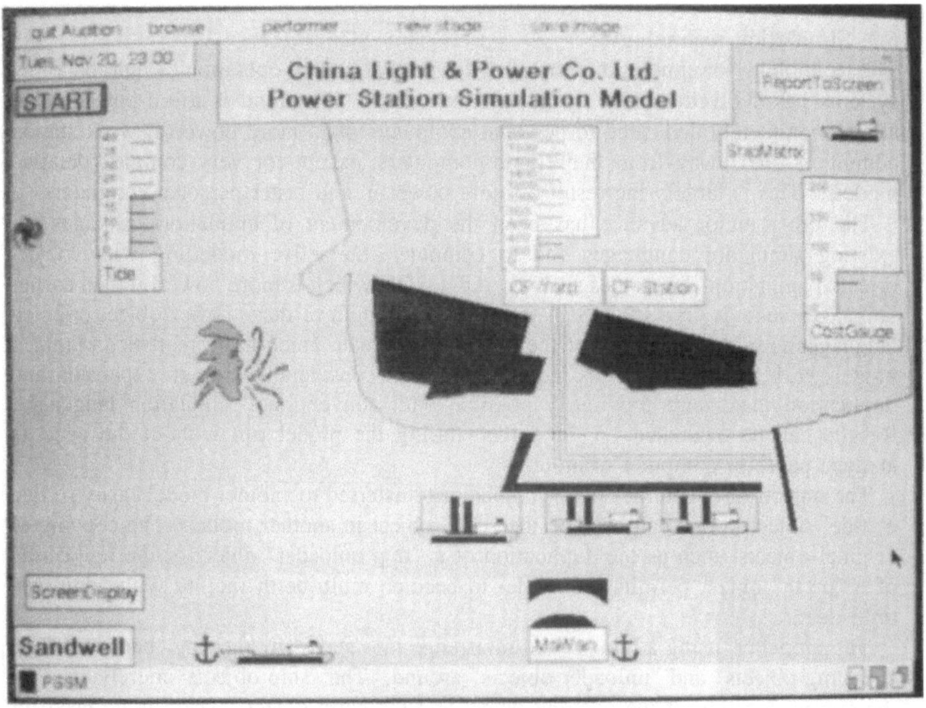

Fig. 5 Bulk Terminal Simulation

The model produces concise tables of berth and equipment usage, throughput-to-date, ship statistics and economic ships costs, stockpile storage levels, power plant consumption and an echo of major input. Using the model operators and managers can undertake for themselves the analysis of operations and expansion options.

The animation of PSSM is best illustrated in Figure 5 which shows an expanded state of the facility beyond the existing 2 berths, 5 ship unloaders and single power station. The simulation configuration was easily altered to the four berth arrangement with two additional unloaders using "prototype" objects, making copies, renaming the copies, shifting the copies and some minor re-configuring. This model has also been used for other coal unloading power stations.

7 Conclusion

The development and operation of a mine focuses on finding the most economic method of recovering material from the ground and converting it into a marketable commodity. Frequently, shipping the finished product to the customer is not part of that process.

The transportation system is subject to many factors which are beyond the control, and frequently the knowledge of the mine operator. There is a temptation to rely on the a competitive free market system to give the most economic cost for shipping the mine product to the customer. This approach may deliver the product to the customer but it is highly unlikely it will be at the most economic cost.

To achieve the lowest shipping cost the method of transport from mine to customer must be subject to the same degree of exhaustive economic analysis used in designing the mine. This cannot be achieved without an understanding of the entire transportation system, including ports and deep sea shipping which are discussed in detail in this paper. With this background the alternatives can be assessed, and the best transportation system selected.

Appendix A

Principle planning factors

- the annual tonnage and grades of material to bc handled;
- the seasonal rates of mine or plant production;
- distance from mine or plant to terminal;
- size of rail cars and unit trains or trucks and truck fleet;
- terminal handling rates;
- size and distribution of shipping fleet and ocean shipping distances;
- applicable freight rates, demurrage charges and dispatch bonuses.

Design variables to be considered for the terminals are:

- availability of land, climatic conditions, physical characteristics, geotechnical conditions, and environmental sensitivity of the area;
- train or truck waiting yard layout;
- train or truck unloading rates, number of cars and cycle times;
- stockpiling and reclaiming rates;
- total storage capacity required;

- number of different products or grades to be handled;
- handling and storage characteristics for the bulk products;
- stockpile layout including number and location of stockpile for each grade and individual storage capacity of each stockpile;
- number of stackers, reclaimers, or combined stacker/reclaimer systems;
- covered storage enclosures and isolation barriers for protection against contamination, emissions and exposure;
- physical arrangements of material and major equipment on site;
- shiploading rates, number and types of berths; and
- ship anchorages.

Appendix B

World Bank environmental requirements

1. Water-related impacts
1.1 Impacts caused by dredging
1.2 Impacts of dredged material disposal
1.3 Construction of piers, breakwaters and other waterside structures
1.4 Alteration of harbour/port ship traffic patterns
1.5 Ship discharges - oily ballast; bilge water; sewage
1.6 Spills: detection and clean-up of spills;
2. Waterfront industry discharges - sanitary and non sanitary
3. Land-related impacts
3.1 Excavation for fill
3.2 Wetland damage and filling
3.3 Loss of usable uplands to expanding waterfront/industrial areas
3.4 Noise from ports and harbourside industry
3.5 Effects of dust and other airborne emissions
3.6 Traffic burden projections
3.7 Handling and disposal of solid shore generated wastes
3.8 Runoff from raw material storage
3.9 Waterfront drainage
3.10 Industrial liquid wastes not discharged to harbour
4. Air-related impacts
4.1 Important background information
4.2 Fugitive emissions
4.3 Gases, smoke and fumes
5. Hazardous materials/cargoes
5.1 Categories - gases, liquids, solids
6. Socio-cultural impacts
7. Review of existing and proposed regulations affecting the proposed port or harbour development and its construction
8. Need for construction or facility operation environmental monitoring

With the exception of hazardous materials most of these items apply to a coal and concentrate shipping terminals.

Mean and lean—conveyor design for the 1990s

S. P. Zamorano
Bateman Ingeniería Chile S.A., Santiago, Chile

Abstract

The modern design of conveyor belts focused on producing a "Mean and Lean" product, in other words, a conveying system with the minimal capital and operating costs and maximum availability, using the latest technology available, is discussed.

Topics reviewed are computer simulations, high speed belts, belt resistances, the use of horizontally curved conveyors, (including their dynamic behavior) and steep inclined conveyors. Also the optimization of conveyor stringers, belting and idlers as well as non conventional designs, such as hanging conveyors and return strand transportation are discussed.

Theoretical background is discussed and real life examples are presented.
Keywords: belting, bulk materials, conveyor belt, design, idlers, transport

Introduction

The depressed bulk commodities market and the worldwide economic recession increase the always present need for economically efficient transport of bulk materials. Traditionally belt conveying has been the best way of transporting bulk materials in distances from a few meters to several kilometers and quantities up to 20,000 tons per hour. However, traditional design methods based on handbooks or "cook books" are very conservative and tend to produce over-sized systems. This over-sizing is required due to the shallowness of the analysis and the high number of unqualified assumptions. An in-depth engineering analysis of a system allows the engineer to reduce the "ignorance" factor and to produce a more efficient system.

During the last couple of decades, belt conveying technology has advanced significantly, mainly due to the use of more sophisticated analysis and design techniques, but also to the use of novel design concepts. This paper intends to give an overview of these two subjects.

Basic design parameters

Capacity

In most cases the design capacity of a belt is determined directly from the operational requirements. "X" tons per hour must be transported from "A" to "B". However, in complex plants, the transport needs are subject to other factors, such as surge capacities, events up and down stream, system availability, etc. Up to a few years ago, these multiple variable problems were approached using "rules of thumb", such as, one shift surge capacity or a design capacity 50% higher than the nominal, etc.

When confronting the design of a complex system or a high cost one, the use of modern computer simulation techniques is a must. Up to very recently, simulation was synonymous with long and boring procedures, whereas modern object orientated software makes the model creation an easy and fast task, with available hardware running complete simulations in a matter of seconds.

As an example, we can mention simulation work done on the Richards Bay Coal Terminal, currently the largest Coal Terminal in the world with a capacity of 53 million tons per year, in order to evaluate the required modifications to the plant for successive capacity expansions. The model had to take into account the almost 150 different routes that the coal can take from rail to ship. The terminal has six rail wagon tipplers, eleven stackers and reclaimers, almost one hundred conveyors and three ship loaders, all of this for eight different users. An additional problem is the high uncertainty on the train and ship arrival times.

The original simulation was written in 1979, when the terminal capacity was 20 million tonnes per year, and used to run on a 8086 PC (XT), with an alphanumeric interface and taking up to eleven hours for a single simulation run. In 1992, the simulations were done on a UNIX work-station, using a graphic interface, and taking a few seconds per simulation run (on a plant more than twice the size).

A computer simulation allows the analysis of the many possible operational conditions, and the determination of the transport capacities required to meet the production or handling goals.

Belt speed

The faster a conveyor belt operates, the more material it can transport for a given width. For a set transport requirement, the faster a conveyor moves, the narrower and therefore the cheaper the system can be.

Traditionally, belt conveyors operated between 0.5 and 3.5 m/s, with latter number still being shown in some handbooks as the maximum recommendable speed for a belt. In some applications the top velocity is effectively limited by operational constraints, typically, the need to avoid the degradation of friable products, but in most cases this maximum speed is restricted only by mechanical and structural parameters as well as the inconvenience of loading and unloading a fast belt.

The longer a conveyor, the cheaper it is to operate the system at high speed. When transporting large lumps or abrasive materials an accelerator conveyor is recommendable, feeding the material to the main belt at a fraction of the main belt speed and reducing the wear and tear of the main belt covers. However, if the system is short, the cost of this extra conveyor and loading point eliminates the advantage of a faster belt.

Figure 1 shows a conveyor system transporting R.O.M coal with 150 mm lumps over a distance of 6 km at a rate of 2,000 tons per hour. The conveying speed is 5.8 m/s (1,140 f.p.m) and the belt width 1,050 mm (42") with an accelerator belt that operates at a speed of 3.5 m/s (690 f.p.m.) protecting the main belt. After more than two years of operation and several million tons of material transported, no wear is visible on the covers of the main belt. The high speed made necessary the use of variable speed drives that allow a maintenance inspection speed of 1.5 m/s (295 f.p.m.).

Figure 1, Koorfontein overland conveyor, South Africa

The design of this conveyor demanded the simulation of the transient behavior of the belt as well as the analysis of possible resonance problems in the return strand and the supporting structure. The use of a "normal" speed belt would have required a wider belt with a significantly higher capital cost.

Another case where high speed conveying is economically viable is the transport of large volumes of material. When moving more than 8,000 m^3 per hour a "normal" speed belt would have to be almost 3 meters wide. Although this is technically possible, the cost per meter of such a system is very high as components are non-standard. Figure 2 shows a conveyor transporting power station coal at a rate of 11,000 tons per hour at a speed of 6.3 m/s (1,240 f.p.m.). The belt is 2.2 meters wide. The low abrasiveness of the material conveyed and the high cost of a wider belt or a dual system made the fast belt the best economical option.

Figure 2, 11000 t/h conveyor, Richard Bay, South Africa

The author led a feasibility study on the upgrading of a 1,800 mm wide conveyor from 5,500 tons per hour of power station coal to 11,000 tons per hour. The first option was to replace the belt for a new one, 2,200 mm wide. However, in order to avoid changing the full conveyor, with the associated down time, a study was done into speeding up the belt to 10 m/s, to handle the extra volumetric capacity needed. The analysis included power requirements, belt tensions, belt covers wear, possible resonant vibrations (of the belt, the idler frames and the structure) and the problems of loading and unloading a belt at such a high speed. The proposed solution included the use of larger diameter idler rolls (229 mm instead of 152 mm), longer loading skirts and specially designed deflection plates. As the current structure had to be replaced due to corrosion problems, the economic advantage of the faster belt disappeared and the wider belt was chosen. However, the basic engineering which was performed did indicate the feasibility of operating a belt at more than 10 m/s, as previously affirmed by Harrison [1].

Idler design

The design and arrangement of idlers is one of the main factors determining the friction resistance of a belt system.

Obviously, the type of bearing and the sealing arrangement will determine the rolling resistance of the roll itself (together with the relation between the shaft and roll diameter). Bearings with small clearances, such as taper roll or regular ball bearings are easily affected by penetration of foreign particles. Some sealing arrangements based on lip seals increase the roll resistance due to the lip friction, although this extra resistance disappears when the lip is worn and the seal is not effective anymore. This is the reason why high powers are needed to first start-up some new conveyors with badly designed idlers. Also, in the case of poorly designed labyrinth seals the drag force of the grease packed in the seal increases the roll resistance significantly, with a similar effect of high friction forces on start-up. Furthermore, if the labyrinth is not effective preventing dirt penetration, the dirt-grease mixture not only increases the friction forces but also wears the seal away, making the seal useless.

Furthermore, in the presence of high loads, the roll shaft is deflected significantly, and in the case of bearings with little tolerance to misalignment, such as taper roll bearings, the result is a considerable increase in the rolling resistance [2] and early failure. The problem is exacerbated by the unavoidable misalignment between shaft and bearing produced during manufacturing, that can very seldom be reduced consistently below 6' of arc. This misalignment is the maximum allowable for a taper roll bearing, so any deflection caused by load will result in early bearing failure and increased rolling resistance.

The best results (the least rolling resistance) are obtained when using large clearance, deflection tolerant bearings, such as deep groove ball bearings, as recommended by leading manufacturers like SKF [3]. Taper roll bearings are not appropriate for conveyor idler applications.

Belt resistance

The other two main components of the rolling resistance of a conveyor are the indentation and the flexing resistances.

The indentation resistance is produced by the energy dissipated by the deformation suffered by the belt in the contact point with the roll, as shown on figure 3.

The diameter of the idler roll is inversely proportional to the indentation resistance, i.e., the larger the diameter the smaller the drag force as the deformation of the belt is reduced. In the same way, the higher the belt tension, the lower the resistance as the angle of contact belt/roll is smaller, and so is the deformation. The load on the belt and the distance between idlers are proportional to the indentation resistance, as they increase the force between roll and belt and as a consequence the deformation and indentation of the belt. All these factors can be evaluated (or at least estimated) using analytical and experimental methods [3,4]. The other important factors in this force are the belt properties, such as cover elastic modulus, hysteresis, rheology (or in vibration terms its damping coefficient) and the belt stiffness.

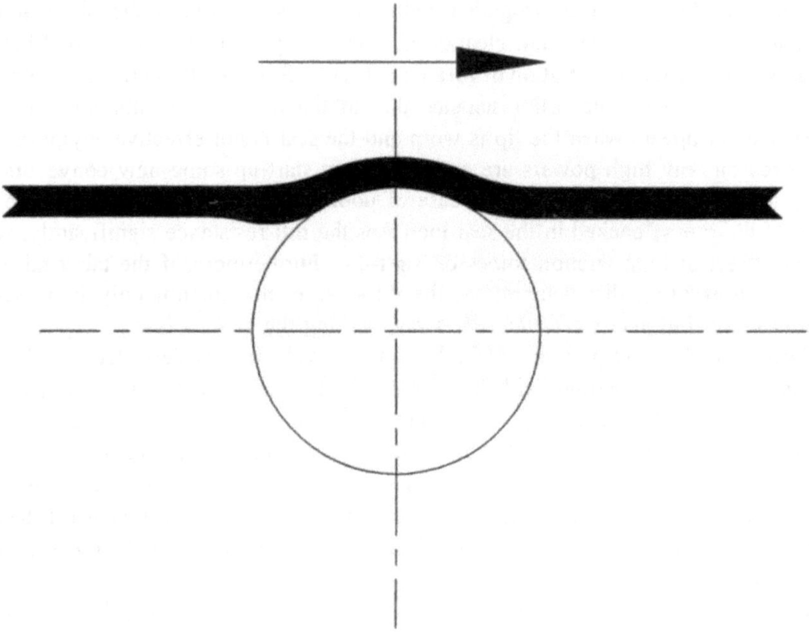

Figure 3, Belt indentation

The flexing of the belt between idlers produces an energy loss due to the internal friction (or damping) in the belt and in the material transported. The resultant friction force is called the flexing resistance, and is dependant on the actual deflection of the belt. In other words it is proportional to belt load and idler spacing and inversely proportional to the belt tension.

The CEMA design method takes into account most of the above factors. However, the actual resistance values calculated are greatly conservative. Other design methods, such as ISO and Goodyear apply artificial factors based on historic data. An accurate design must be based on up-to-date research data, applicable to the system to be designed.

In summary, a conveyor designed for low friction resistance must have idlers with large clearance ball bearings, large diameter rolls and with a high average belt tension and minimum sag. Of course, the best design will be determined by the relevant capital and energy cost, and also providing that any potential saving is higher than the cost of a detailed analysis. Otherwise, a "cook book" design and engineering common sense should be applied.

Belt class and dynamic behavior

Traditionally, the required belt class is determined by the rated tension of the selected belt. This nominal tension involves a safety factor of 10 in respect of static ultimate strength in the case of fabric belting (with vulcanized splices). In steelcord belting this factor is usually between 6.7 and 7.5. In the case of aramid fabrics (Kevlar) this factor is highly dependant on the splices and quite variable as the technology is not yet fully developed. However, it is not rare to find enquiry specifications calling for a safety factor of 10, regardless of the type of belting. Probably a result of specifications being copied from project to project, without real updating.

The above factors are all related to the maximum static tension on the belt, as they assume a dynamic or start-up multiplication factor between 1.6 and 2. This factor is related to the additional forces produced by the drive to accelerate the conveyor or the brake (if any) to stop it.

These simple assumptions are not applicable to long and/or high capacity conveyors. In the first place, an incorrect start-up procedure can produce tension multiplying factors of up to seven, causing, as has happened on more than one occasion, the catastrophic failure of the belt. Secondly, the cost of high strength belting makes the use of huge "ignorance" factors uneconomical. Therefore, the analysis and prediction of transient tensions becomes necessary.

The occurrence of elasto-dynamic forces during starting and stopping, over and above the rigid body dynamic forces proportional to the acceleration rate, is related to the excitation of the natural frequencies of the system [5], or as it is called in solid mechanics the generation of stress waves in the belt. These resonant vibrations or tension waves, as one wishes to call them, can be estimated by simplified methods or calculated by finite element analysis, depending on the complexity of the system and the accuracy of the analysis required. Usually the design approach is to avoid the elasto-dynamic problem by proper starting and stopping procedures, or if this is not possible due to design constraints such as power failures, design the belt and associated structures in accordance with the transient forces calculated. In this way the occurrence of catastrophic failures can be avoided, and more economically viable static safety factors, based mainly on splicing and fatigue considerations (about 5 for steelcord) can be used. This approach was followed in the design of the conveyors shown in figures 2 and 4, and no dynamic related problems have been encountered during their operation.

Figure 4, Majuba overland conveyor, South Africa

Conveyor stringers

The design of the supporting structure of a conveyor belt is not usually considered a critical issue. However, this structure sometimes costs as much as all the mechanical components of a conveyor.

The basic and most repetitive structure in a conveyor is the stringer, and consequently will give the best returns from optimization. Traditionally these stringers or tables, as they are sometimes called, have been designed according to structural codes with the main criteria being usually the maximum deflection and aesthetic considerations. An optimal design must be based on dynamic factors as well.

The design of a stringer must be done in such a way that resonance between the rotation of the idlers and the structure is avoided. In this way, conveyors with the same loads per meter running at different speeds will have different stiffness requirements for the stringer. The same applies for the transverse vibrations of the belt (flapping).

Once again, the possible saving in structural steel must offset the cost of the detailed dynamic analysis, except in the case of fast belts (over 3.5 m/s) where static design methods are not enough.

Advanced applications

Long, single flight conveyors

A cost analysis of a conveyor will show that a large portion of the cost of the system lies in the transfer points. This is applicable both to the capital and the operating cost of the belt. Therefore, a single flight conveyor will in most cases be cheaper to build and operate than a multiple flight system.

Conveyor designers often get asked questions like: how long can a conveyor be? or what is the maximum length for a conveyor? Like most things in life there is no absolute answer. The most important factors in the maximum feasible length of a conveyor flight are the availability of high tensile strength belts and their cost.

Current technology and production costs result in the optimum length for a single flight conveyor being between 6 and 12 kilometers. This is the point where the lowest cost per meter is generally produced. This is of course a general statement.

In practice almost any conveyor above 2 kilometers uses steelcord belting, as take-up requirements for traditional fabric belting become awkward. For a fabric belt the take up length required for a 2 km system is between 30 and 60 meters, compared to the 3 to 6 meters required for a steelcord belt.

In underground coal mines, like the ones in the South African coal fields, steelcord belting is seldom used, due both to the fire resistant requirements of the belting and the problems encountered, performing a hot vulcanized splice underground, in a fiery environment. Another factor is the use of temporary installations, that are repositioned every few months. In these conditions PVC belting with mechanical splices are used. Due to strength limitations, the maximum practical length of a single flight is about 2 km. However, mainly in collecting conveyors, longer belts are sometimes required.

In order to avoid the intermediate transfer stations, a single flight belt can be achieved by using linear or booster drives. A booster drive is formed by an auxiliary belt positioned underneath the main one, which transfers driving power by means of belt to belt friction. In this way the maximum tension in the belt can be reduced as shown on figure 5, which shows the tension profile on a belt with multiple booster drives. The design of this sort of system is based on research data and operational experience [6].

A potential application for this kind of technology is very long overland conveyors, especially when material degradation must be minimized. One of the reasons is the almost exponential relation between strength and price on steelcord belting. A preliminary study done on a 21 km system showed a 30 % saving between the use of a single flight belt, with 2 booster drives, and a conventional system with 2 flights.

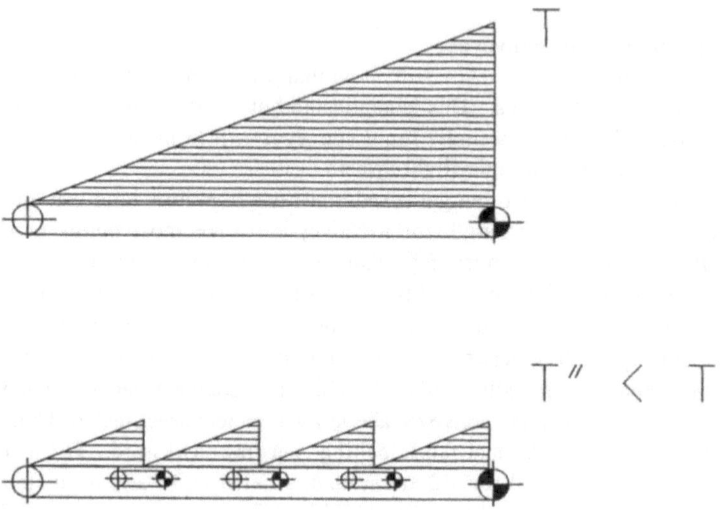

Figure 5, Tension distribution with booster drives

High lift and capacity conveyors

In many applications, for example in-pit crushing operations, it is necessary to lift the material up to several hundred meters in the minimum distance possible. When moving large tonnages, high strength belting and the determination of transient dynamic forces are required.

The largest system of this kind in the world is located in Germany and transports 1,800 t/h of R.O.M. coal over a distance of 3,745 meters and with a lift of 783 meters. The installed power is 7,2 MW and the belt class 7,500 kN/m.

In general, these sort of conveyors do not present many problems on start-up, but they do present problems during stopping. On a power failure or aborted start-up, the tension drops rapidly from the maximum tension to the minimum, and if transient forces are produced, belt tension can become zero. In this condition the belt folds between idlers and snaps up again when the tension recovers, expelling any material that could have been in the belt. If the conditions are right, a non-linear coupling of the longitudinal (tension variation) and transversal (flexing) oscillations can occur. If this happens the consequence is almost surely catastrophic failure.

In this sort of belt the prevention of the occurrence of very low or zero tensions in transient conditions is of paramount importance. If this is not possible due to design constraints, a dynamic simulation must be performed and the conditions necessary to avoid non-linear vibrations found and implemented.

Down hill conveyors

Steeply inclined downhill conveyors are often regenerative, in other words they generate energy when loaded. These types of conveyors are quite common in mountain mining areas like Chile, where the material is often taken from the mountain to the beneficiation plant downhill. Figure 6 shows one such system.

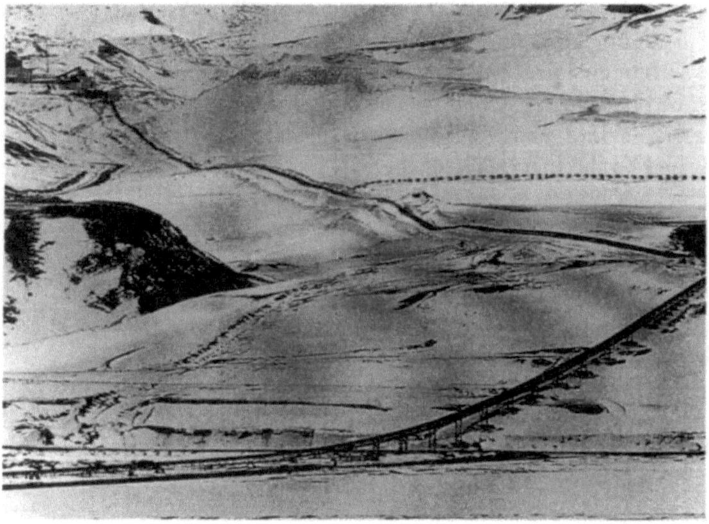

Figure 6, Heap leach conveyors, Tres Cruces, Chile

A downhill conveyor must have enough power to drive the empty belt and enough braking capacity to withhold the loaded belt. The main safety feature of such a conveyor is a failsafe braking system, able to stop the conveyor in a power failure and overload situation. A runaway conveyor will often destroy the drive due to centrifugal forces and sometimes the drive and tail station.

Braking must be applied in a fashion such that no transient forces are created or the system must be designed to withstand them. Sometimes variable speed drives are used to ramp up and down the belt speed and the brake is reserved for locking off the standing belt and in case of power failure.

When designing a system like this, the conservative position is to use a realistically low friction factor for the calculations. Otherwise the regenerative force can be underestimated and the system under designed.

Horizontally curved conveyors

Very often, due to geographic conditions, it is not possible to install a single flight straight conveyor system. Therefore, it is advantageous to curve conveyors and save thereby on earthworks, bridges, transfer stations, and so on.

The first problem to be addressed is the difference in tension between the inner and outer edge of the belt, as the outer edge follows a longer trajectory and tends to elongate more. If the tension in the outer edge is too high, damage can be produced to the belt carcass. If the inner edge tension is too low or zero, the belt will sag excessively.

The second problem is the existence of a radial force towards the inner curve, that will cause the belt to drift off the inner curve rolls. Various systems have been proposed to control this drifting. Probably the simplest one is using Garland type idlers. By lateral shift of the idler configuration an equilibrium between radial and gravitational forces can be obtained. A similar idea uses rigid idler frames, which are pivoted above themselves. The third method is to employ standard idler frames which are essentially super-elevated on the inner edge of the curve and horizontally cambered and twisted. This method has proven to be superior to the Garland and pivoting type due to the mechanical sensitivity of these first two systems. With the right test data, all the relevant frictional. tensional and gravitational forces present in a conveyor curve can be simulated in a computer program, and a satisfactory design produced. This was the methodology used to successfully design the conveyors on figures 2 and 4, as well as a curved conveyor that transports 6,000 t/h at 5.8 m/s [7].

Non conventional designs

Hanging conveyors

In underground mines it is not unusual to hang a conveyor belt from the roof instead of using the footwall to mount it on. This configuration facilitates cleaning underneath the belt and eliminates the need for supporting legs. In some cases of deep hard rock mining, like the South African gold mines, the cost of the civil works required to mount a conveyor can be significant, and much higher than installing roof bolts to hang the belt from.

Totally suspended systems, with garland idlers hanging from a cable stringer that hangs from the roof, are used in coal mines. In hard rock mining, it is more appropriate to hang a conventional continuous steel stringer due to the large loads involved.

Low capacity belts

Figure 7 shows a conveyor designed to handle 180 t/h of gold ore at 2,000 meters depth. The belt is 580 meters long and has a vertical lift of 157 meters. Due to the maximum lump size (300 mm), the width of the belt was chosen as 1,050 mm (42"). The conveyor is loaded by a loading conveyor fed by orepasses.

Traditionally, due to the problems in accurately controlling the feed, the conveyor would have been designed according to volumetric capacity and the belt speed set accordingly. However, in this case that would have resulted in a high class belt, large pulleys and a huge high reduction drive, all of this impossible to handle underground, without mentioning the high associated cost. It was decided instead to load the belt intermittently, utilizing only a fraction of the volumetric capacity. A power transducer on the drive motor activates an alarm when the maximum load has been reached in the belt. The operator stops the loading conveyor, waits for the main belt to clear and starts loading again. If the belt keeps on being loaded, an overload switch trips the belt. In the event of the alarm and the overload switch not working, the integrity of the system is protected by the fact that belt and structures can take the full stall torque of the drive. In this way, the capital cost of the system was reduced significantly and a fully workable system was given to the client.

Figure 7, Inclined conveyor, Kinross Gold Mine, South Africa

Return strand feed
In most mining and industrial plants raw materials are brought in, while product and waste are sent out, often using the same route. Conveyor belts can be used to transport material in two directions. Historically, this has been done by using reversible belts. The disadvantage of this system is that material cannot be transported in both directions simultaneously. An alternative is to use the return strand of the belt for loading purposes. In this way, material can be transported in both directions simultaneously.

Figure 8 shows a new conveyor belt feeding the return strand of an existing one. In this case the client had an existing raw material import conveyor and required a product export conveyor system. A new conveyor was built to collect the material from the product storage building and used to feed the return strand of the existing overland conveyor. At the other end a transfer station and an unloading conveyor can transport material in both directions with virtually no extra power needed. The cost of the project was 40 % of the cost of a system with a new overland belt.

Figure 8, Return strand feed, I.O.F., South Africa

Conclusions

The capital and operational cost of a conveyor system can be substantially reduced by means of proper analysis and the use of novel ideas and techniques. This kind of cost reduction will make a contribution towards maintaining the competitiveness of conveyor belt technology and towards the economic mining of raw materials.

References

1. Harrison, A., Hayes, J.W., Roberts, A.W. (1982) **The Feasibility of High Speed Narrow Belt Conveyors for Bulk Solids Handling**. Mechanical Engineering Transactions, Number 3.
2. Greune, A., Hager, M. (August 1990) **The Energy Saving Design of Belt Conveyors**. Bulk Solids Handling.
3. Stewart-Lord, M. (1991) **Rolling Bearing Characteristics for Maximising the Life of Conveyor** Idler Rolls. International Materials Handling Conference.
4. Spaans, C. (November 1991) **The Calculation of the Main Resistance of Belt Conveyors**. Bulk Solids Handling.
5. Zamorano, S.P. September 1991) **Dynamic Analysis of Conveyor Belts**. The South African Mechanical Engineer.
6. Küng, W. (1989) **Booster Drive for Twistdraai Colliery**. Design Report
7. Zamorano, S.P. (January-February 1992) **The Largest Capacity Horizontally Curved Conveyor in the World**. Technology S.A.

Economic advantages of belt conveying in open-pit mining

H. Lieberwirth
Department of Mining and Materials Handling Systems,
Krupp Fördertechnik GmbH, Duisburg, Germany

Abstract

Transport systems create a large share both in the initial investment costs and in the operating costs of open pit mines. The main competitive transport systems to be used in larger mines - the discontinuous material transport by trucks and the continuous material transport by belt conveyors - are compared. The advantages of the belt conveyor system are shown by applications of some in-pit crushing and overland conveying systems.

1 Introduction

Declining grades, rising costs, falling production rates, serious overmanning, growing environmental protection demands - these are only a few of the problems with which the mining operations are being faced at present.

Partial compensation of the ever-rising costs is possible in the form of advantageous transport technologies. If investment costs alone are considered for the necessary equipment, about 10% is required for removing materials (drilling and blasting), 31% for the loading equipment and 59% for the transport systems (Fig. 1) [1] [2].

The situation is nearly the same with respect to operating costs. 22% is invested for removing materials (drilling and blasting), 18% for the loading and 60% for the transport of materials (Fig. 2) [1] [2]. These two comparisons show quite clearly that material transport is the most expensive factor in the mining process.

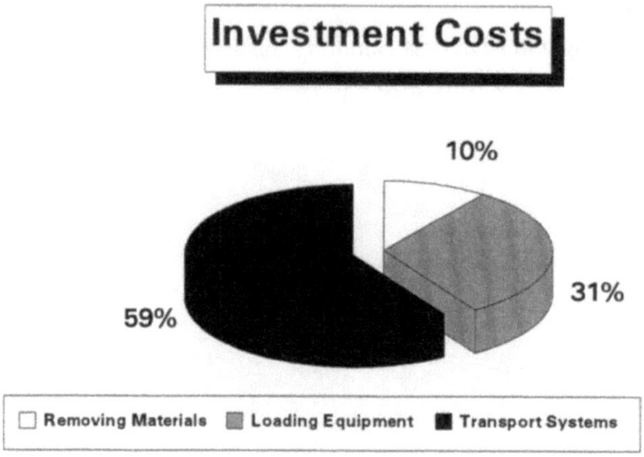

Fig. 1 Distribution of investment costs

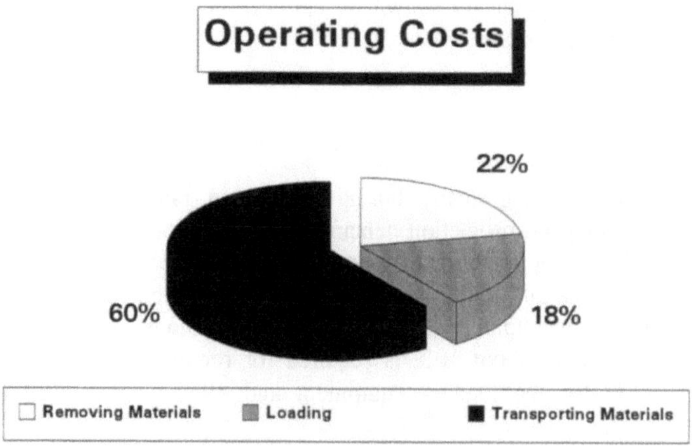

Fig. 2 Distribution of operating costs

2 Transport systems

There are only two main competitive transport systems from many alternatives which have been established for larger open pit mines with high outputs in the last decades (Fig. 3):

- Discontinuous material transport by trucks;
- Continuous material transport by belt conveyors.

Other systems, like ropeways, slurry pipelines or mine railways tend to be used for such mines only under unusual circumstances. They are not widely distributed and consequently have not been further considered in this paper .

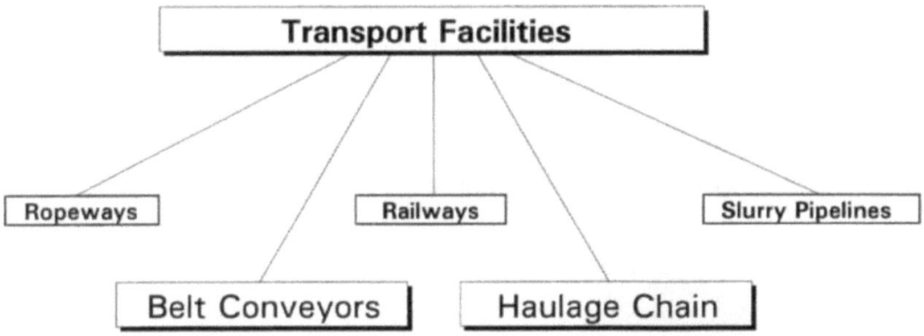

Fig. 3 Transport systems for open pit mining operations

The initial investment costs of a belt conveyor system are often slightly higher than those of a haulage chain depending upon the length of the transport system. But even the argument of lower investment costs for discontinuous transport systems has to be checked for every new installation.

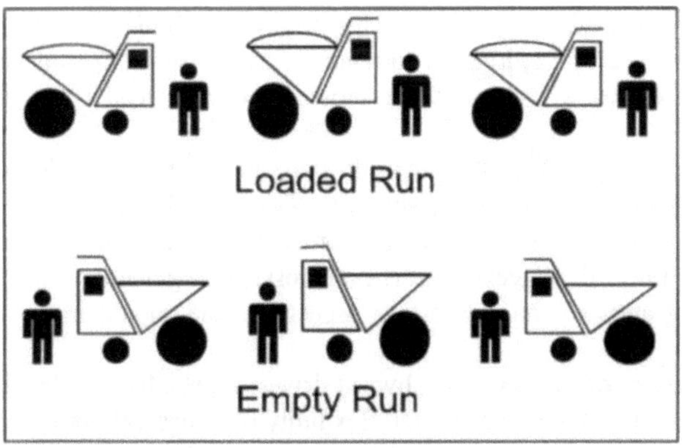

Fig. 4 Workforce for haulage transport

Considering the price for a 150-t-truck of about $1,000,000 it is often a question of throughput, transport distance and the difference in altitude to be climbed, to determine which transport system is more favourable. Already for a mine of 15 m depth investment costs for trucks may be in the same order as for conveyors. However, experience shows that the total production cost per ton of mineral - one of the most important figures in a mining operation - is influenced much more by the operating costs than by the investment costs.

For each truck, one driver per shift is necessary, but the driver runs an empty truck about half the time (Fig. 4)! Moreover, additional staff must be available for repairing and servicing the trucks, maintaining the roads, providing fuel and for the truck-fleet management. The average wage of bargaining unit employees in the Canadian copper mines is about $C38,000 per annum. Including benefit costs this figure even rises to $C50,000 per annum [3]. Therefore, wages, including costs of fringe benefits, account for roughly 1/3 of the total operating costs (Fig. 5).

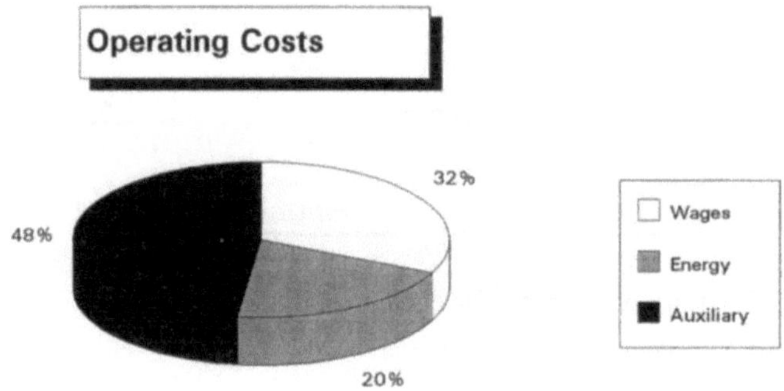

Fig. 5 Share of wages and energy within operating costs

Another important aspect to be considered when comparing the operating costs of different transport systems is the waste of energy. With a truck, about 40% of the total weight is dead-weight (Fig. 6). This means that about 40% of the total energy consumption is necessary alone for moving the truck itself. Considering a complete round trip, the energy consumed for moving the dead load rises even further to about 60% [4].

For a belt conveyor this value amounts only up to 20% [5]. Moreover, in countries without their own mineral oil reserves, diesel fuel has to be paid for in foreign currency, while the electrical power for conveyor systems can be generated locally. By installing downhill conveyors, it is even possible to generate electrical energy by transporting material - an interesting aspect for the mountainous region of the Andes.

The introduction of trucks with hybrid-drives especially in the early nineteen eighties was an attempt to connect the flexibility of a diesel-driven truck fleet within the mine to the economical electro-drive system on the slope, as the energy costs per kilowatt-hour are usually higher with diesel fuel than with electro-energy. But

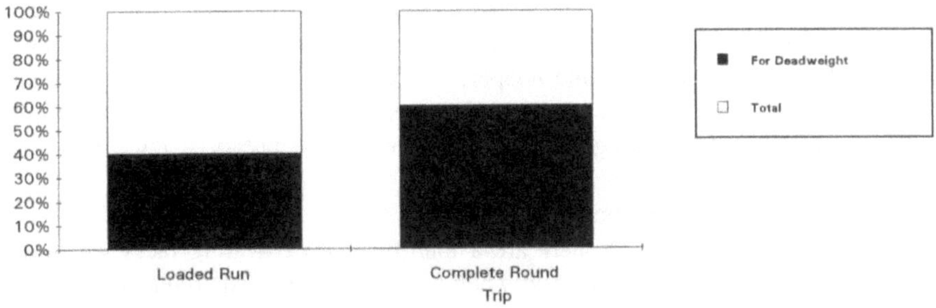

Fig. 6 Energy consumption of a truck

this attempt was only partially successful. In principle the trolley-assisted transport systems were limited to mines with special deposit designs like the Phalabora mine or the Sishen Iron Ore Mine in South Africa.

3 In-pit crushing systems

3.1 General

Whereas trucks can handle almost any material size, crushers are usually required for reducing the size of the material being handled so that it corresponds with the belt width, belt speed, troughing angle and inclination of the conveyor.

Fig. 7 Mobile crusher plant

There are two main locations for the in-pit crushers:

1. At the digging face (Fig. 7)
   Such a plant is fully mobile and is fed directly by an excavator. The plant discharges directly or in most cases via a belt wagon or conveying bridge onto a shiftable conveyor with mobile hopper. Fully mobile plants need no extra site preparation if they use the same moving mechanism as the loading excavator. The location at the digging face is applicable only if one or a few excavating faces exist.
2. At one or more strategic points in the mine
   Such locations apply when there are a multitude of excavating faces which are not each worked continuously and when the excavators need frequent relocation to different benches as practised in most deep open pit mines. The plant is semi-mobile (Fig. 8) and follows the pit advance at appropriate time intervals. It is fed by trucks which haul material from the face to the crusher. Their haulage route is kept as short as practicable and avoids any uphill haulage under load if possible. The semi-mobile crushing plant discharges via mobile and/or shiftable belt conveyors into permanently installed conveyors. It requires preparation of the installation site, mainly excavation and sometimes civil work, particularly for large units.

Fig. 8 Semi-mobile crusher plant

Fig. 9 Crusher at Chuquicamata mine

The maximum capacity of semi-mobile crushing units these days averages 10,000 t/h, reducing the material down to 0 - 250/400 mm. One unit of this kind is to be found in the famous Chuquicamata mine, Chile's largest open pit and the world's largest copper-producing mine with an altitude of approx. 3000 m and a desert climate. Not only the ore is crushed and transported continuously but also the waste and the overburden (Fig. 9).

The problem of steep slopes for material transport was solved in different ways for the waste and ore transporting systems. Whereas for the waste system three ramp conveyors with an inclination of 17° are used to transport the material out of the mine over a total difference in altitude of 422 m, the most suitable solution for the ore system was to build a tunnel through the mountain [6].

The decision for continuous transport of waste and overburden is more remarkable, as for overburden conveying the necessary preparation and dumping costs are an extra, to be offset by savings in trucking and truck haul road costs. But the large quantities of overburden to be transported over fairly long distances at great lifting heights justify the crusher-conveyor-spreader system.

Other difficult climatic conditions are found in Fort McMurray (Fig. 10) where Syncrude Canada began to operate a semi-mobile crushing plant even if the material is frozen down to -55°C.

Fig. 10 Crusher at Fort McMurray

3.2 The Quidashan mine

One of the recent installations of this kind is now under construction in the Quidashan mine in the People's Republic of China (Fig. 11).

It is planned that the production of iron ore will be expanded from the present design production capacity of 8 million tonnes per year to 17 million tonnes per year and the total of ore, waste and overburden from 20 million tonnes per year to 51 million tonnes per year (Fig. 12).

The whole system will consist of two separate parts (Fig. 13):

- the ore crushing system with semi-mobile crusher with belt conveyors for the transport of ore from the mine to some silos at a train loading station and alternatively to a store yard,
- the waste crushing system with a semi-mobile crusher with belt conveyors for the transport of waste and overburden from the mine to the waste dump.

Fig. 11 Location of the Quidashan mine in the People's Republic of China

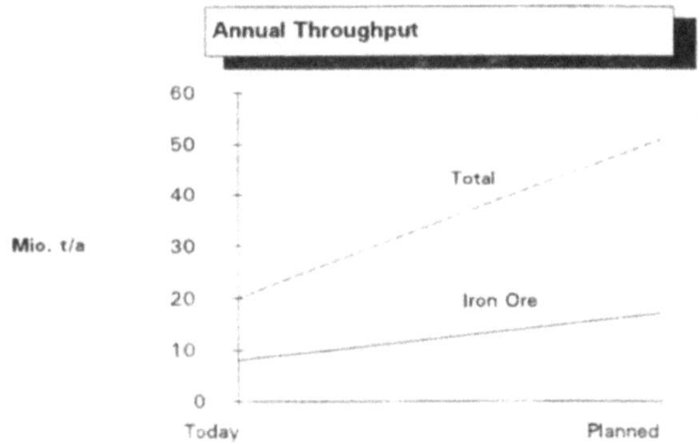

Fig. 12 Annual throughput of the Quidashan Mine

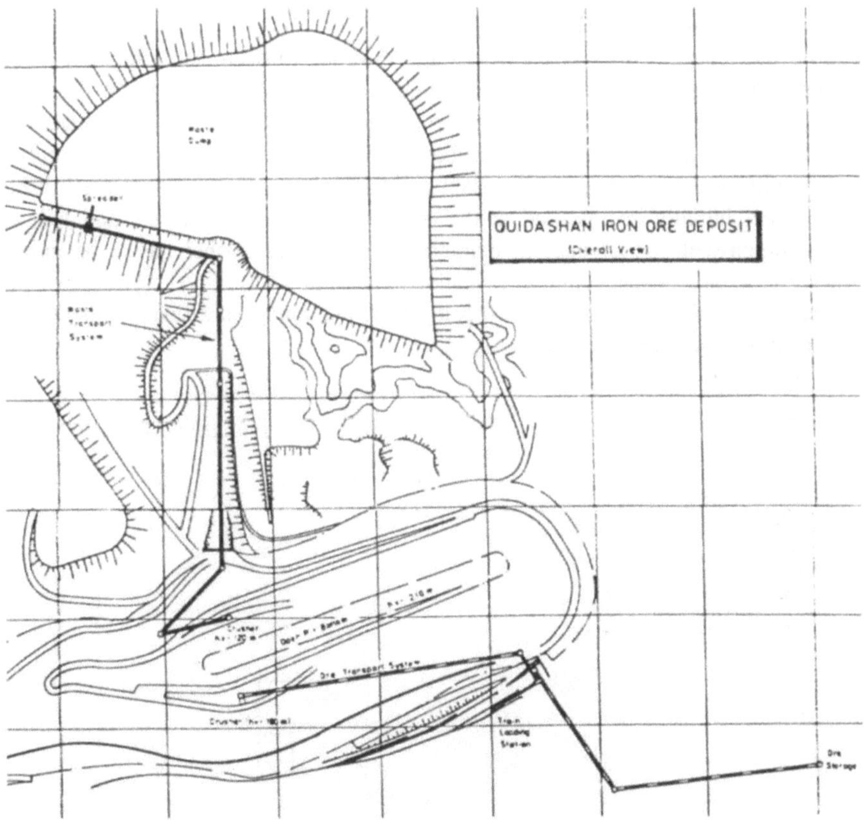

Fig. 13 Quidashan Iron Ore Mine, overall view

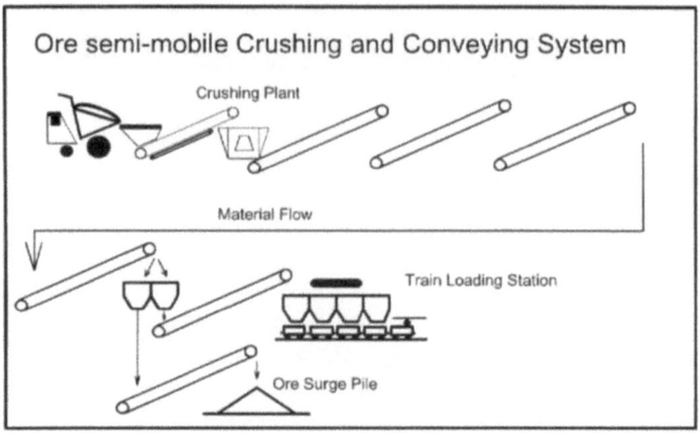

Fig. 14 Ore system - flow sheet

The ore crushing system (Fig. 14)
The ore is transported by in-pit trucks and dumped into the receiving slot of the crusher. The crushed ore is hauled by a belt conveyor to ore storage on the 30 m level. From the 17 million tonnes of ore produced per year, 8 million tonnes will be delivered to the existing Quidashan beneficiation plant while the other 9 million tonnes will be delivered to a new plant at Diaojuntao. The semi-mobile crusher system will initially be installed at the +42 m level on the surface and will be moved down with the advance of extracting work. The crusher has to be moved every one to two benches and will finally be fixed at 180 m level in the open pit. The conveyor system consists of two belts at the beginning, and finally 6 belts with total length of 3.2 km.

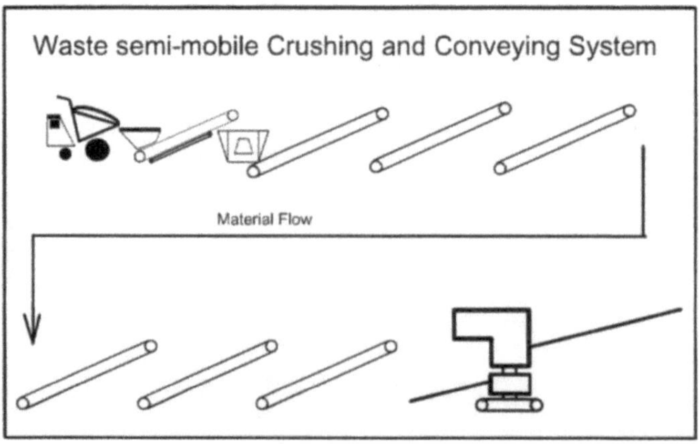

Fig. 15 Waste system - flow sheet

The waste crushing system (Fig. 15)
The waste will be hauled by in-pit trucks to the receiving slot of the crusher. The crushed waste is hauled by conveyor and is distributed by means of a movable conveyor-tripper-stacker system. The waste mobile crusher system will be initially installed at +46 m elevation, then moved down with mining. It must be moved once every one to two benches and will finally be fixed at -120 m elevation in the pit. The conveyor system consists of three belt conveyors at first, and finally five belts with a total length of 3.4 km.

The transport system
Comparison of different transport systems preceding the contract for crusher/conveyor systems have shown, that the continuous systems in connection with the truck-fed in-pit crushers are most efficient. This is even more interesting as it concerns a mine in a country where direct labour costs represent a lower portion of operating costs than in an equivalent western operation. Staffing is usually between one-and-a-half to two times the western level. But the wage rate lowers the total effect to 50% of western costs. The electricity costs appear to be comparable to US prices (a recent study took 6.2 cents/kWh), whereas consumable reagents and fuel can vary considerably between their relative costs in the West and in China [7].

The ore and waste crushing conveyor systems are controlled automatically by means of computers and are operated centrally. The mine will employ an Automatic Truck Dispatching System for the in-pit trucks with a blending system to improve equipment efficiency and to assure supplies of grading balance concentrate for the beneficiation plant. The maximum distance, constantly driven over uphill by loaded in-pit trucks is 1.5 - 2.0 km.

4 Overland conveying systems

As site conditions arc worsening and the preparation of minerals of newly-opened mines leaves a lot to be desired in existing plants, it often becomes necessary to transport the material over long distances.

Although the advantages of belt conveyor transport systems, in particular for longer conveying distances and larger volumes of material, have already been stressed above, there are often obstacles preventing straight conveyor routing. If a belt conveyor is to bypass mountains, valleys or existing buildings, it must be constructed in several flights or of horizontal curved design (Fig. 16). In our experience, a curved conveyor is usually the better solution as the maintenance costs of a single flight system are considerably lower compared with systems of two or three flights; the operational safety increases and the investment costs are often lower. Running cost reductions of as much as 40% (in some cases even 60%) are consequently reached [8].

Some examples for long distance overland belt conveyor systems are shown in table 1.

Fig. 16 Horizontally curved belt conveyor

Table 1. Examples of overland conveying systems

Site	Location	Length (km)	Capacity (tph)	No. of flights
Fosfatos de Bu-Craa	Sahara	100.0	2200	11
Alberta (Fig. 17)	Canada	10.8	700	1
Soma	Turkey	5.3	1000	1
Alabama (Fig. 18)	USA	4.0	1090	2
Optimum	South Africa	5.3	1000	1
Bayswater power st.	Australia	5.6	1000	1
Trichardt [9]	South Africa	10.7	700	1

The overland belt conveyor of the Obed Mt. Coal Co. (Fig. 17) operates at temperatures of down to -45°C. It is driven by three drives with 447 kW (600 HP), each coupled by scoop-tube fluid couplings with the bevel spur gearings.

Fig. 17 Overland conveyor of Obed Mt. Coal Co. (Alberta, Canada)

Fig. 18 Curved conveyor system for limestone (Alabama, USA)

Fig. 19 First horizontally curved overland conveyor (Soma, Turkey)

One of the recent installations of horizontally curved long-distance overland conveyors in Western Turkey economically transports crushed lignite from a mine in a mountainous region to the power station. The first conveyor went into operation in 1987 (Fig. 19). It descends about 320 m and replaced the trucks previously used.

Due to the overwhelming efficiency of the system the power company called for two more conveyers when extending the capacity of the power plant. These two conveyors have now also been commissioned (Fig. 20).

The conveyors include several major features rarely combined in a single flight:

- overland long distance conveyor,
- horizontal curves,
- steep vertical convex and concave curves to match the difficult terrain,
- downhill conveying,
- frequency converter drives with variable belt speed,
- generating operation mode for backfeeding energy to the network.

The conveyors include two 4 km radius horizontal curves, one left-hand and one right-hand. The medium radii of the vertical convex and concave curves are as low as 300 m and 470 m.

In sections of horizontal curves, the idler frames are banked by predetermined angles to horizontal axes. Banking ensures that the belt tracks correctly and prevents spillage of material from the 1000 mm wide belt. The carrying idler frame is of novel design (Fig. 21). The frames are mounted on a central pipe with a bracket, making banking adjustments easier if required. The conveyor frames consist of modules, each 6 m long, containing three carrying idler frames and one return idler frame. The majority of the conveyor modules are anchored in concrete foundations on the ground.

Fig. 20 Horizontally curved overland conveyor Soma 5/6

Fig. 21 Horizontally curved overland conveyor Soma 5/6, conveyor frames

The belts are covered with galvanised corrugated steel sheets for weather protection. The covers are easily dismountable. The conveyor flights include two belt-turnovers each at the head and tail end to keep the return idlers and areas under the conveyor flights clean.

A tried and tested simulation program offers the possibility of optimising belt tracking and the operating system, especially for gentle starting and stopping of the belt.

5 Conclusion

Mines are faced with the problem of ever rising costs. A partial compensation is possible by the introduction of advantageous transport technologies. For mines with large material quantities to be transported these are mostly continuous belt conveyor transport systems.

Technological constraints of belt conveyor transport led to the development of mobile or semi-mobile crusher plants for preparation of the material.

For material transport outside the mine itself, overland conveyor systems are an appropriate solution. Site conditions may require a horizontally curved design of the belt. Even for difficult conditions the best solution can be found with the aid of a tried and tested simulation program.

6 References

1. Franke, R. (1982) Combined Mining Systems for Open Pit Mines. **bulk solids handling**, 1, 75-80

2. Müller, G. (1986) Senkung der Betriebskosten im Festgestein-Tagebau durch Einsatz von Brecher-Band-Systemen. **fördern und heben,** 8, 556-559

3. Kennedy, A. (1991) Highland Valley: B.C. 's World Class Copper Mine. **Mining Magazin**, 3, 132-138

4. Bullivant, D. (1984) An Appraisal of the Correct Transport System for a Surface Mine. **bulk solids handling**, 4, 769-772

5. Cohrs, H.H. (1984) Mobilbrecher im Tagebaueinsatz. **fördern und heben,** 5, 376-379

6. Einenkel, O.; Richter, B. and Pelzer, W. (1992) Das neue Fördersystem J-1 Norte im Kupfererz-Tagebau Chuquicamata, Chile, unter besonderer Berücksichtigung der elektrotechnischen Ausrüstung. **Braunkohle,** 5, 30-40

7. Wesely, R.J. and Turk, T.W. (1993) Recent trends in refractory gold treatment projects in China. **Mining Engineering**, 8, 1016-1018

8. Alles, R. (1993) Horizontal Curve Engineering Above and Below Ground. **Conveyor Belt Engineering for the Coal and Mineral Mining Industries**. Society for Mining, Metallurgy, and Exploration, Inc., Littleton, Colorado, pp. 119-123

9. Geilert, U. (1990) Conveyors - 'Seldom Found Outside Mining'. **South African Transport**, 8, 33

Mining – Technology

Future of pneumatics in percussive drilling—the South African experience

C. W. Hunt
Boart SECO, Roodepoort, South Africa

Abstract

Although for the last twenty years or so hydraulic drills have become more reliable and accepted in the market, pneumatic drills still have some part to play in mining and construction. They are much less efficient and potentially less powerful, but they are easier to maintain and cheaper to run. Their use in unsophisticated markets is assured for many years to come.

Many underground mines around the world still use hand held or airleg mounted jackhammers. Although rig mounted pneumatic drills have been replaced in many areas by the more efficient and sophisticated hydraulic drills, hand held drills still remain pneumatic. There has been a move, particularly in South Africa, to introduce hydraulic jackhammers running on water. The claims for these units are higher penetration rates, higher rotation torque, more efficient use of power, less noise, no fog, etc. Unfortunately the drills are more expensive and complex than pneumatic machines and usually the life between breakdowns is less.

By careful design and the use of modern techniques, pneumatic jackhammers can be produced with a penetration rate comparable to hydraulic drills. Similarly, if rotation mechanisms other than the traditional rifle bar system are used, the rotation torque can also be increased. The lubricant fog, normally issuing from the exhaust, can be significantly reduced by the use of coatings inside the drill and by directing the lubricant required to the correct areas. Exhaust noise can be reduced by muffling; drill steel noise, although difficult to reduce, is also present in hydraulic drills. In short the majority of advantages apparently available with hydraulic units can be largely achieved by pneumatics.

The efficiency of the pneumatic system, when comparing output to input power, is less than hydraulics and will always be so. The power costs will therefore be higher. There are however many other costs involved. If a careful overall cost analysis is done then the result need not favour the hydraulic system.

The current recession in the world and in mining in particular has imposed restrictions on capital expenditure and placed operating costs under close scrutiny. The smaller operations are less likely to be able to afford to change to mechanised mining. They will have to remain with jackhammers and labour intensive operations to survive. These jackhammers will most probably be pneumatic.

1 Introduction

Pneumatic jackhammers have been used in South African gold and platinum mines for narrow reef stoping and development for a considerable time. During this period they have been developed and refined so that, although made to fine tolerances, these drills are rugged enough to survive in the mining environment.

Currently there are some 68,000 pneumatic drills in service in the mines of this co untry. They are manufactured and supplied mainly by three companies: Atlas Copco, Compair and SECO. The first two companies share approximately 10% of the market while SECO has the remainder. There is consequently a large spare parts market which is shared by these OEM companies and several alternative suppliers. Not surprisingly, in today's depressed economy, there is considerable competition for the sale of new machines and even more competition for the sale of spares. Each OEM company has to compete on a total basis with alternative suppliers. The OEM's do so at a considerable disadvantage. Firstly they have to sell a mature product competitively in order to have machines running in the field to generate spares sales. Secondly they have to do some ongoing research and development to ensure that their machine design stays competitive. Thirdly they have to be able to stock and supply any part of the drill at delivery times which suit today's discerning customers. They cannot adopt the strategy of the alternative suppliers of manufacturing and supplying only the fast moving wear parts. All these factors have influenced the selling price of drills and parts such that a pneumatic jackhammer and thrust leg, depending on the type, now sell for about R2900,00.

The underground performance of these machines in terms of holes per shift depends on several factors. The drill itself must have been designed to give a percussive output that is sufficient to drill effectively and also be able to be pushed by the operator using a thrust leg. Performance in terms of penetration or metres per minute is dependent on the pressure and amount of air available. In a mine where the compressor is usually on surface, the reticulation system necessary to get the air to the drill is long and circuitous. Even if carefully designed, this inevitably results in pressure drops and air leakage. If badly designed or, as is more usual in older mines, the system just grew as the mine was extended, then the leaks and pressure drops become serious. The performance of the drill at the end of this system may then only be a fraction of what was intended.

With the invention and ultimate successful use of hydraulic drills, by manufacturers such as Montebert, Atlas Copco, Tamrock and others, the potential advantages of higher efficiency, greater power, lower noise levels, etc. have been fully exploited in mechanised mining. In South Africa the Chamber of Mines Research Organisation (COMRO), in collaboration with various manufacturers, saw the potential of hydraulics. They have done a considerable amount of work since the late 1970's on hand held hydraulic jackhammers. Initially work was started using drills operating on 5/95 oil in water emulsions. Over the last few years machines operating on water only have been developed and are operating successfully in several mines around the country. The most significant of these is probably Northam Platinum Mine, near Rustenburg, which was laid out from the outset as a "water hydraulic mine". No air compressors were installed for drilling purposes.

Some 1500 drills are currently operating at Northam on a production basis. The three main suppliers of water driven jackhammers, at the moment, are Ingersoll Rand, Gullick and Sulzer. They are all represented at this mine. Several other companies either have drills or are working on them to be introduced in the future. By the end of this year the total population of water drills in the country will be approximately 2000. This may seem small compared to the 68,000 pneumatic drill population already quoted, but it is a number which is watched with interest by the major manufacturers of rockdrills throughout the world. Watched is probably the correct way of describing the strategy of some of these companies. It is believed by many that currently having three players in such a small market is too many. This is particularly true if the profits being made are small in relation to the effort needed to support and nurture such a new product. The fortunes of the drills and indeed the mine are undoubtedly being watched with interest.

At this moment no one will deny the inherently more efficient operation of the hydraulic drill and system. When driven by high pressure water from electro-hydraulic power packs, or by the use of the natural head of water available in current deep level mines, the system must eventually prove cost effective. The question is, have the mines currently using water drills shown that the system is cost effective with today's technology, machine prices and reliability? It is the purpose of this paper to investigate some of the claims and perceived advantages of the water drill system and put the case for current pneumatic drills as a viable alternative at this particular time. In addition, it puts forward data on two new generation pneumatic drills which may show the way for extending the life of this type of machine for a few more years.

2 Noise

One of the differences often quoted between the pneumatic and water jackhammers is noise. The water machines are perceived to be much quieter than their pneumatic counterparts. This is true, but what must be investigated is how much quieter and what does that mean in practical terms during an underground working shift.

Rockdrill noise comes from three main sources in the drilling system:-

1. Below 125 Hz by impact between the piston and drill steel.
2. 125 to 2000 Hz by expansion of the air from the exhaust.
3. Above 2000 Hz by transverse drill steel vibrations and between bit and rock.

All percussive drills generate the noise shown in Item 1. They all have an impact piston striking against a drill steel or shank adapter. This noise may be muffled to a certain extent by the fronthead or body housing of the machine. A light aluminium housing will muffle the noise less than, for instance, a heavy iron housing.

Fig. 1 Relative noise levels of jackhammers

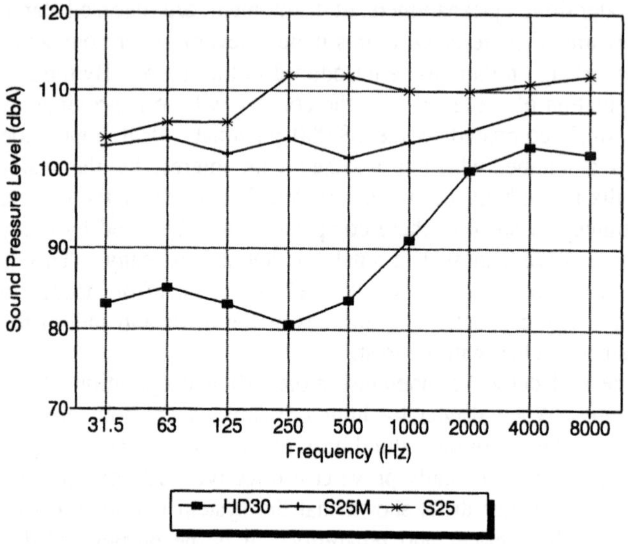

Table 1. Relative overall noise levels

Drill type	Relative noise level
HD30 emulsion hydraulic drill	1
S25M muffled pneumatic drill	4
S25 unmuffled pneumatic drill	16

The noise from the air expansion shown in Item 2 will, of course, only occur with pneumatic machines. Some noise will however be generated by other sources within hydraulic machines in this frequency range.

Item 3 is the noise emitted from the drill steel and between the bit and the rock. When the steel is struck by the piston, a stress wave is transmitted down the steel to the bit and causes the bit to fracture the rock. This wave also initiates transverse vibrations in the drill steel which are one of the sources of noise in this frequency range. The frequency and amplitude will depend on the length and cross sectional shape of the steel. They are a source of lost energy and noise. In addition, more noise is generated between the steel collar and chuck bush if a forged collar is used rather than the rubber one peculiar to the South African mining industry.

The noise levels for two pneumatic drills and an hydraulic drill are plotted against frequency on Fig 1. The S25 pneumatic machine was used in unmuffled and muffled form in the comparison against the HD30 hydraulic drill. It can be seen quite clearly how exhaust muffling has reduced the sound level in the frequency range 125 to 2000 Hz. The HD30 hydraulic machine shows a very dramatic reduction in noise in this range. The overall noise levels for each machine and their relative levels are shown in Table 1. It can be seen that by muffling, the average sound level of the S25 has been reduced from 117 dbA to 111 dbA. Remember that, because a logarithmic scale is used, every 3 dbA drop halves the actual sound level. The drop of 6 dbA therefore reduces the sound level to a quarter of the unmuffled drill. The sound level of 105 dbA for the HD30 hydraulic machine is similarly one quarter of the sound level of the muffled pneumatic drill. Put another way, the muffled pneumatic drill is 4 times noisier than the hydraulic machine and the unmuffled one 16 times noisier!

Work on muffling pneumatic drills has normally been done, as most research work is, to satisfy the requirements of the market. Muffled machines have been available as "off-the-shelf" units for some years, but the demand relative to unmuffled machines has been disappointing. This has certainly resulted in limited resources being allocated to the development of more efficient machine silencing. Although the muffling of jackhammers is more difficult than drifters, because of the limitations on size, better sound reduction can be achieved by more development. An example of what is possible can be appreciated by the values shown on Fig 2. of a modern pneumatic drifter against an hydraulic one.

Fig. 2 Drifter noise levels

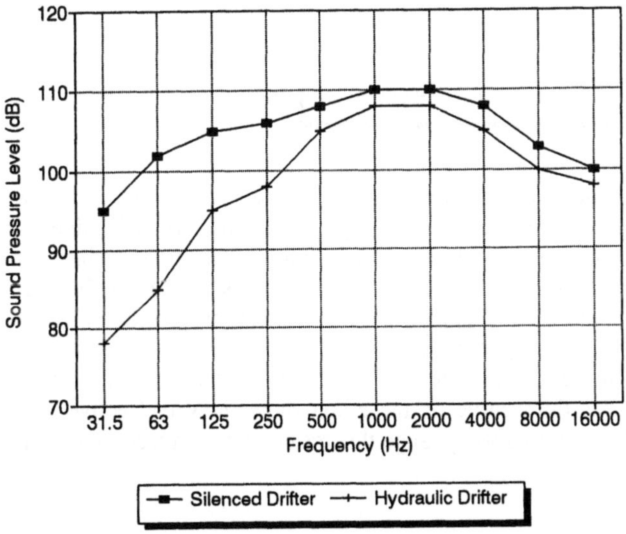

Now let us look at the practical benefits of the quieter drills and whether they can be effectively used. Fig 3. shows sound levels and the exposure time in the work place which are acceptable according to three standards. We have shown the US, UK and SA standards. Of the three, the SA standards are the most stringent. Now checking the three noise levels of 117, 110 and 105 dbA we see that the exposure time for these three drills would be about 5, 10 and 15 minutes before some hearing damage would be expected. In other words hearing protection is required when using all three drills: yes, even with the hydraulic one! The lower sound level of the hydraulic machine is therefore of no real advantage at all.

Fig. 3 Noise exposure standards

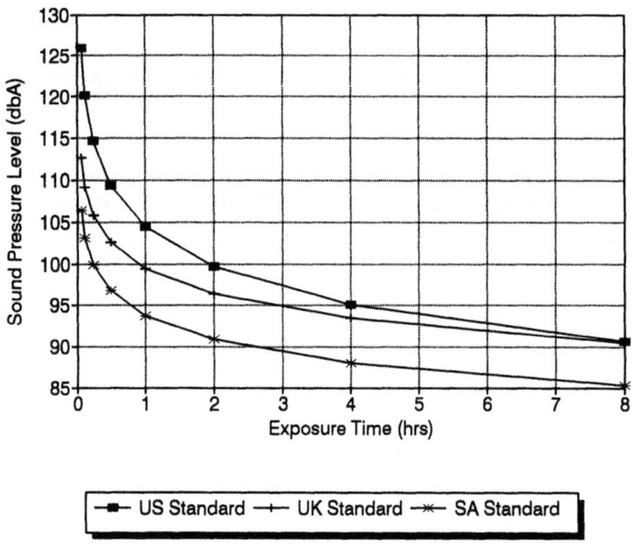

Fig. 4 shows the noise attenuation or reduction to the wearer provided by two different type of ear plugs. The wax impregnated type is slightly more effective than the hard plastic type. Both types do however give more protection against high frequency noise than the frequency noise. Ear muffs worn separately will give a similar amount of protection. If both are worn then maximum protection will be achieved. The use of a muffled pneumatic drill or an hydraulic one reduces the amount of low frequency noise so that protection in the form of ear plugs or ear muffs can be more effective. The third curve on the graph shows how ineffective badly fitting plugs can be. In the low frequency range this results in no protection at all for the wearer.

3 Exhaust emissions

Pneumatic drills exhaust air, water and lubricant into the atmosphere while they are operating. These emissions can cause problems in the working place. The air exhausted drops the temperature and can in fact help the working environment in deep level, hot mines such as those in South Africa. The inclusion of condensed water in that air helps to further cool the surroundings. What is not helpful is the loss of visibility due to the water droplets in the air. The result of this loss of visibility is not normally serious but can hamper the mining operation and result in some reduction in productivity.

Fig. 4 Ear plug attenuation comparison

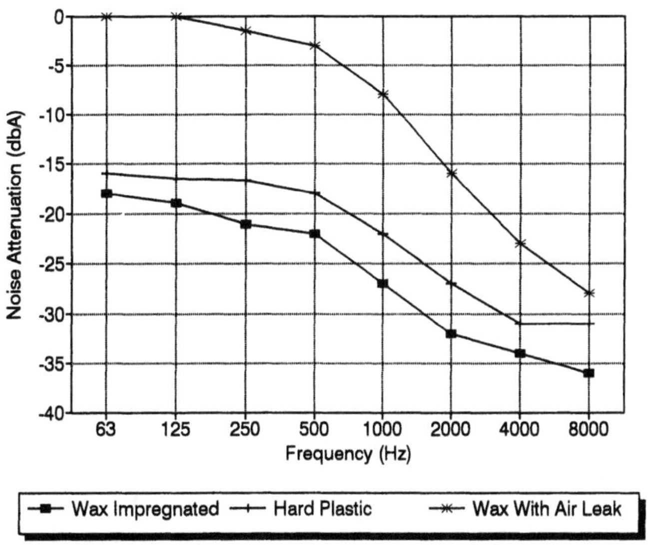

The lubricant that is present in the exhaust air is unpleasant when inhaled and can be regarded as a health hazard. In recent years there has been a significant move to the use of grease lubrication instead of oil in most mines; this comes out of the machine in larger droplets and is very quickly deposited out of the air in the general vicinity of the exhaust stream. Oil, on the other hand, stays in suspension in the air much longer thereby increasing the chances of it being inhaled by the operator.

Current hand held pneumatic drills introduce lubricant to the main input air. Simple methods of directing lubricant out of suspension to the bearings, etc. are used. Inevitably however a large proportion of the lubricant follows the main air flow and leaves the machine with the exhaust. More modern drills are being designed to utilise the lubricant more effectively by directing it to the specific areas of the machine where it is required. As these techniques improve then the amount of lubricant exhausted will be reduced.

4 Penetration rate

A great deal has been made of the power of the water hydraulic machines and the potential benefits in the number of holes drilled per shift. This is achieved by virtue of the percussive output power and rotation torque and power. Generally speaking, the power output of the three main hydraulic drills is around 4 kW and a maximum torque of 80 Nm at the chuck. Torque at the normal chuck rotation speed of 200 to 300 rpm will be less. In practice the torque demanded by normal hard rock drilling conditions is quite low. The high torque available from these machines makes them suitable for drilling in crushed rock conditions often found in deep level mines.

When operating on the water column available in deep level mines, sometimes known as "Hydro-Power", exhaust water from the drills is dumped into the working area. The inlet pressure, depending on the distance from the main water column, will be fairly constant. Because the systems currently operating are new and well controlled, the machine inlet pressure requirement will match the system pressure closely. The resultant power output will therefore be very close to the designed output. This is in contrast to pneumatic systems where pressures at the machine can vary considerably and are more dependent on air demand throughout the mine.

The majority of pneumatic drills in use on the mines, particularly in the stopes, are of the lightweight type with a mass of 22 kg or so, suitable for one man operation. The move to this type of drill has taken place over the last few years as most mines have tried to reduce labour costs. Previous drills required an operator and an assistant. The water drills have in fact reversed this trend: as with a drill mass of around 30 kg, some assistance may be necessary in the stopes.

It is not surprising therefore that higher penetration rates are achieved by the water drills. They normally claim to be able to drill about twice as fast as pneumatic machines and this is probably true. Under normal circumstances penetration rates of 500 to 600 mm/min are common. This would compare with 250 to 300mm/min for the lightweight pneumatic drills. There is also a price to pay for this of course. They usually have to use 25mm hexagonal drill steels instead of the 22mm used by pneumatic drills. This imposes an extra cost for drill steels and poses a transport problem as the steels are heavier: moving them in and out of a stope is not easy. The alternative system of knock on taper button bits would apparently provide a convenient solution to the problem. Unfortunately tests using this equipment at Northam have not proved successful. Even when using 25mm steels the shank wear is very high. Using the knock on bit system means that steels stay underground longer and monitoring the shank wear becomes difficult. The result is that badly worn shanks continue to be used and these damage to drill chuck bushes. Because of this several cases of stuck steels have occurred and maintenance costs have risen. A return to integral drill steels is being implemented.

It is possible to partially or totally counter the performance advantage that water drills currently have. Locally SECO have been experimenting with new concepts in jackhammer design. They have an independent rotation drill called the NOVA which has an output some 75% to 80% greater than the S215 drill which is the current mining standard machine. The NOVA has been tested thoroughly on the test bed and undergone preliminary testing at selected mines. A pre-production batch of these machines will undergo extended underground tests starting in January 1994. A brief specification of the NOVA 70 is shown in Table 2. A performance curve of the

NOVA relative to an S215 is shown in Fig. 5. Also shown in Table 3 is a comparison of some underground results taken with the NOVA against figures reported for emulsion drills. The NOVA results are from a limited period underground. They are only intended to give some indication of the potential for such a drill. The results shown for the emulsion drills were average values for many months of testing. In addition it is assumed that the current water drills have the same output as the earlier emulsion machines. These impressions need to be confirmed by a longer test programme next year. It does however indicate what may be possible when comparing drills having similar outputs.

Table 2. Pneumatic drill specifications

Drill type	Drill rate @ 400 kPa (mm/min)	Drill rate @ 600 kPa (mm/min)	Piston dia (mm)	Stroke (mm)	Impact rate (Hz)	Mass (kg)	Length (mm)	Air cons. (l/sec)
NOVA 70	450	830	70	60	55	26	749	95

(Note: The NOVA 70 was drilled with a 36mm button bit in Norite.)

Fig. 5 Performance comparison of NOVA 70 and S215 pneumatic drills

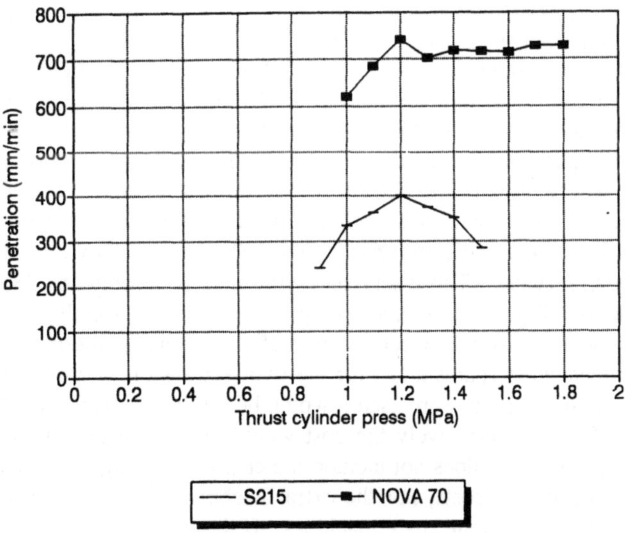

Table 3. NOVA 70 compared with emulsion drills

Production site	No in crew	Hole length (m)	Bit diameter (mm)	Penetr rate (m/min)	Drilling rate (hole/h)	Moving time/hole (mins)	Drilling time/hole (mins)	Total time/hole (mins)
A	2	0.9	31-38	0.53	24.2	0.78	1.70	2.48
B	2	1.2	38-42	0.63	23	0.70	1.90	2.61
C	1*	1.2	27-38	0.57	19	1.05	2.11	3.16
NOVA	1	1.2	38-42	0.48	18.8	0.7	2.50	3.20

The NOVA 70 pneumatic machines can operate on standard mine stoping 22mm integral steels. If preferred of course they can operate with the knock on button bits currently supplied by many companies. Shank wear should be no more than any other pneumatic drill because, in common with them, some lubrication is directed to the chuck bush.

5 Maintenance costs

Because pneumatic drills have been in existence for a considerable length of time the manufacturing costs have been trimmed down to a minimum. Currently the selling price of a pneumatic jackhammer is +/- R2300 and a thrust leg +/- R600. The cost of spare parts is similarly very competitive. In addition these prices are held down by the existence of many alternative manufacturers in addition to the OEM's.

In order to check on the real maintenance costs of SECO machines in the field, an exercise was done at Harmony Gold Mine comparing SECO costs with those of an alternative manufacturer of spare parts. Two sections North and South were chosen: the former using alternative parts and the latter using SECO parts. The results of a 12 months test from August 1992 to July 1993 are shown in Figs 6 and 7. Fig. 6 shows the average number of metres between machine breakdowns for each region. This shows quite clearly that where drills are serviced with SECO parts they last on average 480 metres between services. Those serviced with pirate parts only last 280 metres between breakdowns. Fig. 7 shows the cost per metre drilled for the two regions. Using SECO spares the cost is 21,6 cents versus 25,9 cents using alternative spares. The cost of maintenance in spares alone for the two sections on Pirate parts would be R135752. If only SECO parts were used it would be R113214. This would represent a saving of R22538 or alternatively the cost would be 83% using SECO parts against using Pirate parts. This does not include the cost of bringing the drills to surface, the rockdrill shop labour costs, nor the extra machines in the system to cater for the lower reliability of the pirate spare machines. If these were included on a realistic basis then the savings would be even more significant.

Fig. 6 Metres drilled between services

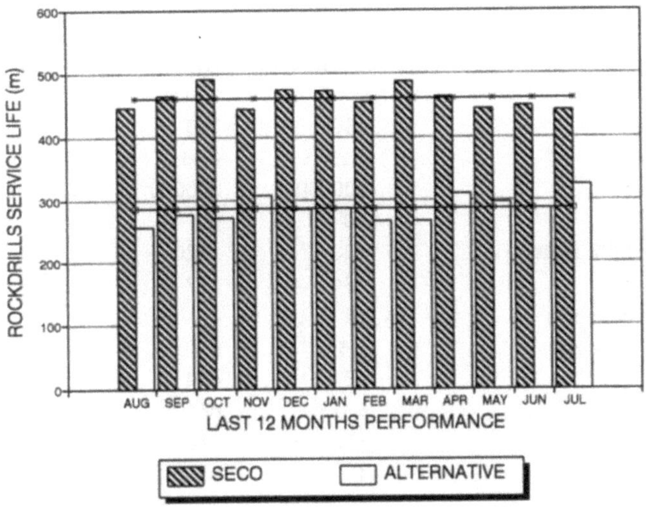

Fig. 7 Rockdrill spares cost per metre drilled

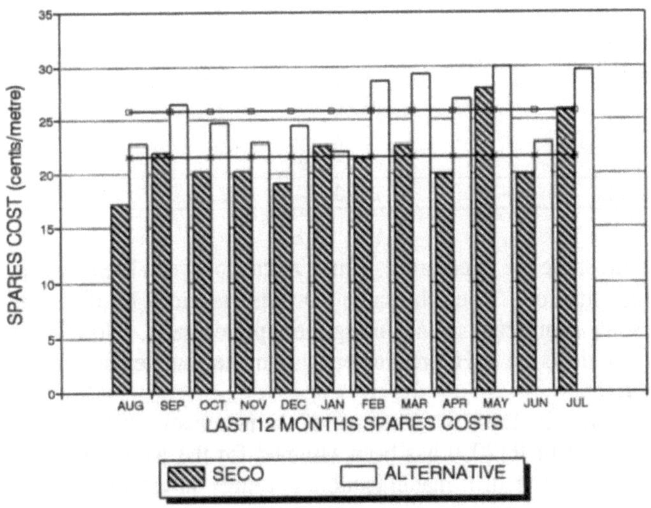

Table 4. Cost comparison of pneumatics versus water drills

	Pneumatic Drill	Water Drill
	1 Man crew	2 Man crew
Cost of drill and leg	2900	12000
Metres/breakdown	450	250
Cost per metre in Rands	0.22	0.91
Metres drilled per month	300000	300000
Metres drilled per shift	30	60
Total metres drilled/month(22 shifts)	660	1320
No of drills	454	227
Average No of days to breakdown	15	4
No spare for every drill operating	0.5	2
Number of spare drills	227	454
Total number of drills	681	681
No of machine services per month	667	1200
Time per service(hours)	1	1.5
Total capital cost of drills & legs(Rand)	1974900	8172000
Total cost of spares /month	66000	273000
Cost of drill service @ R75 ph	50000	135000
Transport cost/metre @ R50 per machine	0.11	0.2
Lubricant cost/metre	0.05	0
Spares cost/metre	0.22	0.91
Maintenance labour cost/metre	0.17	0.45
Drilling labour cost/metre	1.60	1.60
Power cost/metre	0.23	0
Power cost to pump water/metre	0	0.1
Total cost per metre in Rands	2.38	3.26

This case study serves to illustrate the importance of high metres between breakdowns for rockdrills in a mining situation. Even the same machines serviced by different manufactured parts can show significant savings using the superior parts. Now if we consider the case of water drills on a Hydro-Power system the maintenance costs become even more important.

A brief comparison is shown in Table 4. The cost of the pneumatic drill and leg was mentioned earlier. A cost of R12000 has been assumed for the water drill and leg combination. Of the three water drill manufacturers currently in the market and at places like Northam this cost is probably nearer the lowest. A back up of two machines for every one water drill drilling is assumed on the basis of what is happening in practice. The back up of one machine for every two pneumatic drills drilling is again based on actual conditions. A cost to the mines of 8,9 cents per kWh for electric power was given by Eskom. The service cost of R75 per hour is based on commercial fitting rates in industry, it may be a little high for rockdrill shop labour.

The figures come out in favour of the pneumatic drills at the moment. It obviously depends who compiles the figures and where the information comes from as to which has the lower cost. This is not intended as the definitive cost exercise. Its intended to highlight some of the areas of difference between the two systems.

Any mine or section must then do their own sums and make their decision on those facts. Until the price of the water drill comes down considerably and its reliability increases it is doubtful if it really is more cost effective to run, particularly if higher performance pneumatic drills are used. Eventually however the water drill costs and reliability will be improved. How long that will take is not clear at the moment.

6 Summary and conclusions

When comparing two machines or systems, such as pneumatics and water hydraulics in this case, one must take care to compare them under the same conditions. If this is not possible, as is often the case with mining, then one must be sure to have enough information or data to reduce the variables. What has happened in this case is that firstly new hydraulic machines and reticulation systems which have been designed using today's technology are being compared with mature pneumatic products and established air reticulation systems. In addition a heavyweight, high performance water machine is being compared with a lightweight medium performance pneumatic drill.

Over the last ten years there has been a policy on most mines to change to lightweight pneumatic drills suitable for one man operation. These machines have a mass of approximately 22 kg instead of the previous 30 kg machines requiring an operator and an assistant. At best two of these drills could be used with two drillers and one assistant shared between them. The water hydraulic drills are in fact reversing this policy by using drills having masses +/- 30 kg. In the case of both pneumatic and hydraulic drills there is some correlation between the drill mass and its output power. Thus while pneumatic drills, at the request of the mines, have been reduced in mass and power over recent years the hydraulic machines maintained the high mass and have opted for increased power. It should come as no surprise therefore, that one of the advantages claimed by the advocates of water drills is high penetration speeds. Again comparing like with like becomes difficult.

If the lower power trend of the pneumatics is reversed and a powerful pneumatic drill designed, then one of the main selling features of the water drill is immediately nullified. As discussed this has been done recently by SECO with the NOVA. The NOVA machines compete favourably with the hydraulic ones on mass, being +/- 26 kg. If suitable air pressure is supplied then at 550 to 600 kPa their output power will be close to 4 kW. Furthermore, SECO has the possibility of producing a NOVA with a 90mm bore to give 3.5 to 4 kW at a lower operating pressure more in keeping with those normally achieved underground. These machines stand every chance of being as reliable as the current range of pneumatic drills. Their cost and therefore spares cost may be higher than for those machines currently being used. It is unlikely however that they will be more than half the current price of water drills. These developments may therefore see pneumatic drills competing on a more equal basis with hydraulics. Reliability will be at least comparable with current pneumatic equipment. It may however be able to improve on that!

Drill manufacturers will continue their efforts to improve overall efficiencies and environmental acceptance of their products. As already seen major improvements have been made in the areas of noise reduction, oilmist pollution and overall efficiency. With further development, improved penetration rates and higher rotation torque for crushed ground can be achieved. The materials necessary for air drills can normally be of lower strength and less corrosion resistant than those required for water drills. The mass and cost of such drills can therefore be much lower. Water machines will always require the use of stainless steels and will always cost more to manufacture than air machines. Because the capital cost of new pneumatic machines is less, the maintenance costs will also be lower. There is no reason for the life between breakdowns to be any less than their hydraulic or water competitors. In fact, in the short to medium term, there is every indication that life between breakdowns will be higher than water hydraulic drills. Thus all relevant costs associated with air drills, except energy costs, will be lower than hydraulic drills. The capital outlay and enormous replanning of resources and systems necessary to install Hydro-Power in a mine or section where air is currently used make it impossible to contemplate. New mines, or areas of existing mines which require drills and do not have the pneumatic infrastructure, are ideal candidates for water drills to be considered.

In a large number of mines the current recession will mitigate against large purchases of capital equipment. Thus those mines with existing compressor installations will be forced to retain them for at least the short to medium term. Pneumatic drills will therefore remain the dominant choice in those areas. In conclusion, there is plenty of life left in the pneumatic drill. The major manufacturers will continue to design and develop new machines. This will result in further improvements which will benefit the market at large. The customer will of course ultimately decide which way the market moves and this is as it should be.

7 Acknowledgement

This paper was previously published under the title; The future of pneumatics in percussive drilling, at a seminar **Drilling and Blasting in Narrow Reefs and their Effect on the Profitability of Gold Mines** by The South African Institute of Mining and Metallurgy on 26 and 27 October 1993.

Aplicación de un modelo de fragmentación de rocas en la reducción de los costos de voladura en minería

Osvaldo Aduvire P.
Escuela Técnica Superior de Ingenieros de Minas de Madrid, Spain
Hugo Aduvire P.
Universidad Nacional Jorge Basadre Grohmann de Tacna, Peru
Carlos López Jimeno
Escuela Técnica Superior de Ingenieros de Minas de Madrid, Spain

RESUMEN:

Las constantes innovaciones tecnológicas experimentadas en el campo del arranque de rocas con explosivos han mejorado la calidad de la fragmentación de éstas, lo que unido a las mejoras de los equipos de carga, transporte y manipulación de materiales, hacen posible que en la actualidad se puedan maximizar los rendimientos de la maquinaria minera y reducir los costos de producción. Una de las metodologías que permite optimizar los parámetros de voladura en minería a cielo abierto es la predicción de la fragmentación del material volado mediante la función de distribución del tamaño medio de los bloques.

Con este trabajo se pretendía implementar un modelo de cálculo que permitiera estudiar y comparar la influencia que tiene la granulometría de material arrancado mediante explosivos sobre los rendimientos y los costos unitarios en las operaciones mineras.

El trabajo se realizó con datos tomados en la Mina Tintaya, que se encuentra ubicada en la cordillera occidental de los Andes del Sur de Perú, entre los 4.000 a 4.600 m.s.n.m. La corta tiene aproximadamente 1.000 m de largo por 600 m de ancho y una profundidad de 270 m. En la actualidad la producción promedio oscila entre 35.000 a 40.000 t/día con una ley mínima de corte de 0,85 % Cu.

El objetivo principal del trabajo, consiste en elegir y aplicar un modelo semi-empírico de predicción granulométrica de fragmentación de rocas por voladuras, al diseño de las mallas de perforación. Para ello, se analizaron varios métodos que permiten evaluar la fragmentación y algunos modelos de distribución granulométrica del material volado, teniendo en cuenta la energía de los explosivos, la longitud del barreno y las características del macizo rocoso.

El estudio permite ver la influencia que tiene el diseño de la voladura sobre la granulometría del material volado y consecuentemente en la reducción de los costos. El grado de la fragmentación de rocas con explosivos a su vez influye directamente sobre los rendimientos de las operaciones de carga, transporte y trituración.

AGRADECIMIENTO: Los autores desean hacer llegar su agradecimiento a los ingenieros y técnicos de la Empresa Minera Tintaya por sus aportes oportunos y apoyo desinteresado.

1. INTRODUCCION

El estudio se realizó con datos tomados en la empresa MINERA TINTAYA, que se encuentra ubicada en la cordillera occidental de los Andes del Sur de Perú, entre los 4.000 a 4.600 m.s.n.m.

El yacimiento que se explota está mineralógicamente controlado por un *skarn* en rocas calizas, observándose una disminución del contenido de cobre cuando la existencia del skarn es pobre. Este depósito está constituido básicamente por venillas de calcopirita y cuarzo, diseminaciones de calcopirita y abundante magnetita y molibdeno. Cuenta con más de 52 Mt de reservas probadas entre sulfuros y óxidos, con una ley media de 2,3 % de Cu.

La corta tiene aproximadamente 1.000 m de largo por 600 m de ancho y una profundidad de 270 m. En la actualidad la producción promedio oscila entre 35.000 a 40.000 t/día con una ley mínima de corte de 0,85 % Cu.

El método de explotación empleado en la mina es el cíclico convencional y lo constituyen las operaciones de perforación, voladura, carga y transporte. Se utilizan como equipos de producción perforadoras eléctricas BE 45R, palas P&H 1900 (4,5 y 9 m^3), volquetes WABCO de 85 t, perforadoras Track Drill, y además como unidades auxiliares cargadoras frontales, tractores, motoniveladoras, camión cargador de ANFO, etc.

2. FORMULACION TEORICA DEL MODELO KUZ-RAM.

Aplicando éste modelo se puede ver la estrecha correlación entre las ecuaciones empíricas y los resultados de fragmentación de rocas por voladuras en el terreno, teniendo en cuenta la energía de los explosivos, la longitud del barreno y las características del macizo rocoso.

2.1. Función de distribución estándar aplicable al modelo

La representación de los tamaños de partículas de una voladura se ha resuelto por aceptación de relaciones matemáticas, desarrolladas a partir de estudios de fragmentación, la función de distribución que proporciona el mejor resultado es la propuesta por **Rosin-Rammler (R-R)**.

$$F(x) = 1 - e^{-(\frac{X}{Xo})^n}$$

Donde:

a) F(x) = Porcentaje en peso, cuyos fragmentos tienen dimensiones inferiores a X(50%)
b) Xo = Tamaño característico, y corresponde a la dimensión X
c) X = Tamaño de referencia o abertura de malla.

Con la fórmula de Kuznetsov obtenemos el valor medio de "X". Para un valor de F(x) = 0,5 la ecuación queda reducida a:

$$0{,}5 = [\,e\,]^{-(\frac{X}{Xo})^n}$$

De donde resulta:

$$Xo = \frac{X}{(0{,}693)^{\frac{1}{n}}}$$

d) n = Indice de uniformidad. Está relacionado con los parámetros geométricos de la voladura. Este valor "n" determina la forma de la curva granulométrica y varía comúnmente entre 0,8 y 2,2. Un valor alto indica una fragmentación uniforme, mientras que valores pequeños reflejan cantidades importantes tanto de finos como de tamaños gruesos.

El algoritmo para calcular "n" según Cunningham (1987) es el siguiente:

$$n = [\,2{,}2 - 14\,\frac{B}{D}\,] \cdot [\,1 - \frac{W}{B}\,] \cdot [\,1 + \frac{R-1}{2}\,]^{0{,}5} \cdot [\,\frac{L_c}{H}\,]$$

Donde:

B = Piedra (m)
D = Diámetro del barreno (mm)
W = Desviación de la perforación (m)
R = Relación espaciamiento/piedra (S/B)
Lc = Longitud de carga (m)
H = Altura de banco (m)

Si se emplea un diseño de perforación triangular o al tresbolillo, n debe ser incrementado en un 10 %, de igual forma si el encendido es secuencial.

2.2. Modelo de predicción granulométrica

Sobre la fragmentación influyen un gran número de variables, y entre éstas las propias características del macizo rocoso. La optimización de cualquier operación de arranque con explosivos pasa por la medición de los tamaños de la roca volada en determinadas condiciones.

No existe un método o fórmula de predicción exacta, pero en los últimos años se llegó a aplicar desde modelos simples hasta otros más complejos. Se ha optado por utilizar el algoritmo de **Kuznetsov** que tiene la siguiente expresión:

$$< x > = A \cdot [\frac{V}{Qo}]^{\frac{4}{5}} \cdot [Qo]^{\frac{1}{6}} \qquad (1)$$

Donde:

a) <x> = Tamaño medio del fragmento cm (aproximadamente alrededor del 58 % de la distribución está bajo el tamaño <x>)

b) A = Constante que depende de las características de la roca.

- Rocas de dureza blanda	A = 5;	f = 5 - 8
- Rocas de dureza media	A = 7;	f = 8 - 10
- Rocas duras altamente fisuradas	A = 10;	f = 10 - 14
- Rocas muy duras débilmente fisuradas	A = 13;	f = 12-16

El valor de f mide la dureza de la roca de acuerdo a la escala de Protodiakonov.

Para determinar el valor del coeficiente experimental "A" se ha empleado la siguiente expresión:

$$A = 14\,(BL) - 9 \cdot e^{[-\frac{(SC - 67ff - 400)}{630}]^{3,125}} \qquad (2)$$

A = 1 x 14 - 9 x exp[-(1195 - 67 x 0,5 - 400)/630]^{3,125} = 12,5

Donde:

BL = Parámetro que especifica el tipo de voladura
- Voladura en bancos : BL = 1
- Voladura en túneles: BL = 2

SC = Resistencia a la compresión simple de la roca.
- En nuestro caso SC = 1195 kg/cm^2

ff = Frecuencia de fracturas por pie de testigo.
- Roca débilmente fisurada: ff = (0 - 1)
- Roca medianamente fisurada: ff = (1 - 2,5)
- Roca fisurada : ff = (2,5 - 4)
- Roca altamente fisurada : ff = (4 - 6)

c) V = Volumen de roca volada/barreno (m^3/barreno) :
Espaciamiento(S) x Piedra (B) x Altura banco (H)

d) Q = Carga explosiva dentro de un barreno (246,46 kg)

e) Qo = Cantidad de TNT equivalente a la carga de explosivo/barreno (kg)

f) PRP = Potencia relativa en peso del explosivo: (ANFO = 100, TNT = 115)

$$Q \times PRP = Qo \times 115$$

$$Qo = \frac{Q \cdot PRP}{115} \qquad (3)$$

Reemplazando (3) en la ecuación (1), el algoritmo de Kuznetsov se convierte en:

$$< x > = A \left[\frac{V}{Q}\right]^{\frac{4}{5}} \cdot Q^{\frac{1}{6}} \cdot \left[\frac{PRP}{115}\right]^{-\frac{19}{30}}$$

Para los cálculos se ha considerado el PRP del explosivo = 125

3. DATOS UTILIZADOS EN LA APLICACION DEL MODELO

- Los explosivos utilizados en la voladura son los indicados en el cuadro 1.

PROPIEDADES DE EXPLOSIVO	ANFO-AL 10%	SLURREX 40%
Densidad (g/cm^3)	0,86	1,15
Potencia relativa en peso	134	107
Calor de explosión (kcal/kg)	1.303	
Presión de detonación (kilobar)		58 - 78
Velocidad de detonación (m/s)	3.500	5.000

Cuadro 1. Explosivos utilizados en mina Tintaya

- El diámetro de perforación empleado fué de 9 7/8 pulgadas.
- Se ha considerado una altura de banco (H) de 10 m y una longitud de barreno (L) de 13 m.

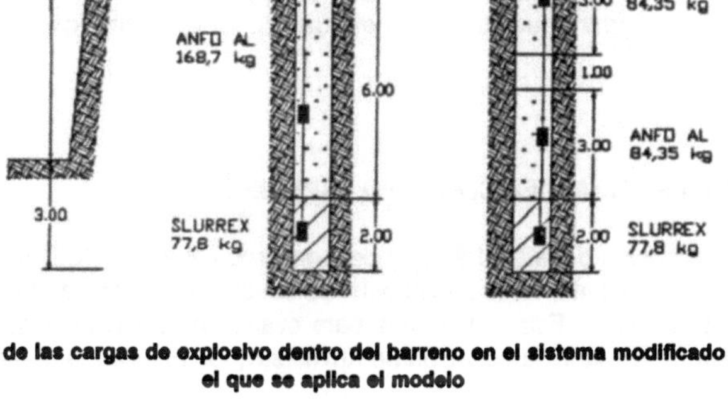

Figura 1. Colocación de las cargas de explosivo dentro del barreno en el sistema modificado planteado con el que se aplica el modelo

- El esquema de voladura puede ser de tipo rectangular o al tresbolillo. Este último da mejores resultados de fragmentación, pero el mallado rectangular se ajusta mejor a las condiciones de operación existentes. La secuencia de encendido empleado es por filas.

- Se ha considerado un solo retacado superior de 5 m para la primera fila de barrenos, mientras que para los barrenos de la segunda fila este se reduce a 4 m y se utiliza un taco intermedio de 1 m, tal como se ve en la figura 1.

- El tamaño crítico de los bloques se ha considerado en función de la abertura de la parrilla que regula el paso del material a la machacadora primaria.

- La relación espaciamiento-piedra se tomó S/B = 1,25 por considerarla la óptima.

- Las desviaciones medias observadas en la mina para longitudes de barrenos de 13 m fue de un 2,3 %, equivalente a 30 cm, incluido el error de emboquille.

- La constante de roca (A), según criterios utilizados en las formulas Kuznetsov, Protodiakonov y por observaciones hechas en el terreno "A" varía entre 12,5 - 13, asumiéndose: A = 13; f = (12-16).

- Para voladuras con cargas generalmente menores de 220 kg de explosivo por retardo no se aprecia daños aparentes en las labores adyacentes, mientras que al disparar cargas mayores a 340 kg simultáneamente se observa fracturación y caídas de rocas. Por consiguiente, la carga disparada por retardo debe estar dentro de este intervalo.

- La longitud de la carga queda definida por la altura del barreno menos los retacados: Longitud de carga = longitud del barreno - retacados y Longitud de carga = 13-5 = 8m

4. ELECCION DEL TAMAÑO DE FRAGMENTACION

Con el objeto de cuantificar y comparar los resultados de las distintas voladuras se eligen diferentes grados de fragmentación representados por el tamaño medio, para posteriormente seleccionar el más conveniente desde el punto de vista técnico económico.

Los tamaños medios elegidos para la determinación de los costos y sus correspondientes parámetros de diseño son:

$$\langle x \rangle = 30;\ 35;\ 40;\ 45;\ 50;\ 55;\ 60;\ 65 \text{ y } 70 \text{ cm}$$

La descripción de las distribuciones de tamaño nos permite observar el rango de variación de los parámetros de voladura que se desean evaluar, como la piedra, los factores de carga y la granulometría.

5. CURVA DE LA PIEDRA Y EL CONSUMO ESPECIFICO

Los valores de la curvas de la piedra (B) y del consumo específico (CE) se obtienen variando el tamaño medio del material volado, manteniendo constante la carga de explosivo por barreno (Q = 246,26 kg/barreno). Esta curva sirve para evaluar el rango de piedras en función del diseño de voladura mediante la ecuación de Kuznetsov, los resultados obtenidos se presentan en el **cuadro 2**.

$$B^2 = Q \cdot \frac{\left[\dfrac{<x>}{A \cdot Q^{\frac{1}{6}} \cdot [\frac{PRP}{115}]^{-\frac{19}{30}}}\right]^{\frac{5}{4}}}{R \cdot H}$$

TAMAÑO BLOQUE X (cm)	PIEDRA B (m)	ESPACIA-MIENTO S (m)	VOLU-MEN (m^3)	TONE-LAJE (t)	CONSUMO ESPECIF. (kg/m^3)	FACTOR POTENCIA (kg/t)	n	Xo (cm)
30	4,36	5,45	238	666	1,036	0,370	1,64	175,1
35	4,80	6,00	288	806	0,856	0,301	1,63	175,3
40	5,22	6,53	341	954	0,725	0,259	1,62	175,6
45	5,62	7,02	395	1.106	0,626	0,223	1,61	175,9
50	6,00	7,50	450	1.260	0,546	0,195	1,59	176,2
55	6,37	7,96	507	1.420	0,486	0,174	1,58	176,5
60	6,73	8,41	566	1.585	0,436	0,156	1,57	176,8
65	7,07	8,84	625	1.750	0,394	0,140	1,56	177,2
70	7,41	9,26	686	1.921	0,359	0,128	1,54	177,5

Cuadro 2. Variación de los parámetros de las voladuras en función del tamaño medio de fragmentación

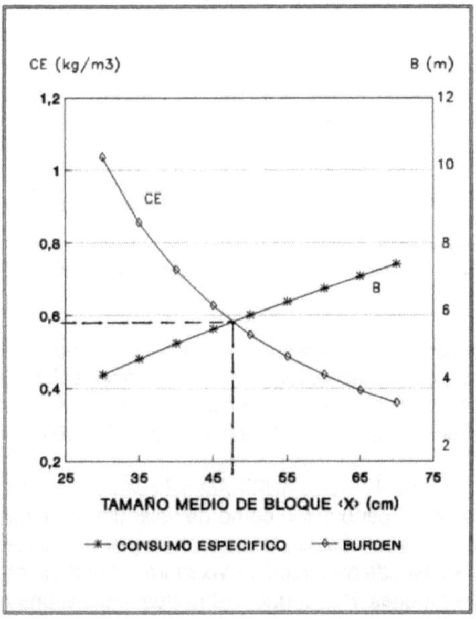

Figura 2. Determinación del Tamaño Medio Optimo a partir de las Curvas del Consumo Específico y Piedra

La curva de distribución se obtiene variando la piedra para un tamaño de referencia determinado, obteniéndose un tamaño máximo característico para cada diseño de voladura. Mediante las ecuaciones de Cunningham y Rosin-Rammler obtenemos los valores de **n** y **Xo** para un tamaño de referencia **X** igual a 140 cm, tal como se ve en el cuadro 2.

En la figura 2 se representan las curvas del Consumo Específico y de la Piedra en función al tamaño medio de los bloques. A partir de estas curvas podemos determinar el Tamaño Medio Optimo <X> de la fragmentación que es igual a 48 cm, a cuyo valor le corresponde una Piedra Optima (B) igual a de 5,85 m y un Espaciamiento de 7,38 m.

Para valores de la Piedra mayores a 5,85 m el Indice de Uniformidad "n" desciende, lo que indica que la granulometría a obtenerse será bastante irregular y mal fragmentada, con gran cantidad de bloques de tamaño grande, que incrementarán los costos de perforación y voladura secundaria, aunque aparentemente los costos de perforación y voladura primaria descienden según se incrementa el tamaño del bloque, más no así el costo de las operaciones de conminución del material volado.

6. COSTOS DE PERFORACION Y VOLADURA

Si consideramos constante el costo unitario de perforación primaria de 8,613 $/m perforado (0,133 $/t) se ve que existe una reducción del 29 % frente a los costos obtenidos con el modelo propuesto.

Con los costos de voladura primaria sucede lo mismo, logrando reducciones del 8 % entre el costo sin aplicar el modelo y el costo obtenido con el modelo de predicción granulometrica.

OPERACION ARRANQUE	B (m)	S (m)	VOLUMEN m^3	TONELAJE (t)	CARGA EXPLOSIVA (kg)	CONSUMO ESPEC. kg/m^3	C.PERF. PRIMA.	C.VOL. PRIM.	COSTO P1+V1
							$/t	$/t	$/t
S.M.	5,00	6,00	300	840	246,46	0,822	0,133	0,308	0,441
C.M.	5,85	7,31	428	1.197	246,46	0,576	0,094	0,283	0,377

Cuadro 3. Comparación de valores aplicando y sin aplicar el modelo de predicción granulometrica, considerando el costo de explosivo en 2,458 $/kg. (S.M. = Sin modelo, C.M. = Con modelo).

De igual forma, si comparamos los costos conjuntos de perforación y voladura primaria del cuadro 3, esta diferencia se mantiene, logrando reducciones del 15 % entre los costos sin el modelo y los obtenidos aplicando el modelo KUZ-RAM. También hay que tener en cuenta que cuanto más uniforme sea la fragmentación menores serán los costos de voladura secundaria.

Con los valores obtenidos con el modelo KUZ-RAM del cuadro 2 podemos determinar las curvas de costos primarios, tanto de perforación como de voladura, en función al tamaño medio del material volado, figura 3. Estos costos descienden según se incrementa el tamaño de los bloques, mientras que el costo de perforación y voladura secundaria crece según se incrementa el tamaño medio de los bloques. Por lo que, habrá que llegar a una solución de compromiso sobre el tamaño medio del material fragmentado de tal forma que se trabaje con un tamaño adecuado intermedio como el determinado con el modelo.

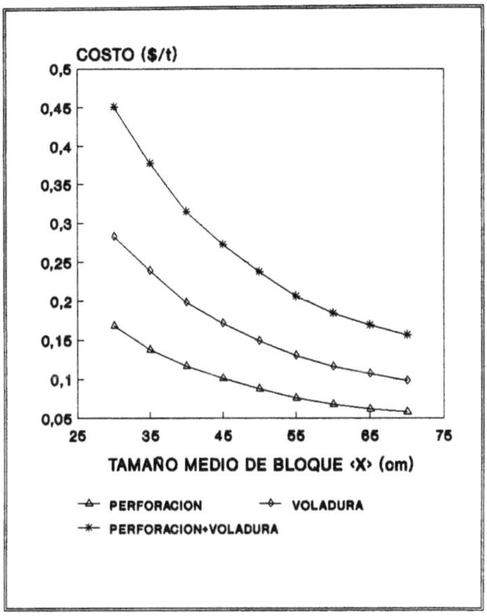

Figura 3. Variación de los costos de perforación y voladura en función al tamaño medio de fragmentación

7. CONCLUSIONES

El algoritmo descrito en el trabajo presenta una gran sensibilidad al costo base de perforación y voladura, reduciéndose ambos costos en un gran porcentaje al diseñarse adecuadamente los parámetros geométricos, y empleando los explosivos y accesorios de voladura adecuados.

Para el caso estudiado la granulometría óptima para reducir los costos de perforación y voladura, es de 48 cm y la piedra puede variar entre 5,5 a 6 m, para valores mayores el índice de uniformidad desciende, lo que no es recomendable porque se obtendría una granulometría muy variable.

Un diseño adecuado de la malla de perforación y una buena voladura permite controlar la estabilidad del talud final de la mina, reduciendo la velocidad pico de partícula referida al grado de vibraciones que producen los disparos en el macizo rocoso.

BIBLIOGRAFIA

- ADUVIRE, H.: (1992). Determinación de un Modelo de Predicción Granulométrica en Voladura de Rocas en Mina. Tesis UNTAC - Facultad de Ingeniería de Minas. Tacna (Perú).
- BAUER, A. et al. (1983). Principles and Aplications of Displacing Over in Strip Mines by Explosives Casting. SME-AIME, October.
- BERTA, G. (1985). L'Explosivo Strumento Di Lavoro. Italexplosivi.
- CUNNINGHAM, C. (1983). The Model for Production of Fragmentation from Blasting. Lulea.
- FERNANDEZ, L. et al. (1993). Predicción de la Fragmentación de Rocas. Rev.INGEOPRES Nº8
- LOPEZ JIMENO, C. et al. (1987). Manual de Perforación y Voladura. Ed. ITGE. España.

Quality excavation with instrumented and automated drilling equipment

Pertti Koivunen
Tamrock Oy, Tampere, Finland

Abstract

The quality concept is now much more widely understood than it used to be a decade ago. Then it was limited more to the production and technical finishing of a certain machine or piece of equipment. Today quality management comprises all the chain from recognizing the user's needs, designing, manufacturing and delivering the equipment up to task planning and end use. Excavation which is done with an awareness of quality and safety is also economic and effective.

Drill holes which are correctly positioned, have the correct alignment and proper length are the basis of good-quality blasting operations. Here the benefits of instrumentation and automation have been proved during recent years. Profitability analyses of controlled contour tunnel excavation and surface blasting as well as underground stoping have proved that the profits gained by precise drilling are very significant. The instrumentation of drilling equipment is accordingly giving a good return on investment. In practice it has been shown that in a road tunnel drilled by a computerized drilling jumbo, the saving in concrete lining was 0.26 million USD for each km of tunnel. In addition, the pull-out was increased by 10 %. In an open pit mine, with 6 Mt/y output, an increase of drilling accuracy was found to improve the economy by 0.35 million USD annually, due alone to reduced drilling and blasting costs.

In rock excavation using the drill and blast method the quality concept means optimized blasting, minimum damage of the surrounding rock, maintaining the required contour tolerances and correct positioning of the tunnels, ramps and other underground caverns. From the ore excavation point of view it means low dilution and minimum ore losses. In bench blasting the marks of quality are, for example, correct fragmentation and control of level, casting and toes.

The fast development of information technology and availability of electronic components for difficult conditions have made the instrumentation of drilling equipment a standard tool for today's miners and contractors. Drilling instrumentation comprises a full series of equipment starting from gravity based angle indicators and ending at modern computer controlled units. In addition to angle measuring, modern instruments perform many other tasks, such as giving the hole direction, measuring the hole depth, recording penetration rates, collecting operation data and drilling parameters as well as collecting production rates. The computer-controlled drilling jumbos perform pre-programmed drilling patterns. Today's automated long hole drilling units drill holes up to programmed depths, taking care of both coupling and uncoupling of the rod string and control of the drilling itself.

Quality awareness is increasing the use of instrumentation and automation. The level of instrumentation and automation is chosen in accordance with the operator's needs in order to optimize the return on investment in each particular case. Future enlargement possibilities will include rock identification based on the drilling parameters. Drilling of total fans independently will be the next step in long hole instrumentation and automation. Advanced positioning and navigation systems are in active use in both underground and surface applications. The drilling equipment will be increasingly integrated into the mine planning and control systems with the overall goal of a more optimized excavation and total mining process.

1 Introduction

Hard rock drilling has gone through a period of intensive mechanization during the last twenty years. Both the penetration rates and shift capacities as well as long term rates have increased. Each generation of hydraulic rock drills and drilling systems have brought a new step upwards with regard to production capacities. Today, net drilling rates are not a bottleneck in rock excavation operations.

Besides capacity considerations, other important aspects have gained in importance; such as work safety and quality of work. They have a parallel effect on the excavation result and economy. Excavation which is done with quality and safety awareness is generally also economic and effective.

The quality concept is now understood much more widely than it used to be a decade ago. Then it was limited to the production technical finishing of a certain machine or equipment. Today quality management comprises all the chain from recognizing the user's needs, designing, manufacturing and delivery of the equipment up to task planning, use of the equipment and end product.

2 Quality concept in rock excavation

In rock excavation using the drill and blast method, the quality concept means optimized blasting, minimum damage of the surrounding rock, maintaining the required and ordered contour tolerances and correct location of the tunnels, ramps and other underground caverns. From the ore production point of view it means low dilution and minimum ore losses as well as fluent total process from drilling to milling. In bench blasting the quality identification marks are, for example, correct fragmentation and control of casting, level and toes. Drill holes which are correctly positioned, have correct alignments and proper lengths are the basis of good-quality blasting operations. Here the benefits of instrumentation and automation have been proved in recent years.

3 Technology background

Hydraulics and electronics today offer practically unlimited possibilities to control the drilling process. Long-hole drilling rigs, tunnelling jumbos, mechanized rock bolters and raise borers include high technology components. Some years ago one could still doubt the reliability of electronics in underground or open pit mining conditions. Now there is a population of hundreds items of electronic equipment in use in mining and civil engineering and the most advanced computer controlled drilling units have gained a stable foothold in the market. Practice has proved that the performance and robustness of electronic hardware and sensors is now at a completely different level than a decade ago, and this really has been of major importance in the mining environment. Component development together with the rapid advance of information technologies have made the instrumentation of drilling equipment a standard tool for today's miners and contractors.

4 Underground production drilling

A blast hole can be exactly determined by giving the collaring point, the length, the side angle and the tilt angle. The practical conditions, where the operator has to align the holes are very demanding in terms of the instruments' reliability, handiness and ease of use. Especially difficult, but common, are wide drilling drifts, crossings, inclined long-hole raises and parallel hole lines which are in curved drifts. Sometimes there may not be as deep as possible awareness on the part of the drill operator of the real meaning of drilling accuracy. All this means that the actual holes deviate compared with the plans. The total deviation consists of the set-up error, alignment error, the error caused by bending in rock and the depth error.

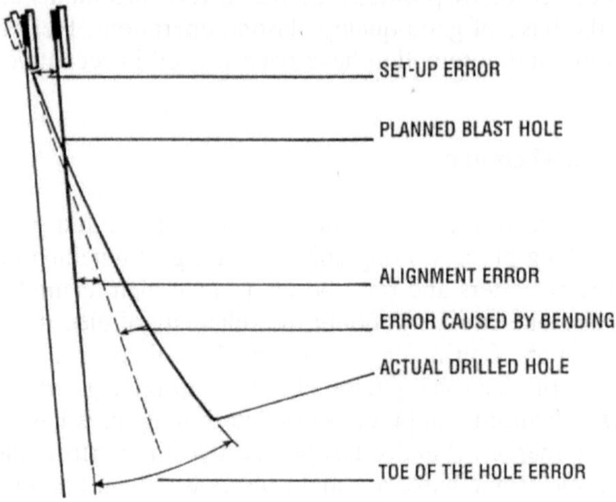

Picture 1. The elements of drilling error.

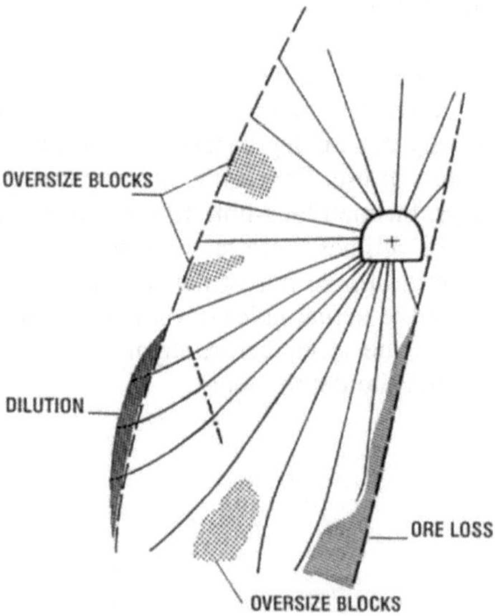

Picture 2. The meaning of drilling accuracy in stope economy.

Improved accuracy may have tremendous economic meaning in terms of fragmentation, reduced dilution, less ore losses and less need to redrill, as basically can be seen from picture 2. Picture 13 shows how big an affect drilling accuracy can have on both specific drilling and specific charge. The diagram is based on a research work in open pit mining and will be discussed in more detail later in the paper. The result is, however, worth noting here also, when underground mining is concerned.

4.1 Instrumentation of long-hole rigs

The longitudinal position and hole line or ring or fan direction is normally found by means of wall marks. These can be supported by painting the centre line on the roof. Today's miners have effective pieces of equipment to maintain drilling accuracy in all circumstances. The latest laser type positioning devices fitted on long-hole rigs have very sharp and extending beams. The accuracy of the position then largely depends on the wall marks of the drilling place, which should be no problem to the mine surveyors. Of course the human factor, i.e. the operator's skill and especially motivation, is of vital importance, but it can be ensured simply by having the quality aspect as a part of the wages base.

The modern aligning devices normally give two angles in two planes at right angles. The indication is based on precision sensor techniques. Depth measuring and indication is a useful option which can easily be incorporated as a production follow-up function. The angle accuracy is 0.3 degrees and the depth accuracy is in the centimetre range. This means a big improvement compared with earlier practice based on the operator's counting of the drilled rods or tubes. Tamrock TMS, picture 3, is a good example of a modern, handy angle indicator. It can be specified with different options depending on the needs.

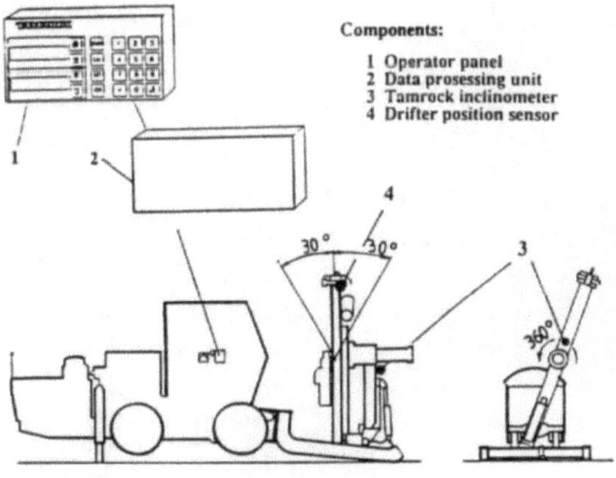

Picture 3. The components of a modern angle indicator, Tamrock TMS, for long-hole rigs.

From present-day surveying techniques could be found even more advanced instruments for the alignment of blast holes. For example, computer-supported total survey stations with continuous measurement and automated following of the target. Certainly they give a very high accuracy but, on the other hand, they put much more demand on care and use. So at the moment their application in relation to stope drilling is not practical.

4.2 Automation

The main goal of automation in long-hole production drilling is to increase effective drilling time and thus achieve more drilled meters per shift. Automation also results in accurate control of the hole length. Once the operator has collared the hole, the rig is left to drill the hole to the preprogrammed depth by itself. When the correct depth is reached, the rig automatically shuts off drilling, retracts the drill string, uncouples the rods and places them in the rod cassette. The increase in shift drill meters through automation can exceed 25 %.

Picture 4. Tamrock Solo 1000 Sixty, computerized long-hole drilling rig.

5 Drifting and tunnelling

The quality of tunnel and drift excavation is usually measured as the surface quality of the excavated underground space. High quality and precision tunnelling features can be listed as follows:

Excavated cross section accurately follows the theoretical profile
Tunnel, drift or ramp is correctly located
Fracture zone above the theoretical profile is not excessive
Pull-out of rounds is maximized even with longest rounds
Specific drilling and specific charge are minimized

The rock conditions as well as type of explosives and ignition systems together with the design of drilling patterns have their effect on the blasting results and excavation quality. Besides them there is one common factor: the precision of the drilled holes in practice. Only correctly positioned and aligned drill holes, with correct length, finally make the quality possible.

Different position and direction marks, put by the surveyor, combined with marking of the required profile with paint, together with careful working practice, can result in reasonable quality. But it still leaves too much scope for human errors or routines, like excessive look-out angles of contour and bottom holes or excessive number of holes. Modern instruments with manually operated drilling jumbos or computer-controlled drilling jumbos are nowadays available, when controlled and continuous quality is wanted.

5.1 Instrumentation of tunnelling jumbos

Simple gravitational angle indicators have been in use for about ten years. With precision sensoring and some more microprocessing on the unit, the measurement is more accurate, can be referenced to the actual tunnel direction, and also other important parameters can be measured.

The Tamrock TMS is an example of an advanced angle indication system for tunnelling. It has optional features, such as hole depth and penetration rate measurement and display. All this can be done with a relatively simple operator panel and data processing unit.

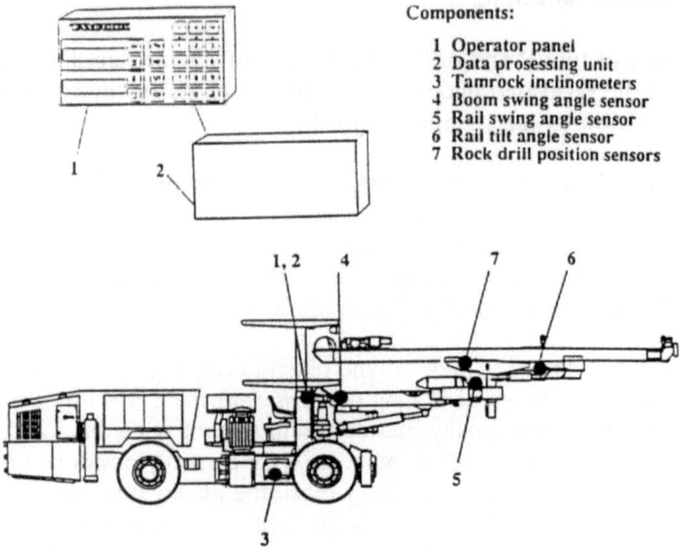

Picture 5. The Tamrock TMS jumbo measuring system.

Picture 6 presents the quality of tunnel profile produced by inexperienced operators using the TMS angle indicator.

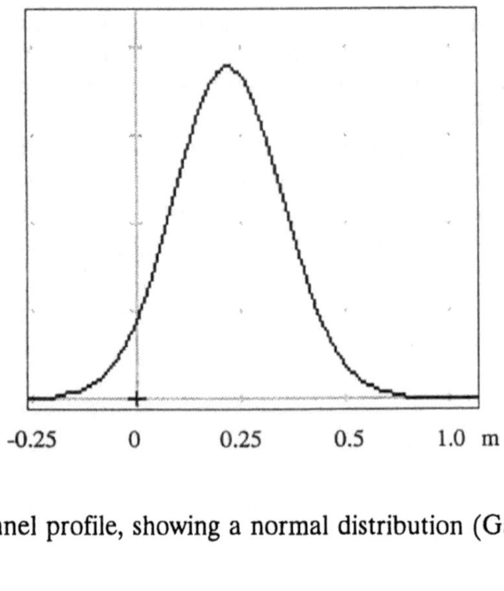

Picture 6. Tunnel profile, showing a normal distribution (Gauss) curve.

In mining the evenness of ramps and transport levels can greatly affect the maintenance costs of LHD and transport equipment. The maintenance of an even bottom itself is also easier and can happen at lower costs. The bottom of the tunnel as shown in picture 7 is obtained using a jumbo equipped with the TMS angle measuring system, when the tunnel direction and position is taken from a laser beam.

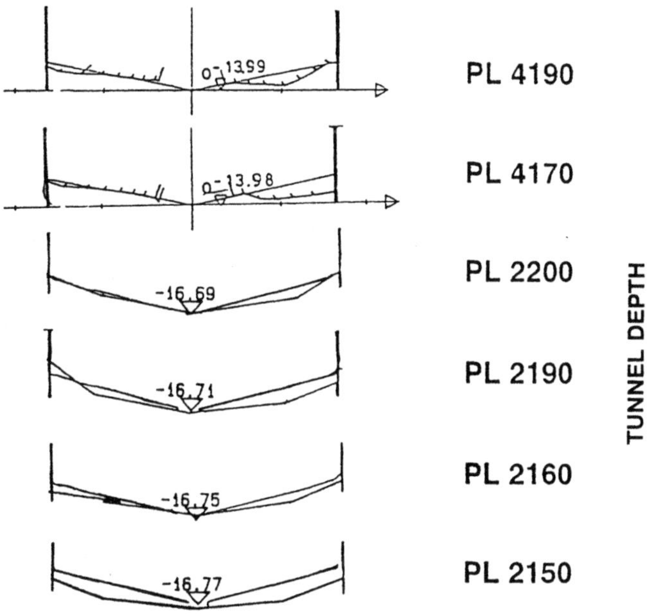

Picture 7. Tunnel bottom controlled by the TMS angle measuring system.

In this paper it is only possible to indicate roughly what the correct or incorrect location of a long-hole drilling drift means in terms of dilution and ore losses, picture 8.

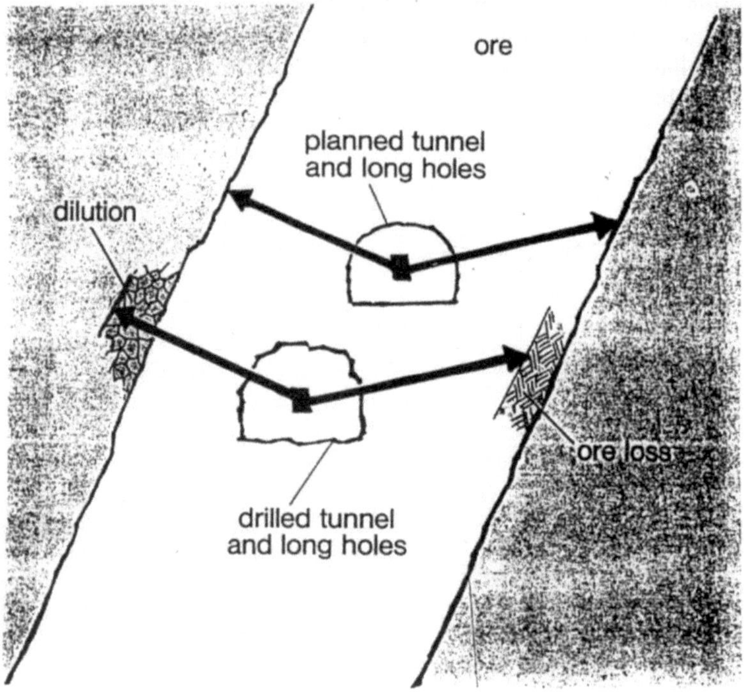

Picture 8. An indication of the importance of the location of a long-hole drilling tunnel.

5.2 Automation of tunnelling jumbos

The ultimate step of instrumentation is the computerizing of tunnelling jumbos. Although unmanned operation is not feasible today, a major part of each tunnel round drilling can nowadays be performed automatically according to a preprogrammed pattern. The operator supervises the process and interferes using manual override when it is necessary to settle a difficult drilling situation. So we can say that computer control is a tool to make good quality. The first computerized jumbos were introduced in Europe and Japan as early as 1985. Today there is a second generation of these machines on the market. There are twelve computerized Tamrock jumbos in operation. So far all experience has been gained at construction worksites. Profitability analyses and user reports have stated that the reduction of overbreak, and so the reduction of lining concrete volume, as well as increase of pull-out factor are receiving remarkable rates. According to Castano (Torino 1992) savings in concrete costs alone are USD 260000 per 1 km of 87.5 m^2 road tunnel. An important additional factor was the shorter project time, because the pull-out with computerized drilling was found to be 97 % compared to 87 % with manual drilling. In practice it has been possible to reduce the number of holes in patterns by 10-20 %.

6 Surface drilling

Collaring, aligning and hole deviation errors influence the actual drilling pattern by increase or decrease of the spacing and burden from the planned pattern. Depth error also causes toe problems.

Too short a burden and spacing leads to local overcharging, resulting in excessive throw, excessive ground vibrations and increased amount of fines. Too long a distance between holes causes oversize blocks, back break and, again, vibrations.

Also on the surface quality drilling is a prerequisite of successful blasting. When the holes are drilled exactly in the planned locations inside the rock mass, the blasting process can be optimized. The goals of quality benching can be listed:

Muckpile fragmentation (no oversize boulders or too much fines)
Muckpile shape and looseness resulting in good loading conditions
Minimum environmental disturbance (noise, ground vibrations, airblast)
Quarry face quality
Final wall quality (smooth quarry walls, no loose boulders)
Maximum safety (minimum flyrock, stable quarry face and walls).

Quality drilling is characterized by the following:

Collaring accuracy
Aligning accuracy
Hole straightness
Hole length accuracy

The aligning error has two components, direction of inclination to the blasting direction and across the blasting direction.

The hole length errors can occur for two reasons:

The hole length is measured erroneously during drilling
The needed hole length from the surface to the required bench bottom level is measured incorrectly

In the most unfavourable case both error elements are additive.

The collaring points on the surface are measured and marked manually. In cases where the blasting company is paying special attention to the collaring precision, the locations are measured by total stations and marked on the bench. When the positioning is instrumented with on-board systems, different satellite, inertia or total station systems are needed.

Hole straightness is a question of drilling technique. Good results are achieved when the correct combinations of drill rods and bits are used. Drilling tube and guide tube systems also improve hole straightness.

6.1 Instrumentation of surface drilling rigs

Advanced measuring techniques today give the capability of controlling hole alignment and hole depth to within required tolerances.

In addition to alignment, modern instruments allow the operator to define the aligning with reference to a certain direction. This is the direction to which the blasting takes place, picture 9.

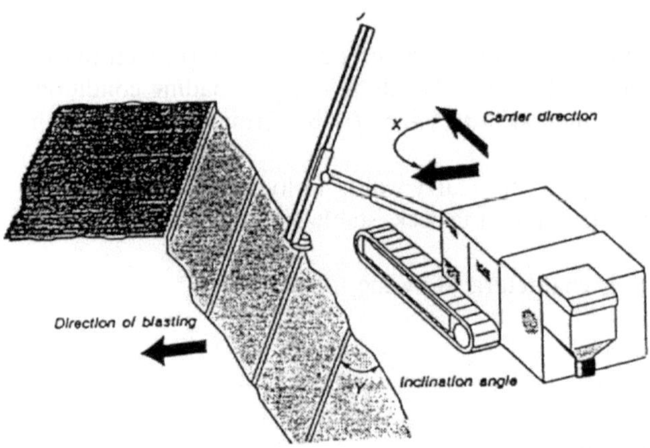

Picture 9. Inclination angle and direction of blasting.

Hole depth can be measured and displayed during drilling.

In picture 10 the Tamrock TIM 2000 measuring system is presented. By using TIM the operator can accurately align the feed to the correct inclination and direction of blasting. An optional depth indicator shows the drilling depth, total drilled meters and penetration rate.

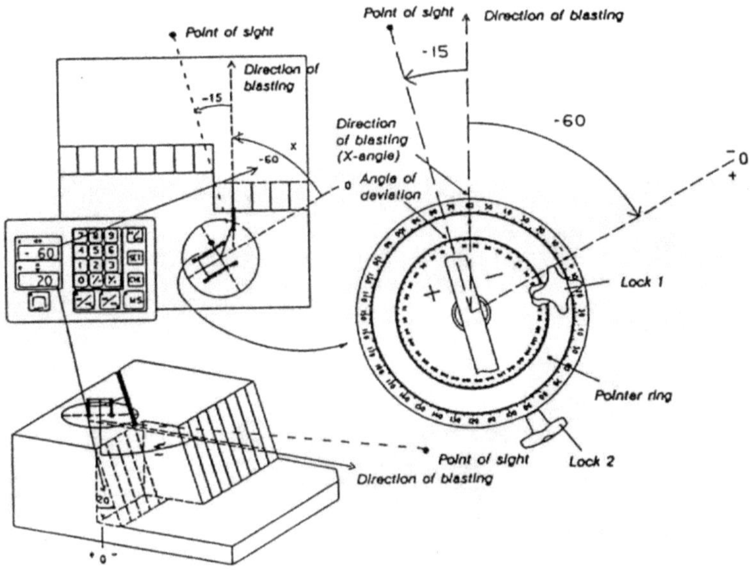

Picture 10. Alignment of feed with Tamrock TIM 2000 system.

The affect of uneven drilling surface can be avoided by the use of level planer, pictures 11 and 12.

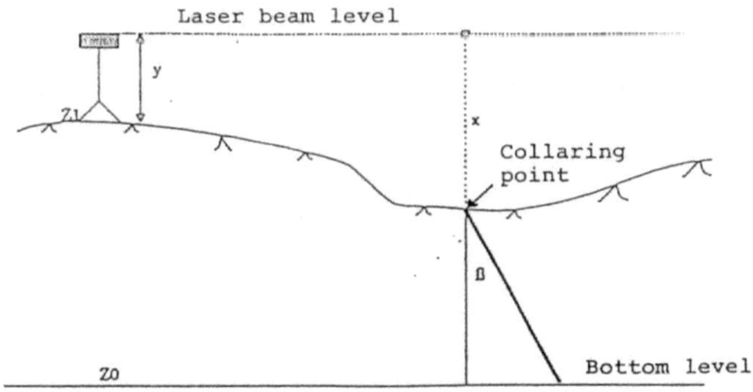

Picture 11. Principle of collaring height detection with a level planer.

Picture 12. Practical application of a laser type level planer.

The big economic advantage of aligning accuracy can be illustrated by the diagram in picture 13 (Kainulainen 1991). When the original 3.0 m burden can, thanks to improved accuracy, be increased by 20 cm, the specific drilling and charging reduces 15 %. In another example the existing burden is 2.5 m. Increasing the burden 30 cm means 25 % less drilling and charging. All this can be achieved with equipment already existing today. An open pit mine with 6 Mt/y production, using 115 mm holes, has found that an increase of 30 cm in actual burden will reduce the cost by 0.35 million USD annually (2 million FIM).

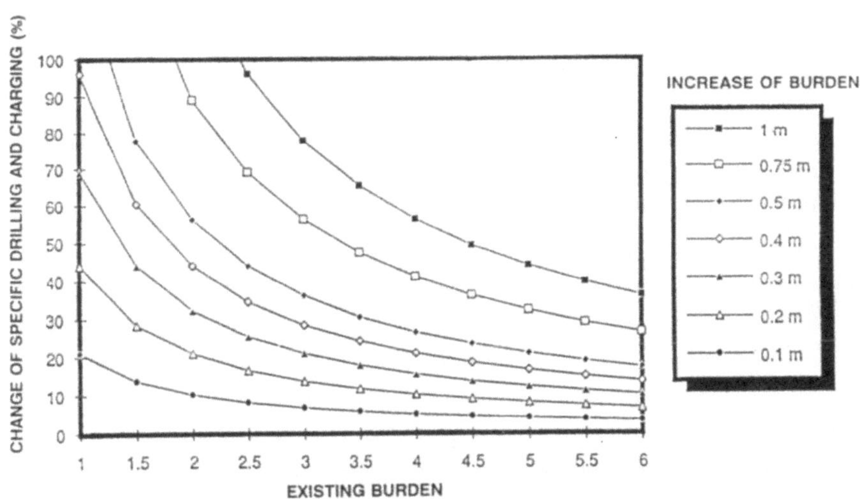

Picture 13. The reduction of specific drilling and specific charge related to existing burden and increase in burden due to improved accuracy. E/B ratio is considered constant at 1.25.

7 Conclusions

The benefits of instrumentation and automation in mining and civil engineering have been proved during the last decade. Today the key question is to utilize the existing technology to its full potential. Quality awareness is increasing the use of instrumentation and automation. The level of instrumentation and automation is chosen as per needs in order to optimize the return on invest in each particular case. Instrumentation offers interesting new possibilities, such as collecting data about drilling parameters and identifying rock properties on the basis of them. Drilling of complete fans independently will be the next step of long-hole automation. Advanced navigation systems for accurate location of equipment both underground and on the surface have been developed. Drilling equipment will be increasingly integrated into mine planning and control systems with the overall goal of a more optimized excavation and total mining process.

8 References

Kainulainen, V. (1991) Drilling Accuracy of Finnish Open Pit Mines, Thesis for Masters of Science in Mining Technology, Helsinki University of Technology (in the Finnish language)

Castano, D. (1992) Analisi della Precisione di Abbattimento in Galleria, Thesis for Masters of Science, Politecnico di Torino, Facolta di Ingegneria (in the Italian language)

Lower costs and higher productivity by use of mechanization in Chilean underground mining

Andrzej Zablocki
Atlas Copco Chilena S.A.C., Santiago, Chile

Abstract

Chile is well known as one of the leading mining countries in the world, particularly in terms of copper production, and has also become well known for low production costs. Today the mining companies have proved that modern mechanized equipment plays a major role in improving productivity and in maintaining the leading position of Chilean mines.
A strong movement to mechanization started in the mid 1980's. This paper describes the mining methods at selected underground mines as examples of the application of mechanized equipment. Productivity, production costs, equipment unit costs, and capacity comparisons are given. The importance of proper maintenance is stressed, and examples of the influence on unit costs and availability are mentioned. The partnership between the mine and the equipment supplier is not an uncommon practice in this respect.
Keywords: Mechanized underground equipment, stoping methods, costs and productivity.

1. INTRODUCTION

The mining industry in Chile has a long history going back to early Spanish colonialism. Today, this industry plays an even more important role in the Chilean economy. The country ranks high in the world's production of many minerals. (1,3,5).
The bulk of copper production comes from the huge Codelco mines and the La Escondida mine. Gold production is also growing fast, increasing from only 3.500 kg in 1979 and reaching 39.500 kg in 1992(3).
More than half of the total ore production comes from open pit mining, and the rest from underground operations, of which the major contribution is from Codelco's block caving mines. Codelco is a state owned company.

An interesting and relatively unknown aspect is the production of copper and gold from the smaller underground mines of the private mining sector.
The most common mining method in this sector is sub-level open stoping, (Figure 1).

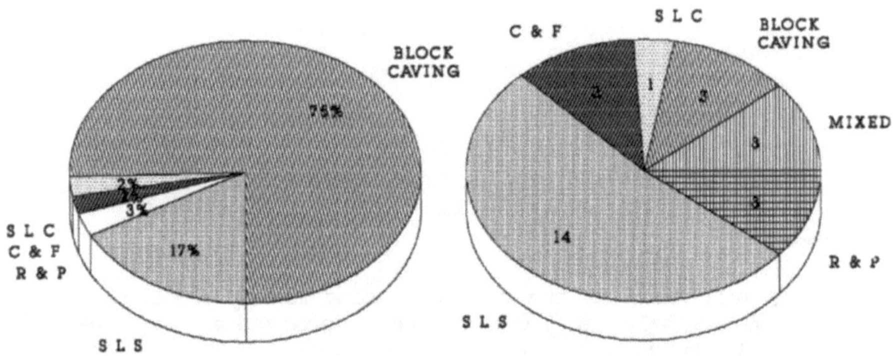

Figure 1. Underground stoping methods in Chile, showing the distribution of annual output (left), and the number of mines (right).

Geological potential, management expertise, a skilled work force, existing infrastructure, easy access to the orebodies, and favourable climate are helping even small mines to survive during the period of low metal prices. In addition, for several years now, many mines have invested in new technology and highly mechanized mining machinery to further improve their productivity and reduce costs.
Degree of mechanization is a very relative subject. Following up on the equipment performance and cost distribution is one of the ways some mines are trying to fit it into a certain level in order to be able to compare with the results of others.
This paper describes four examples of relatively smaller mines employing mechanized stoping methods on orebodies that are not conducive to bulk block-caving methods. A brief summary of these deposits is shown in Table I.

TABLE I. Main data for selected underground mines in Chile

Mine	Mining method	Grade	Annual production Mt
Punta del Cobre	Advanced sub-level stoping	Cu 1.7 % Au 0.6 g/t Ag 10 g/t	1.0
La Cascada	Post-pillar cut-and-fill	Cu 2 %	0.9
El Soldado	Large hole sub-level stoping	Cu 1.5 %	2.4
Mantos Blancos	Sub-level stoping (post-pillar cut-and-fill)	Cu 1.35 %	1.9

2. PUNTA DEL COBRE

Punta del Cobre mine[1] situated near the city of Copiapó in northern Chile, went all-electric in 1985. Until then, pneumatic equipment had been used, with production running at 325.000 tons per annum. After 1985, three major steps forward in terms of production and economy were achieved.

- After the switch to electric hydraulic equipment, the production more than doubled to 720.000 T per annum.
- up to 1985, consumption of energy was 3,79 kilowatts per ton, and the consumption is now 1,59 kilowatts per ton. The total energy costs are lower despite the doubling of production.
- increased capacity enabled the mine to aim for a production target of 1,1 Mt. per annum in 1992.

The mining method employed consists of advanced sublevel stoping and ramp hauling.

Ore from the volcanic sedimentary deposit is mined from sub-level horizontal galleries measuring 4 x 3,6 m. in cross-section. Undercut holes are drilled from the galleries in either a fan or ring pattern, and in special cases vertical holes are drilled on free faces.

Stoping starts from slots at the end of the sub-levels, which are formed by raises mined by drop raising method.

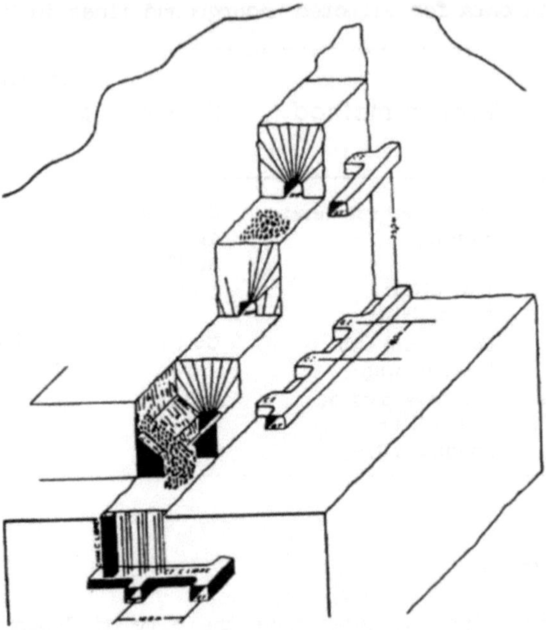

Figure 2. Sketch of the advanced sub-level stoping method used at the Punta del Cobre mine (5).

The stoping method is known as advanced sub-level stoping because each level has its own drawn points and transport drifts for independent, and immediate extraction. Single or multiple drill drifts are driven at a spacing of 18 m, depending on the width of the mineralization.
An example of the relation between total production costs and development costs is shown in Figure 3.

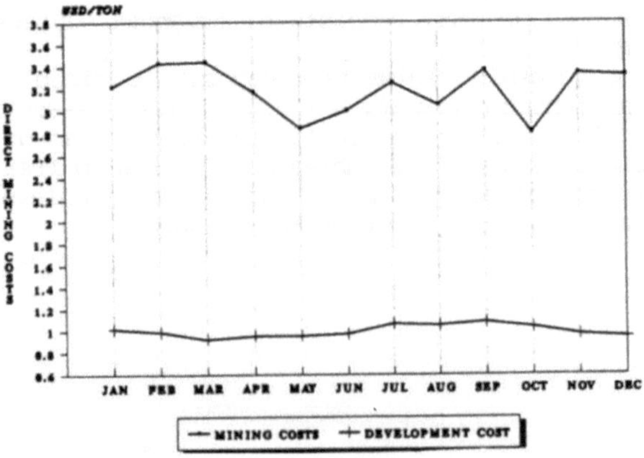

Figure 3. Relation between total underground production and development costs.

For long-hole drilling a Simba H254 drill rig is used. The holes are 18 m. long and 64 mm. in diameter, the burden is 1,8 m. and the hole spacing is 2,5 m. Performance is 7 tons of ore per metre drilled. The rig has an average capacity of 115 m. drilled per shift.
For drifting, both single-and twin-boom Boomer type hydraulic rigs are used. Cost distribution for a typical $18m^2$ drift is shown on Figure 4.

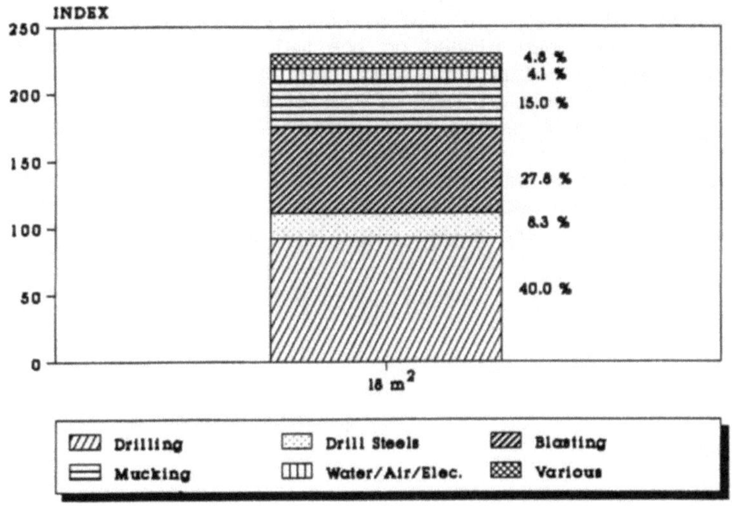

Figure 4. Cost distribution for a drift of $18m^2$

Figure 5 shows the distribution of drilling costs for longhole and drifting rigs. The costs are closely related to rock hardness and abrasivity. An idea of the rock hardness can be gained from the button-bit life in drifting, which is 650 m, and the penetration rate with a COP 1032 rock drill, which is 0,95 m/min.

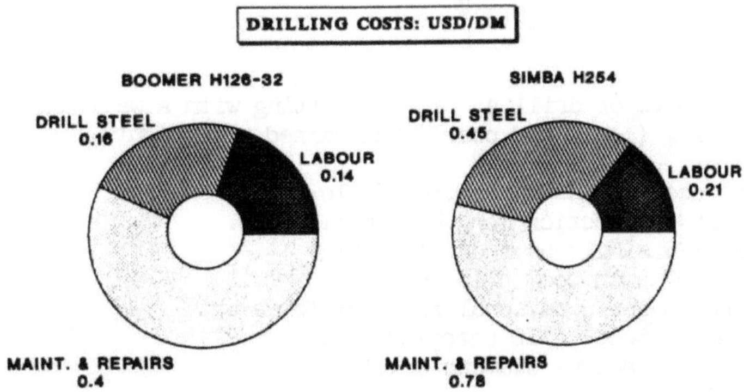

Figure 5. Drilling rigs cost distribution, drifting rig (left) and longhole rig (right).

Standard front-end loaders and articulated dump trucks are currently used for loading and transportation. However, the mine was seriously considering changing to low-profile Wagner equipment for the new mining sections in order to reduce the drift sizes or, alternatively, to increase the loader and truck capacities. Practical tests have shown that a reduction in drift size, from 18 to 15m^2 for example, would reduce the direct excavation costs from 230 to 190 US dollars per metre of tunnel (without considering a possible reduction in rock reinforcement). The final decision depends however also on other factors such as: difference in capital cost, ramp inclination, etc.
Special emphasis has been placed on the maintenance of equipments, and the economic life time of them has been established. For example, a study was made in 1989 of the effect of good maintenance on drilling costs for a period of 4 months time. The results are illustrated in Figure 6. The first case shows the cost of drill steel equipment when a well-maintained feed and a stable boom are used. The costs in the second case are more than double only because a badly maintained feed and a less rigid boom were used.

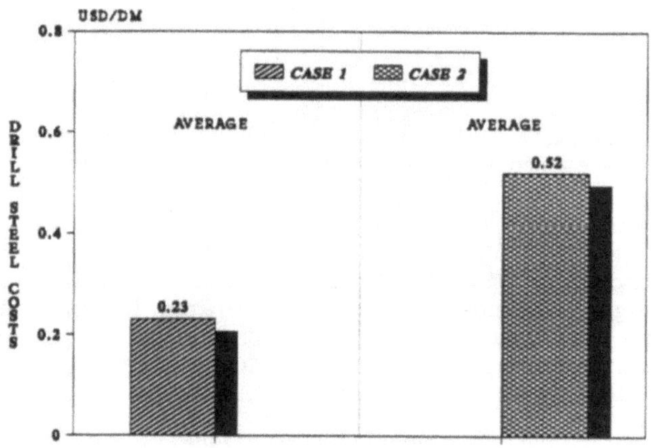

Figure 6. Costs of drill steel for drifting with a well-maintained rig (left) and poorly-maintained rig (right).

Equipment used in the mine is as follows:
2 Simba H254 production long hole drill rigs.
1 Boomer H126 single-boom drifting rig.
1 Boomer H127 twin-boom drifting rig.
Swellex bolts for occasional roof reinforcement
1 PT-61 Nitro Nobel ANFO charging truck.
1 Diamec 262 Craelius diamond drill.
4 Volvo L120 loaders.
4 Volvo A25 trucks.
1 Volvo A30 truck.

The mine has no compressed-air installations. Limited secondary drilling is done with air-driven Puma jack-leg drills supported by XA 125E portable compressors equipped with electric motors. These can be transported to the required sites.

3. LA CASCADA MINE

La Cascada mine is located at the north extremity of the country, close to Iquique. Its Sagasca exotic type of orebody is the only one of its kind in Chile[1]. It is located in an almost horizontal stratum covered by 140 m of overburden. The main feature of the ore is its very low compressive strength value of only 280 kg/cm^2.
The mine was designed to use the post-pillar cut-and-fill method (Figure 7).

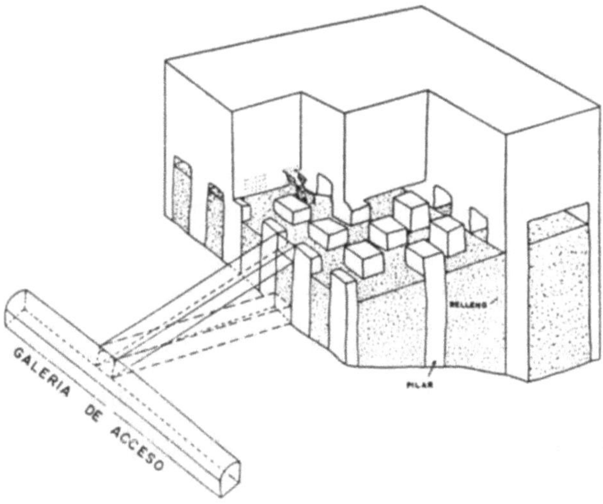

Figure 7. Principle skech of post pillar cut-and-fill method, la Cascada mine.

The room dimensions are 10 m x 5 m, and the square-shaped pillars are mostly 10 m x 10 m but in some cases 8,5 m x 8,5 m. Before full production was started at the end of 1989, 3600 m of ramp and 1900 m of development drifts was completed. The first cut of the rooms was done in the contact zone at the base of the orebody. Once this ore had been mined out, backfill material was introduced to give access to a second cut. This continues until the roof of the orebody is reached. The backfilling factor is 0,46 t of backfill per tonne of ore.

The development and production drilling is done with electro-hydraulic single-boom Boomer H134 and twin-boom Boomer H135 rigs. A comparison between the monthly drilling capacity of the single-and twin-boom rigs is shown in Figure 8.

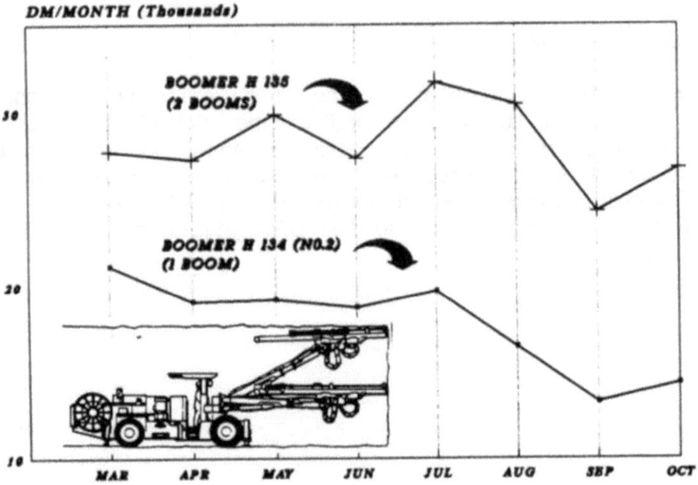

Figure 8. One versus two boom hydraulic face drill performance (1990).

Because of the plastic behaviour and very low compressive strength of the Sagasca ore, the rigs are equipped with a high-impact-frequency rock drill (6000 blows per minute), which is a special version of the COP 1238. This was proposed by the supplier and has yielded very good results.

Of interest is the introduction of water-mist flushing and a feed beam that allows drilling with rods up to 5,4 m long to achieve an effective hole length of 5 m (which is not so common in the mining industry) and so match the pillar dimensions. The drilling parameters in this type of orebody are as follows:

Drilling length	= 4,8 to 5,4 m
Drilling diameter	= 45 to 51 mm
Bit life	= 1800 m
Drifter rod life	= 3700 m
Practical penetration rate	= 2 m/min
Water consumption	= 4 lt/min
Energy consumption	= 1 KWh/per drilled meter
Drilling performance	= 1,8 ton/per drilled m
Total drilling cost	= 1,16 USD/per drilled m

Charging is done with electro-hydraulic driven Nitro Nobel PT-61 ANFO charging trucks. These trucks are very popular in Chile, and are in operation in several mines. The net charging capacity is up to 100 kg of explosives per minute. For roof and pillar stability, special

techniques such as smooth blasting are applied. The blasting factor is 1,39 t of ore per kilogram of explosives, and the cost is 1,0 US dollar per tonne of ore.
Because of the relatively short estimated life of the mine (4 to 5 years), loading and transportation are done by front-end-loaders and standard on-road trucks. The loading capacity is 110 t/hr, and the transportation capacity 60 tons/hr. The productivity is 28 tons per man-shift for direct production employees.

The following equipment is used in the mine:

2 Boomer H134 single-boom drilling rigs
1 Boomer H135 twin-boom drilling rig
2 Nitro Nobel PT 61 charging trucks
3 Volvo L120 front-end loader.
10 Standard 12 m3 on-road trucks.
1 Atlas Copco Brokk 3000 scaler.
1 shotcreting unit.
1 electric compressor with several jack-leg drills.

4. EL SOLDADO

El Soldado mining complex is located 110 kms north of Santiago1.
Since the purchase of the mine by Exxon in 1978, mechanization of the mine and modernization of the mill have been pursued. In 1988, the capacity of the trackless sub-level stoping mine was increased from 3,3 to 11,5 kt per day. The total capacity reached 16 kt per day, most of which came from underground operation. The stoping productivity was increased from 14,5 t per man-shift in 1978 to 28,5 t per man-shift in 1984, reaching 32 t per man-shift in 1988. Today, the mines underground production represents less than half of the total(4). The mine was a pioneer in Chile in applying big hole stoping operations and the results are very interesting.
By consideration of the several periods of the mechanization process, interesting comparisons can be made between various types of equipment and capacities. Figure 9 shows the development of drifting equipment from jack-leg drills in combination with pneumatic overhead loaders, through pneumatic drilling, up to electro-hydraulic drilling in combination with 8yd^3 scooptrams. The development headings average 18,5 m^2 in cross sections. The face rounds typically contain about 55 holes 41 mm in diameter drilled 3,85 m deep.
The relatively large orebodies and good ground conditions allow the use of sub-level large blasthole method (Figure 10).

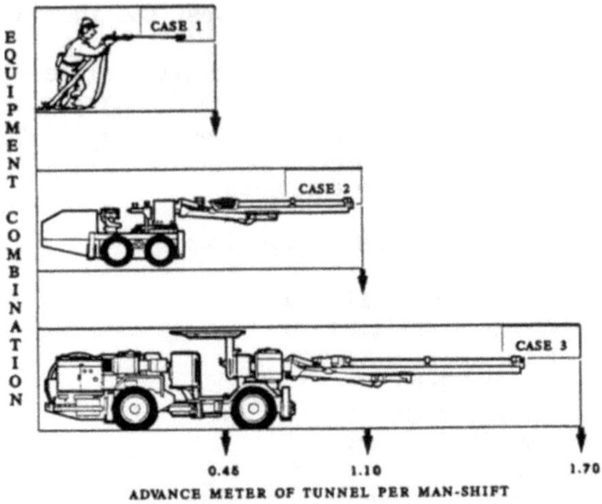

Figure 9. Equipment combination versus tunnel advance.

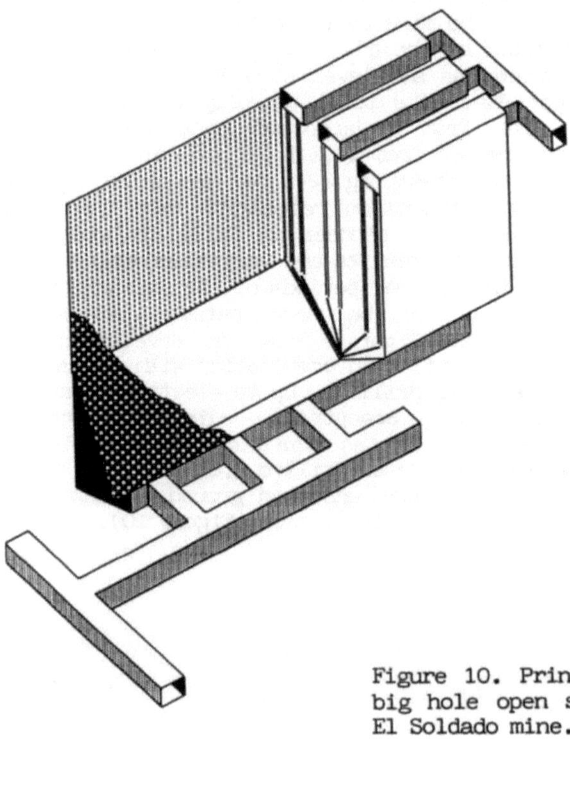

Figure 10. Principle skech of big hole open stoping method. El Soldado mine.

The nominal stope dimensions are 100 m long x 35 m wide, and up to 90 m between sub-levels, with a pattern of load-haul-dump draw points 5,5 x 3,7 m developed at the base of the stope.
Undercutting is accomplished with a fan pattern of 64 to 76 mm diameter holes up to 25 m in length loaded with ANFO. A study was made of the use of Simba top-hammer long-hole hydraulic rigs and their capacity hourly in relation to hole depth is shown in Figure 11.

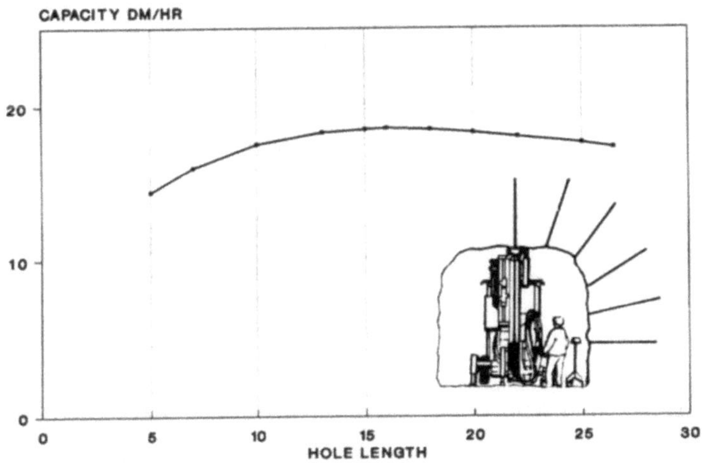

Figure 11. Relation between the capacity of long hole drill rig and hole length (T38 rods, hole diameter 64 mm - upwards drilling).

The change from pneumatic to hydraulic drilling increased the productivity from 56 to 125 drilled meters per man-shift using 64 mm bit, and the bit life improved from 250 to 400 m.
Blastholes 165 mm in diameter and up to 70 m in length are usually drilled vertically on a 3,5 m x 4,5 m pattern, and occasionally in underhand fans of 5 to 7 holes.
Simba 264 rigid down-the-hole rigs using an air pressure of 18 bar for COP 62 hammers permit high-capacity drilling with acceptable precision (with a deviation of 0,5 per cent to a maximum of 3 per cent). Figure 12 shows a cost distribution when drilling with DTH equipment.
The primary blasthole powder factor is about 280 gr/t. The fragmentation is normally excellent, and the stope waste-rock dilution is about 5 per cent. The biggest single shot broke 1,1 Million tons of ore in a moss blasting in April 1993.
Production from the stopes is loaded by 10 m^3 load-haul-dump units and trammed to ore-pass dumping points, end-dump highway trucks of 60t capacity being used for transporting the ore out of the mine.

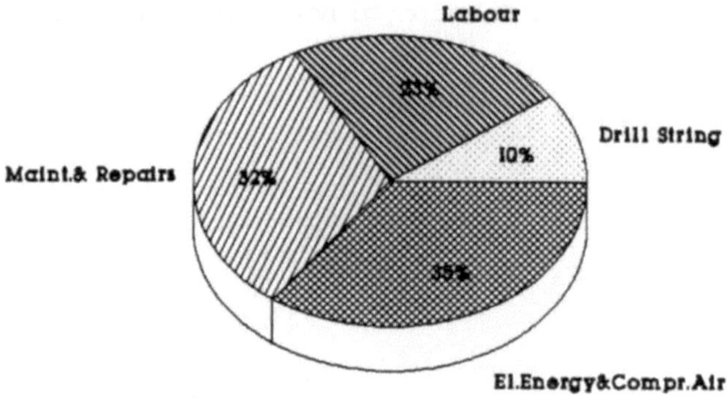

Figure 12. Cost distribution for semi-hydraulic long hole drilling.

The primary mining equipment used at El Soldado is as follows:
4 x Boomer H127 drilling rigs
1 x Simba H252 undercut drilling rig
1 x Simba H221 undercut drilling rig
3 x Simba H264 DTH production drills.
2 x Boltec H321 mechanized bolting rigs.
3 x Nitro Nobel PT61 ANFO charging trucks.
2 x $6m^3$ load-haul-dump units
4 x $10m^3$ load-haul-dump units
10x 60t trucks
2 x 40t trucks
2 x Service trucks

5. MANTOS BLANCOS.

Mantos Blancos mine, is a major private contributor to Chile's copper exports, extracting the ore from the foothills of the Andes 1.000 km north of Santiago close to Antofagasta [1].
The orebody, consisting of both copper oxides atacamita and chrisocole and copper sulphides like bornite and calchosine, was discovered in 1913, but an economic process for oxide ore was not devised until mid 1950's.
Underground Mantos Blancos mine was put in operation in 1970, using cut and fill and sublevel caving methods. From 1975 to 1992 cut and fill was very important method, working big stopes of 15 m. wide 7,5 m height with a gap of 1 m at the bottom to the backfill material.

The mine's long experience in the application of post pillar cut-and-fill method, (Figure 13) served as an example for other mines in Chile.
Face hydraulic drilling rigs were used for both production and development.
Today only sublevel stoping method is used.
For the next 7 years, underground sulphide production will be of 1.9 million tons of ore of 1.23% of ins. Cu.
In 1975, Mantos Blancos used pneumatic mechanized drilling equipment for post- pillar cut-and-fill and sub-level stoping methods. In 1982, the mine started changing to hydraulic rigs.

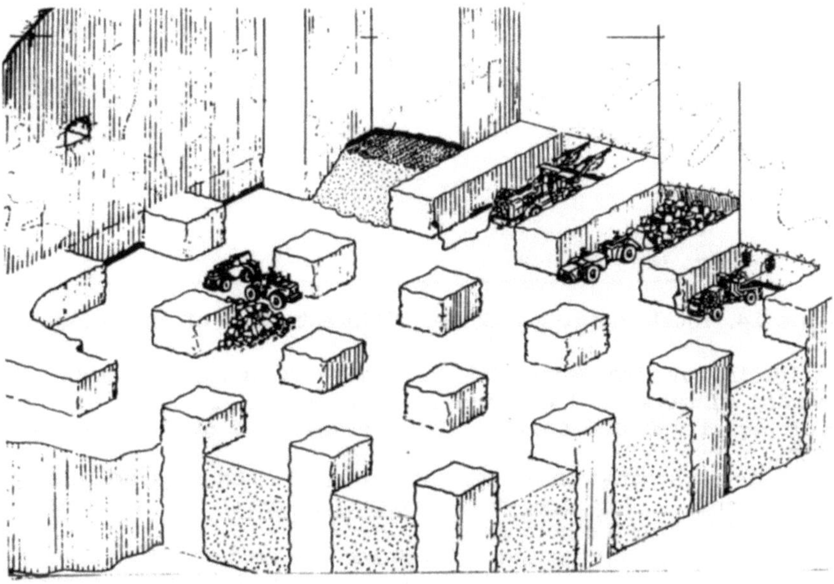

Figure 13. Sketch of the post-pillar cut-and-fill operation at mantos Blancos.

Two-boom hydraulic rigs are used to drill 3,2 m long holes of 51 mm diameter. The capacity per rig is 450 to 500 m drilled per shift (compared with 200 to 250 m with pneumatic equipment). The performance amounts to 4,4 tons per drilled meter. An indication of the rock hardness and abrasivity is that the life of 51 mm button bits is 250 m and the rod life 1220 m.

Use is made of 2-boom Nitro Nobel PT100 ANFO charging trucks. The consumption of explosives in development is 450 g/tons, while in cut & fill was 150 g/tons.
Lately, a third Nitro Nobel's PT-61 explosive charging truck was added to speed up a development work for sub-level stoping operations. This unit is equipped with a newly developed ANOL CC vessel and the difference in charging times can be seen in Figure 14.

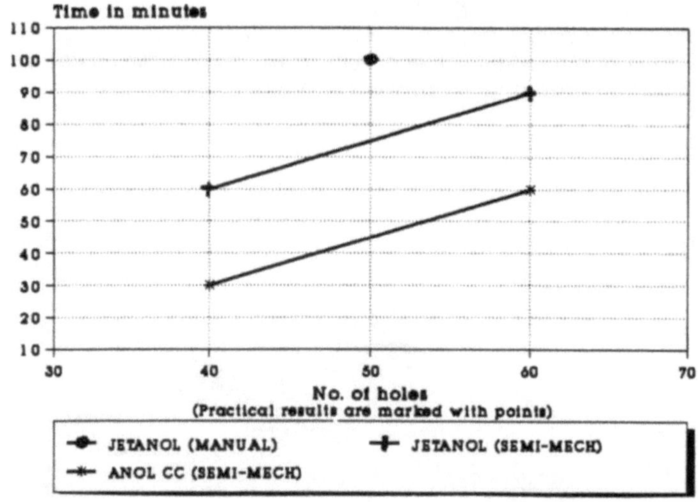

Figure 14. Total charging time for different types of Anfo vessels.

For loading the ore Wagner ST8 load-haul-dump units are used. At a tramming distance of 150 m and inclinations of up to 10 per cent, their capacity is 120 tons/hr. The average utilization is 450 h per month.
In sub-level stoping which dominates today, Simba rigs are used for undercutting. For long-hole production drilling, a contractor operates two down-the hole rigs.
The method of servicing and maintaining the mechanized equipment is worth mentioning since the maximum capacity and mechanical availability of mechanized equipment are of vital importance. When the new equipment arrived in 1989/1990, the mine looked for a partner for service, technical back-up, and support. Atlas Copco was asked to guarantee that the productivity targets would be fulfilled. A service contract signed between the two companies guarantees the minimum acceptable performance of all the equipment (drilling rigs and scoops), leaving the mine management free to concentrate on achieving maximum production at lower costs. Atlas Copco is, guaranteeing 85 per cent availability for Boomer H128 face rigs, and 80 per cent for ST8B scoop trams. The availability achieved for both of these units in 1991 was over 91 per cent.
Figure 15 shows the influence of the mechanical availability of the 8yd^3 LHD on the number of units required for a certain tonnage considering tramming distance of 150 m, tramming capacity 8,2 tons and the tramming speed of 6 Km/h.

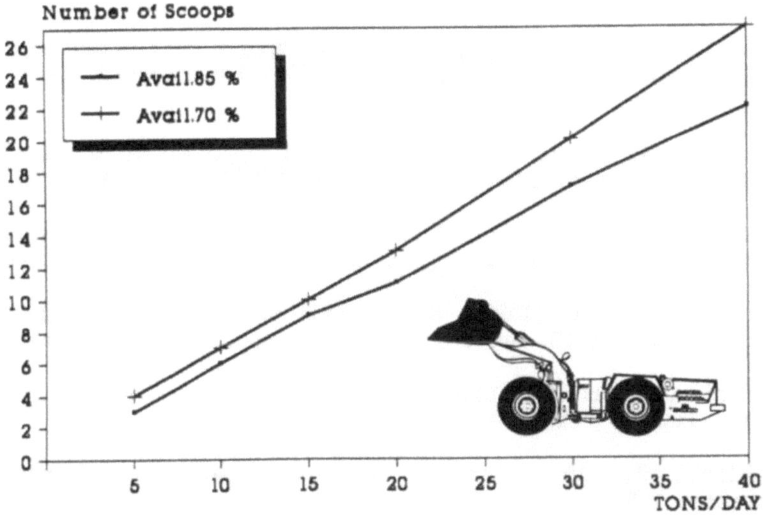

Figure 15. The influence of mechanical availability on the number of units required for a certain tonnage in case of 8yd^3 LHD loaders.

The following equipment is used in the mine:
1 x Boomer H135 face drilling rig
2 x Boomer H128 face drilling rigs
2 x Simba H353 long hole drill rigs
3 x Scalers
2 x Boltec H323 bolting rig
2 x PT100 Nitro Nobel ANFO charging trucks
1 x PT61 Nitro Nobel ANFO charging truck
1 x PT100 Nitro Nobel utility truck
1 x Diamec 262 Craelius diamond drills
7 x ST8B Wagner Scooptrams
6 x ST8A Wagner Scooptrams

6. REFERENCES

1. Zablocki, A. (1992) **The Application of Modern Mechanized Equipment in Chilean Underground Mines.** Massmin 92, Johannesburg, SAIMM 1992 pp. 351-356.

2. Zablocki, A. (1993) **Survey of Chilean Underground Mines; Ore Type, Mining Methods and Production.** October 1993 (Not published).

3. Latinominería (1993) **Gold Comes to Shine Once More in the Mining Future of South America.** August 1993, pp. 15 - 16 (In Spanish).

4. Minería Chilena (1993) **El Soldado Changing Into Open Pit Operation.** Year 13. nr. 145, July 1993, pp. 17 - 21 (In Spanish).

5. Minería y Desarrollo (1993) **Chile and Its Copper - Projection Into 2010.** January / March 1993 (In Spanish).

6. Zablocki, A. (1993) **Personal Discussions with Mine Executives.** December 1993 - January 1994.

Tunnel-boring machine for Magma Copper Co. mine development

H. A. Janzon
The Robbins Company, Kent, Washington, U.S.A.

Abstract

The Board of Directors of the Magma Copper Company has recently approved investments for the development of a new orebody. Mine Management's decision to go in for excavating the new main drifts by means of a Tunnel Boring Machine (TBM) is expected to lead to a new era in fast mechanical rock excavation in hard rock mines: from the moment a machine was ordered until the completion of a tortuous 10.5 km long drive no more than 30 months will elapse. The project is certain to be followed with great interest by the mining industry the world over. This paper describes the background of the investment decision, the geological and geotechnical conditions and presents the TBM to be used on the project.

1 Introduction

Mechanical rock excavation by means of TBMs has made significant progress within the construction industry for driving tunnels for a large range of purposes, such as water supply-, sewage collection-, highway- and rail road tunnels. However, to date the method has not seen a large scale breakthrough in the mining community.

There are a number of reasons for this:

- few mine managers want to be the "first to stick their neck out";
- in most cases, mine plans are laid out to suit existing equipment;
- the concave shape of a bored tunnel's invert is seen as a drawback, as mine haulageways generally require a flat floor;
- TBMs are usually considered to have poor curve-ability and not to perform well in hard rock, etc.

Nevertheless, the majority of mine managers understands that the technique – if it can be made to work satisfactorily – does offer advantages for the working environment, as well as promises of better total efficiency through improvements in the fields of haulage, crushing, ground reinforcement and also with a view to personnel recruitment. From time to time, one therefore sees efforts being made to utilize a TBM – or other means of mechanical excavation such as roadheaders – for development also in hard rock mines.

The Magma Copper Company's development plans for a new orebody present such a case.

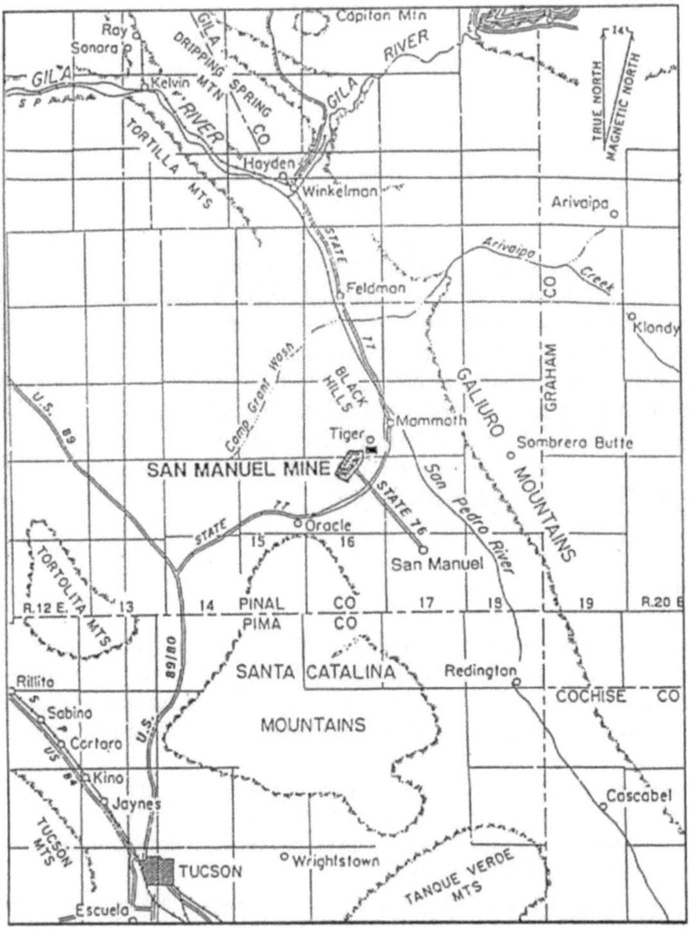

Fig. 1 Location of the Magma Copper Company San Manuel mine

2 Project background

The Magma Copper Company (MCC) is located in San Manuel, a town some 80 km NE from Tucson in the state of Arizona, USA, see Fig. 1. The company's San Manuel Mining Division currently produces some 56 000 short tons/day of sulphide copper ore – which makes it one of the largest underground mines in the world – from the San Manuel and Upper Kalamazoo orebodies.

The reserves in these orebodies will be depleted by 1998, however. To secure future production, it therefore was necessary to quickly develop the Lower Kalamazoo orebody, with a layout suitable for a modern 50 000 short tons/day block caving operation.

In the autumn of 1991, MCC started planning the mining layout of the new orebody and undertook an in-depth evaluation of the feasibility of using a TBM for rapid drifting.

First production from the newly developed orebody is planned for the last quarter of 1996. Full production will be achieved in the year 2000 and then sustained until 2009, when diminishing reserves will – again – lead to reduced production rates.

3 Geological aspects

The Kalamazoo orebody is the hanging wall segment of an orebody that has been bisected and displaced laterally over a distance of some 2.5 km along the San Manuel Fault, see Fig 2. The San Manuel orebody – which has been mined since 1956 – forms the foot wall segment. The San Manuel Fault curves around the northern and eastern sides of the Kalamazoo orebody and dips from 15° to 45° southward to southwestward.

As stated above, the mining method to be used in the Lower Kalamazoo orebody is block caving. The base of the mass to be caved – about 1 100 m below surface – is to be accessed by drifts on two levels from an existing mine shaft, the San Manuel # 5 Shaft. The upper drift – on the 3 440 ft level – provides access on the grizzly level, the lower one – on the 3 570 ft level – is on the haulage level, see Fig. 2.

Based on the outcome of a 16-month comprehensive feasibility study – concluded in February 1993 – which served to define the new orebody (9 277.5 m of new cores were produced and analyzed!), to lay out the mining area and to determine the drifting method to be used, the decision was taken to excavate both drifts in a continuous operation by means of a TBM, starting from the # 5 Shaft on the grizzly level.

On the grizzly level, about 4.5 km of 4.6 m diam drift is to be bored in a large loop, with a number of curves with 150 m radius, see Fig. 3. The machine will thereafter bore a curved 5.5 % decline around the # 5 Shaft to the haulage level, where two loops – in all some 6 km – will be bored. Here the plan calls for curve-radii of 110 and 150 m, see Fig. 4. The total boring time has been calculated to span 18 months, working 3 shifts/day, 355 days/year.

The two parallel drift sections on the haulage level – clearly visible in Fig. 4 – will serve as a conveyor haulage drift and – interconnected therewith at regular intervals – as a service drift, see the layout presented in Fig. 5.

4 How geology and rock quality will influence mechanical excavation

The San Manuel mineralization lies in the Sonoran Section of the so called Basin and Range Province – west of the Rocky Mountains. The orebodies are the result of a porphyritic intrusion of granodiorite at the end of the Cretaceous period (this – by the way – is the same mineralization which created the majority of the large copper orebodies in Chile). The granodiorite intruded a pre-Cambrian quartzmonzonite. Both rock types belong to the granite family.

The mineralization took place in the contact zone and originally had the shape of a very large, hollow, elliptical cylinder. Later on it was split and displaced by faults and intersected by dykes of dacitic, andesitic and rhyolitic composition.

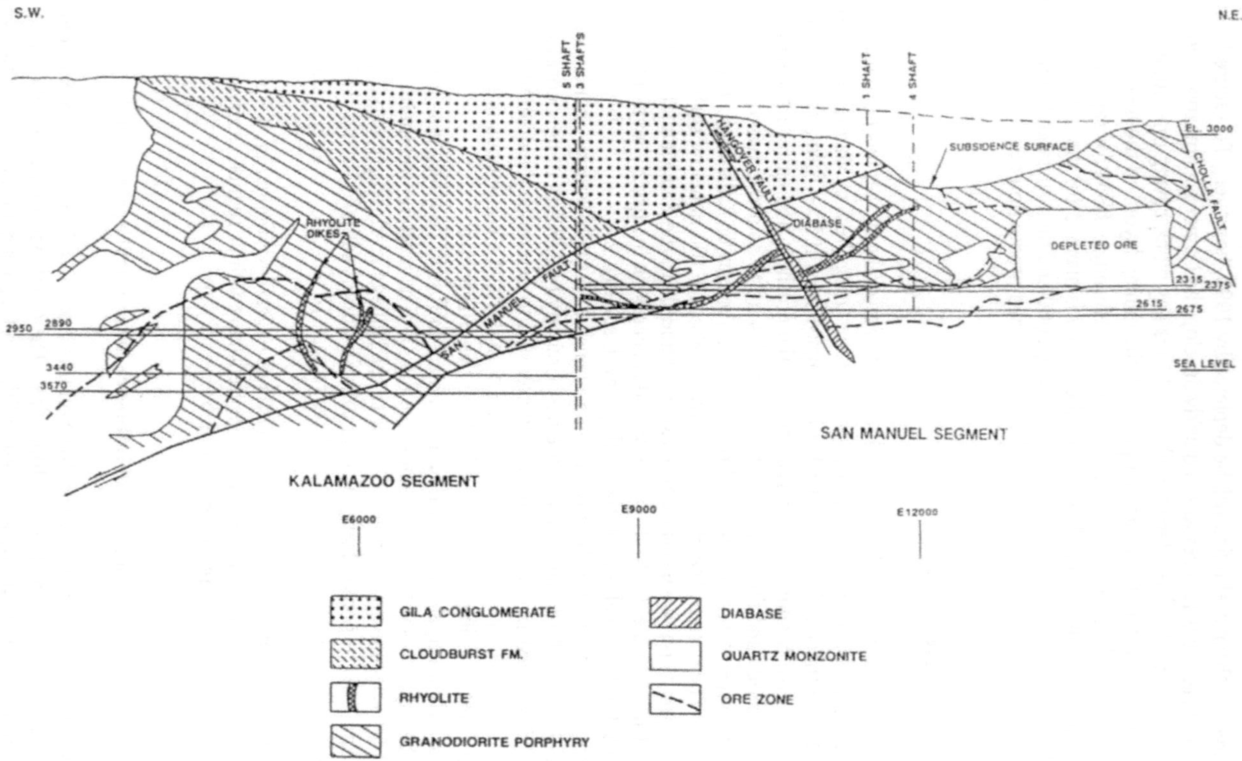

Fig. 2 Diagrammatic geological profile of the San Manuel Mine

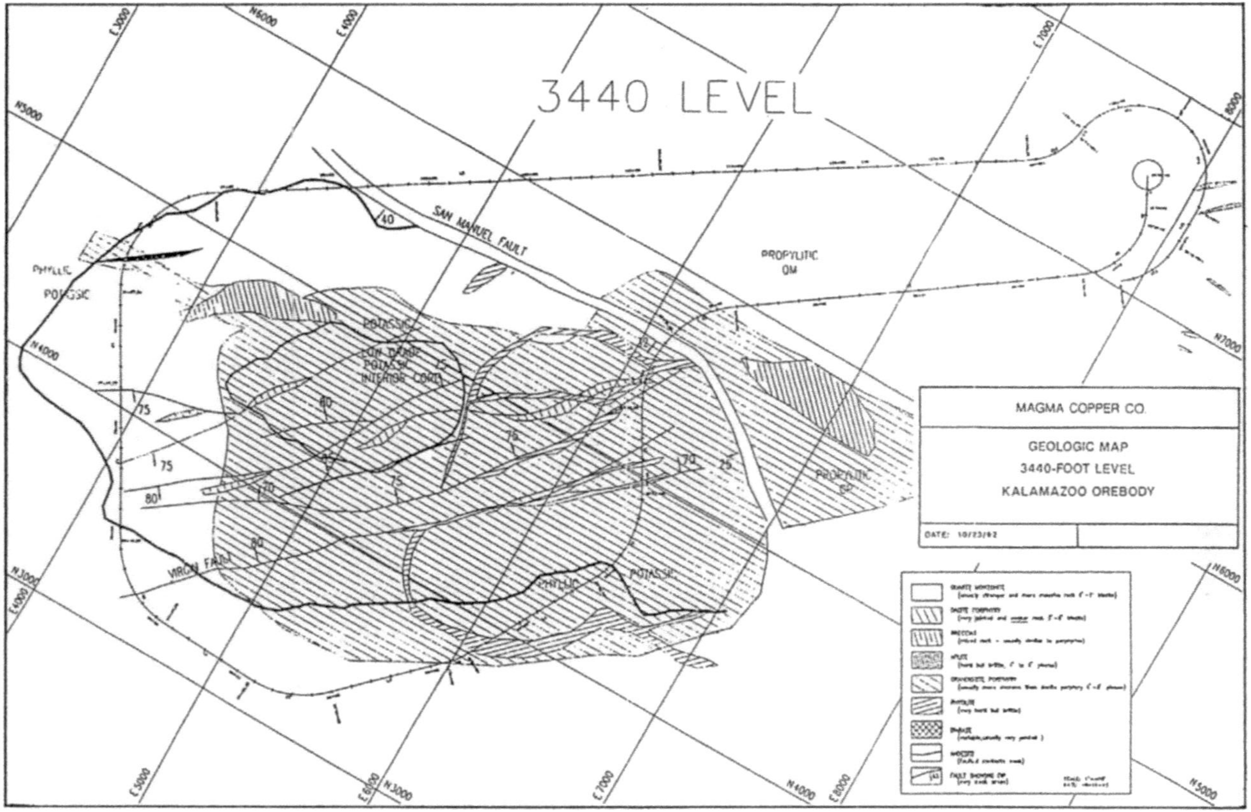

Fig. 3 Boring loops on the grizzly level in the Kalamazoo orebody

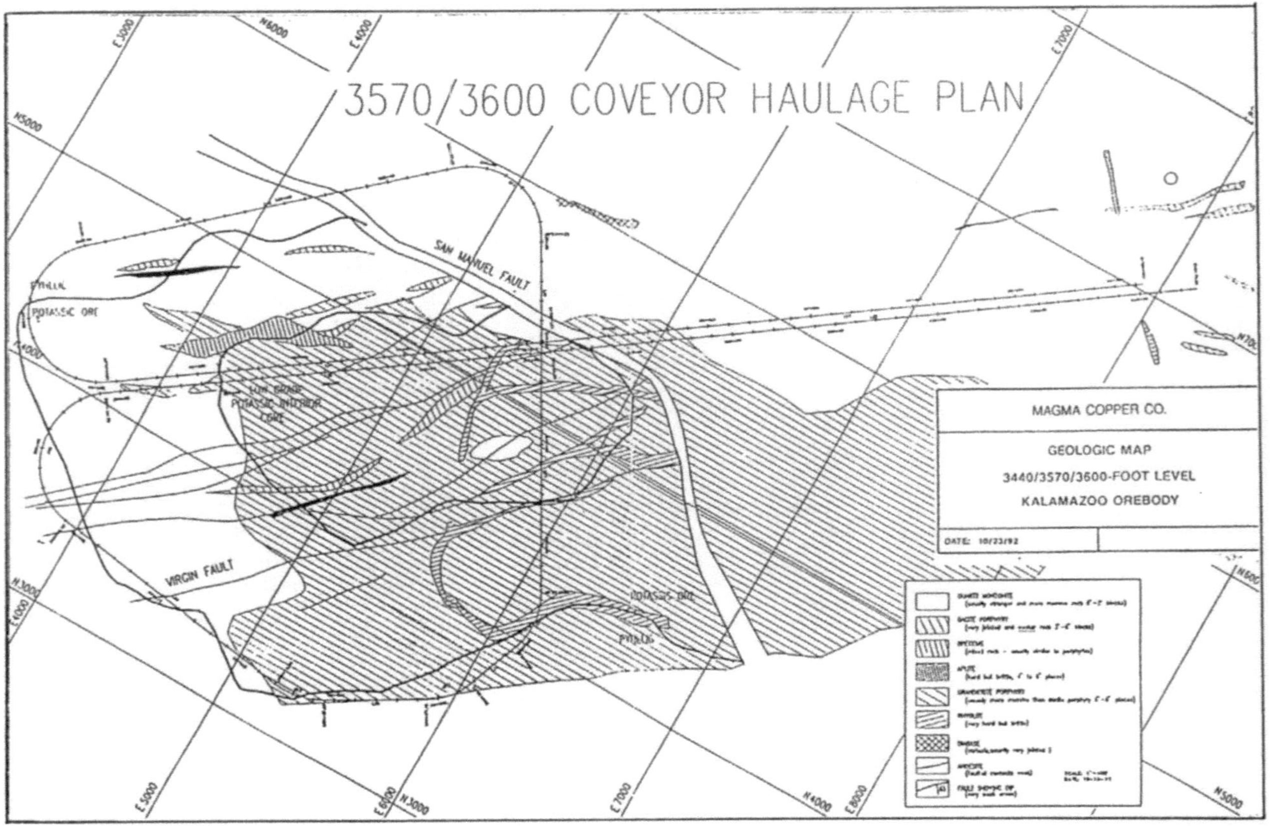

Fig. 4 Boring loops on the haulage level in the Kalamazoo orebody

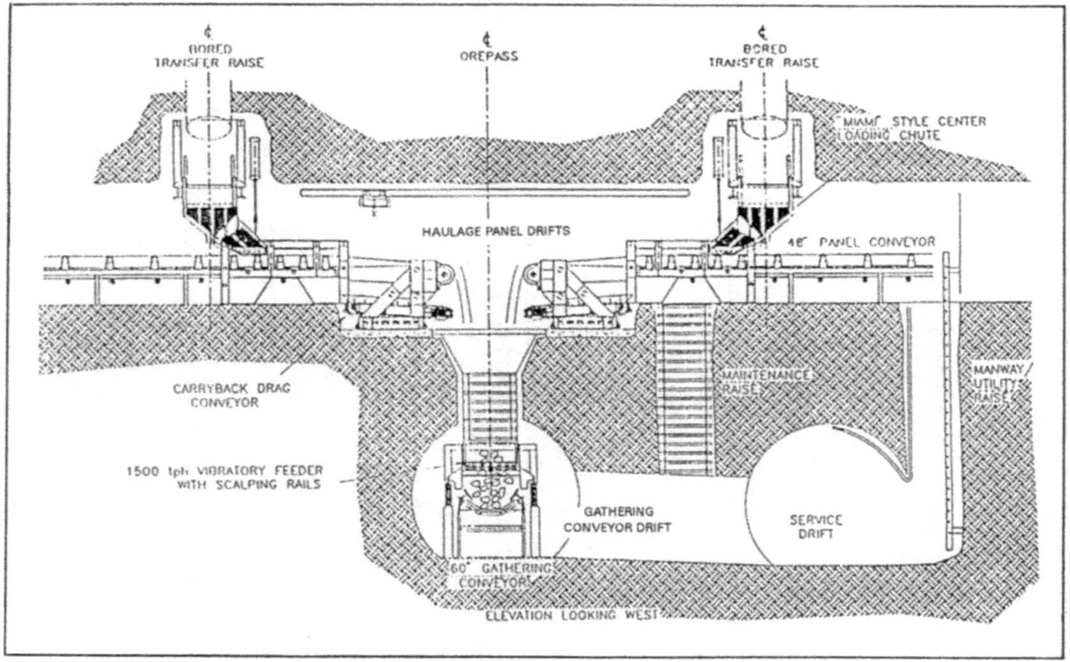

Fig. 5 Haulage level layout in the Kalamazoo orebody

5 The need of rock reinforcement

The structural condition of the rock which the TBM will meet is thus quite complex, ranging from very weak to strong. The area to be traversed contains:

- a stable pre-Cambrian quartz-monzonite with an unconfined compressive strength (UCS) of 150 - 180 MPa;
- two mineralized haloes;
- a mineralized core;
- the San Manuel Fault with a 1 m wide clay zone; because of its flat dip it stays with the bore for some 30 m at a time and influences the bore for some 190 m in all;
- the Virgin Fault zone which dips steeply, but because of a series of related minor faults results in poor rock conditions for about 500 m total;
- the dykes' contacts with the granodiorite and the quartz-monzonite; weak zones some 0.2 - 0.6 m wide will influence the bore for some 180 m in all.

All types of rock have been subject to hydrothermal metamorphosis. The temperatures and pressures of infusion have almost universally weakened the rock by infusion-induced veining, fracturing and jointing. Coatings of the hydrothermal alteration products on rock discontinuities weaken the rock fabric by reducing friction factors and adhesion.

In the San Manuel and Virgin Faults, stand-up time of the far greater part of the rock is short, probably ranging from below 30 minutes to 4 hours. Here it will be necessary to install ring beams and welded wire fabric as soon as possible behind the cutterhead.

Stand-up time in the remaining ground varies from days to months or years. Irish [1] – MCC's geotechnical consultant – recommends systematic rock bolting with a 1.2 m spacing between rows and four 1.8 m long bolts per row.

In all, Irish draws the following conclusions:

"Although none of the rock masses ... is considered to be an excellent environment for tunnelling with a TBM, rock conditions not only should permit effective machine boring, but favor that approach. ... stand-up time, even through the weakest ground of the San Manuel Fault Zone, should be sufficiently long to allow passage of a TBM and for the installation of ground support; and the moderate abrasivity potential of the rock should not inordinately escalate cutter wear ... the structural stability of the rock should be little impaired by the low level vibrations imparted to it by the TBM. In contrast, shock waves imparted by blasting would impair the stability of the hydrothermally altered rock of the orebody especially ... Of course, tunneling with a TBM should be substantially faster than by a drill-and-blast operation".

6 Time schedule

The planned duration of the boring project – a total of 10 363 m, to be completed within 18 months – makes it necessary to advance an average of some 20 m/day, operating on a 3 shifts/day basis. Advance rates in sound rock have been calculated to reach approximately 35 m/day, but may deteriorate to perhaps not more than 3 m/day under the worst conditions.

The tight curves – 110 and 150 m radius – will lower the daily advance; the mine's planners count on 50 % of the normal TBM capacity. Fortunately, no curves are situated in fault zones.

7 TBM equipment to be used

MCC have selected a Series 150 TBM from The Robbins Company for excavating the drifts described above for the development of the new orebody. It will be operated by Frontier – Kemper Constructors Inc. in a joint venture with Deilmann – Haniel GmbH.

Table 1 presents the main characteristics of the machine.

Table 1. Specifications of the TBM to be used at MCC

▸ Cutterhead	diameter	4.62 m
	operating thrust	7 340 kN (748.4 metric tons)
	installed power	1 258.7 kW
	speed	12.156 / 4.052 rpm
	cutters	33 only, 17″ diam, max 222.4 kN/cutter
▸ Minimum turning radius		107 m
▸ Boring stroke		1.575 m
▸ Belt	width	762 mm
	speed	0 - 2.54 m/s
	capacity	552 m^3/hour
▸ El. system	primary power	12 470 V / 60 Hz
	secondary voltage	600 V - main motors
		120 V - control circuit / lighting
		480 V - supply to Back Up Equipment
▸ Hydr. system	high pressure circuit	289.5 bar max pressure
	low pressure circuit	103.4 bar „ „
	auxiliary circuit	241.3 bar „ „
	total power	149 kW

The 240 ton, "open gripper" type machine has a number of features for efficient tunnelling in the mine's blocky – and sometimes weak – ground conditions:

- the cutterhead can be reversed to prevent jamming;
- cutters are back-loaded (installed from behind) for greater safety of the changing crew;
- the machine incorporates roof drill fixtures and a ring beam erector and is equipped with a "finger roof" to support blocks of rock until they can be secured by roof bolts;
- the sound-proof, air-conditioned operator's cab is located on the Back Up Equipment with closed circuit TV monitoring of essential areas;
- ample space is provided in the area between the TBM and the Back Up Equipment for additional rock support work.

8 Project progress

At the time this paper was prepared for submission - early October, 1993 - the following progress had been made:

- Existing underground workings which will be utilized for the Lower Kalamazoo mining operation, have been rehabilitated.
- The mine dewatering pumps have been moved down to in order to lower the mine water level to 3 860 ft below surface.
- Time-wise critical new mine development work has been started as early as in September 1992.
- The underground erection chamber for the TBM at the 3 440 ft level close to the # 5 Shaft has been prepared by MCC.
- The Robbins Company delivered the machine to the site on September 1, 1993, split into parts of suitable dimensions and weights to fit the # 5 Shaft hoisting facilities.
- Site assembly was in progress and the machine is scheduled to start boring by mid-October, 1993.

9 Conclusion

MCC Management has shown aggressiveness in developing plans for mining operations in the Lower Kalamazoo orebody, not in the least when they commissioned the feasibility study to determine - amongst others - the viability of using a TBM for drifting purposes.

They are to be congratulated upon their bold decision to indeed "go for mechanical rock excavation", in spite of the partly difficult ground conditions expected en route.

We are convinced that the TBM selected will reach a high net rate of advance in the hard rock and that it will be flexible enough to tackle the poor rock in the fault zones and the short curve radii required by MCC's mine development plans.

We fully expect to be in a position to submit a favourable interim report when presenting this paper and to report on the successful outcome of this venture at a later date still.

Reference

1. Irish, R.J. (1992) **Geotechnical Design Summary Report**, 3 440-ft level Grizzly Drift and 3 440/3 570/3 600 Conveyor Haulage Drift, Kalamazoo Orebody, San Manuel Mine, Pinal County, Arizona.

Mining – Environmental Issues

Environmental management in the mining sector

M. C. Athié Lambarri
J. Cochrane
International and Policy Services, Environmental Resources Management, London, England

Abstract

Mining is one of the industrial sectors with greatest potential impacts in terms of land-use, environmental pollution and social implications. To illustrate the environmental situation of the mining sector in Latin America, three case studies in Ecuador, Bolivia and Chile were chosen. As the case studies show, the introduction of the environmental dimension into mining programmes and projects is closely related to the political and economic situation of the country, the market tendencies and the international price of the minerals, among other factors.
The mining sector in these, and most other Latin American countries, is beginning to consider environmental protection as part of the design, construction and operation of their installations. As the costs involved are much greater, only a few mining institutions are looking at remediation work and at the provision of productive alternatives to already polluted sites.
Keywords: Mining, environmental management, Ecuador, Bolivia, Chile.

1 Introduction

Mining is one of the industrial sectors with greatest potential impacts in terms of land-use, environmental pollution and social implications. For years, government institutions and private companies in Latin America have been subject to strong pressures to introduce the environmental dimension into mining activities.
International development institutions have consistently required an environmental assessment of new mining projects since the early 1980's, and increasingly, corporate offices of trans-national mining corporations insist on the use of clean technologies. in their foreign plants.
Also, due to the tendency of Latin American countries towards more democratic governments, public awareness and concern in environmental issues is increasing. Authorities are now more receptive to public opinion, which increases the

involvement of pressure groups and interested individuals in the development of environmental policies and legislation.

In some countries, such as Chile, Costa Rica and Panama, the mining sector was the first to introduce environmental regulations and has had a leading role in the development of the environmental legislation and institutional framework. In Latin American countries where there is a substantial mining sector, the environmental implications of mining activities have been tackled with varying degree of success, sometimes under the requirements of an integrated environmental legislation, more often in response to a loose package of laws related to environment, health and social concerns.

There are a number of factors influencing the response of the mining sector to environmental policies and legislation. These include foreign investment and trade agreements, political will, value of environmental protection for public relations and the market position of the mining sector in the country and abroad. Changing circumstances may produce a change of policy regarding the extraction and refining of minerals with direct effects on investment in environmental protection measures. Fortunately, in spite of the economic pressures, there is a trend in most Latin American countries towards a greater consideration of the environmental dimension in the decision making process.

2 The Case Studies

To illustrate the environmental situation of the mining sector in Latin America, we have chosen three case studies in the southern hemisphere as follows:

- Ecuador: Mining Development and Environmental Control Project.
- Bolivia: The Mining Sector and the Environment.
- Chile: Ventanas Environmental Project.

These case studies are particularly interesting because they depict the various attitudes and responses of the mining sector to the environmental challenge.

In Ecuador the mining sector is underdeveloped, representing only 1% of the GNP. The present trend is to encourage the development of the sector but to introduce environmental considerations from the outset. The task however, seems too big for the limited resources of the National Directorate for the Environment.

Bolivia has a long mining tradition and the environmental impacts of this activity are serious and widespread, particularly in the Altiplano. There is a strong process of privatisation of the mining sector taking place since 1985 and there are encouraging signs that the environment will be duly considered in response to the Environmental Law enacted in 1992. Environmental control actions are in recess at the moment while the new government organises itself.

Mining is a very important sector in the Chilean economy and it is technically well developed. Although a proposed umbrella legislation on the environment has not been enacted yet, very significant advances have been taking place in mining centres throughout the country in response to a genuine preoccupation and a significant coordinating work by the Ministry of Mining. A good example of this is the Ventanas Environmental Project located in the V Region near Valparaíso.

As the case studies show, the introduction of the environmental dimension into mining programmes and projects is closely related to the political and economic situation of the country, the market tendencies and the international price of the minerals, among other factors.
However, the mining sector in these, and most other Latin American countries, is beginning to consider environmental protection as part of the design, construction and operation of their installations. As the costs involved are much greater, only a few mining institutions, such as in the Ventanas case, are looking at remediation work and at the provision of productive alternatives to already polluted sites.

3 Ecuador Mining Development and Environmental Control Project [1]

3.1 Background

Mining activity is currently undeveloped in Ecuador and accounts for less than 1% of the GDP. Current mining activity is predominantly small-scale with minimal attention to environmental and health and safety concerns and minimal contribution to tax revenues. There are some large international mining companies exploring for metallic minerals, but private sector activity is on the whole extremely limited. New mining legislation has removed some of the constraints on mining activity by reducing the scope for speculation in mining concessions and liberating land for exploration. However, major concerns such as the lack of basic geological information and inappropriate environmental regulations still remain.

3.2 The Project

The Government of Ecuador, with the assistance of the World Bank, prepared a new mining law which was passed in 1991 and which contains the principles for technical and economical development, as well as environmental control, in the mining sector. The Government is now planning to implement a five year "Mining Development and Environmental Control Project" in order to boost activity in the mining sector and attract foreign investment.
The project consists of a range of support measures designed to increase mineral production in Ecuador, attract foreign investment, mitigate environmental impacts of existing mining activity and develop an appropriate environmental control framework for the future.

A number of environmental measures are incorporated in the project. These are mainly directed at small-scale mining aiming to establish the extent of environmental impact through a pollution and health monitoring programme and to resolve problems of contamination from mercury and cyanide used in processing and from tailings disposal. These are important measures but will need to be backed up by adequate enforcement procedures if they are to be effective.

While it is difficult to predict with accuracy the extent to which mineral production will be increased by the project, there are strong indicators that significant mineral potential exists in Ecuador and that with a combination of support measures, production can be stimulated.

Preliminary estimates indicate that incremental mineral production to a value of US$55 million per year over the period 2000-2010 would yield benefits in terms of tax revenues sufficient to give a 10% return on the project. This would be less than half of the value of current informal gold production. Estimates based on the World Bank projections of future metallic mineral production suggest an internal rate of return as high as 40% even taking into account possible growth in the without-the-project situation. Other less direct or less quantifiable benefits will also result from the project such as spin-off employment from mining activity and improved environmental quality in small scale mining areas.

3.3 Environmental Measures Incorporated in Project Design

The project design incorporates 4 sub-components that relate to the environmental aspects of the project.

Assistance to DINAMA. This sub-component consists of provision of technical assistance in the following:

- development of an appropriate organisational structure for the National Directorate for the Environment (DINAMA);
- preparation of work programme and field procedures for effective inspection of mining operations;
- training on mining-specific environmental control matters;
- establishment and enforcement of standards and criteria that are consistent with existing legislation.

DINAMA will have to increase significantly its resources in order to provide the enforcement necessary to control the environmental impact of both industrial scale and small scale mining, as proposed in the project.

Environmental Baseline Data and Monitoring Pollution and Occupational Health Issues to Mining Activities. The monitoring programme is to be designed to investigate the extent and nature of the levels of pollution that already exist as a result of small scale mining. The work will be carried out in the river catchments and surrounding areas of the three main mining districts, Portovelo-Zaruma, Ponce Enriques and Nambija-Chinapitza. It will include sampling for heavy metals and domestic effluents in ground and surface water, stream sediments, aquatic biota, soils, vegetation and agricultural products as well as human health. Carried out over a period of five years this should provide a very good understanding of the extent of environmental pollution as well as recording improvements due to the implementation of other environmental management actions.

Feasibility of the Containment, Neutralisation and if Necessary Detoxification of Mining Wastes. This sub-component will look at methods of containing mine wastes that would be technically feasible on a small scale. The necessity of, and methods for, the neutralisation for cyanide wastes will also be examined. The concept is that, having come up with a suitable design, a trial containment project would be built both to test the design and to show the small scale miners how it could be done. The detoxification element of this sub-component is aimed at existing areas where a high level of pollution already exists. These would be established by the monitoring programme and cost effective methods for clean-up operations would be examined. This sub-component will be centred on the Ponce Enriques mining district, where the risks to the economic activities are the greatest.

Assistance to Small-Scale Mining. This component is aimed at upgrading mining activities, improving health and safety conditions and reducing the release of mercury and cyanide to the environment. The starting point will be a review of the existing technical, economic and social situation in the mining communities on the basis of which detailed terms of reference for the programme will be developed. It is envisaged that this will be implemented by the Corporation for Geological-Mining-Metallurgical Development and Research (CODIGEM) technical services department with supervision from an international consultant.

4 Bolivia: The Mining Sector and the Environment [2]

4.1 Background

Mining has been the totally dominant industrial activity in Bolivia since the Spanish arrived in the 16th century. Many hundred of deposits have been worked, containing a wide variety of metallic elements. Some of the first deposits to be discovered are still being exploited. There is still considerable potential for future mining both in the Andes (hard rock mining) and in the Amazon (alluvial gold mining).

Mineral production in Bolivia declined from a peak of US$ 765 million in gross metal value in 1980 to a low of US$ 139 million in 1986. Since then a recovery has occurred to US$ 367 million in 1991. In spite of all problems, including the present low metal quotations on the world market, the export of the mining sector amounted to more than 40% of the nation's total export value in 1991.

Hard Rock Mining. In 1952, the government nationalised all major private mines under the new Corporación Minera de Bolivia (COMIBOL). Thus, the private sector was reduced to a number of smaller operations and small scale mining. This private sector has since been subdivided into Medium Mining and Small Mining. These designations are still in use even though nowadays, as a result of the liberalisation of the Bolivian economy which started in 1985, the Medium Mining has surpassed COMIBOL in size of individual operations as well as total output.

Alluvial Gold Mining. The gold originates from the bedrock, from which it has been liberated through glacial or weathering processes. Gold has been exploited in Bolivia since pre-Columbian times. Presently, intense exploitation of gold from alluvial deposits occurs in two main areas within the Amazon basin: along the Tipuani and adjoining rivers draining the Eastern Cordillera; and along the Madera and Madre de Dios rivers close to the frontier with Brazil, the northernmost part of Bolivia.

These areas largely lack even basic infrastructure and are outside any public control to a great extent. Health service is rudimentary and serious accidents are very common. By and large, environmental quality and workers' health and safety are unknown concepts. A particular concern is the use of mercury for the amalgamation of concentrates; this mercury is "heated off" into the atmosphere for the recuperation of gold and in this way enters the soil and water. Another serious consequence is heavy degradation of the landscape by dredging, diversion of rivers, cutting trees, etc.

4.2 Policies, laws and institutions for environment

In 1992, Bolivia adopted a new umbrella environmental law and began developing the Bolivian Environmental Action Plan. The law covers all aspects of environment, including environmental management, and advocates sustainable development and the use of economic incentives. Its requirements are very general and so implementation will require more detailed regulations. In many cases the law is unclear about institutional responsibilities.

The national environmental law created an environmental coordinating body that has the status of ministry, the National Secretariat for the Environment (SENMA); it also created regional counterparts to SENMA that will operate in each of the nine departments, and called for all ministries to establish environmental units. SENMA had broad responsibilities, much broader than its current resources can accommodate. Last year the Congress started drafting new laws on water, forestry and biodiversity but, with the change of Government in July 1993, SENMA started a re-organisation process and both its own activities and the legislative work on the environment came to a pause. SENMA became the Ministry of Sustainable Development with a higher status but, apparently, broader responsibilities still.

4.3 Mining and the Environment in Bolivia

Bolivian mining has caused, and continues to cause, a major negative impact on the natural as well as the human environment. However, data on specific impacts tend to be scattered and not well organised. The main causes of concern, which are common to most mining activities in Latin America, are the following.

- The consumption of scarce and depletable natural resources, such as the use of land, the removal of ores and minerals and the use of water.
  The mining industry in Bolivia consumes approximately 31.5 million m^3 of water each year. This amounts to about 2% of the total water resources available in the Altiplano, where the great majority of the mines are situated.
- Transformation of the landscape, involved in open pit exploitation, dredging activities in alluvial mining, tailing dams, waste rock disposal and waste piles from heap leaching.
- Contamination of surface waters, groundwater and soils from mine water, process water, tailings in process water suspension and leachates from old waste dumps or other sources.
  Effluents from mining operations cause widespread contamination of natural soils and waters. Wastewater often contains extremely high contents of heavy metals, as well as process chemicals such as xanthates and cyanide. The frequent lack of proper tailings dams or deficiencies in their management, result in 30-50% of all waste waters with their contamination being discharged to nature.
  Another major problem is "acid rock drainage", which is caused by oxidation of sulphides in tailings and water bearing rock, and the subsequent formation of acid, metal rich solutions through leaching by rain water. All these sources of pollution pose major potential threats to flora and fauna.
  The spread of mercury to nature is a major concern associated with alluvial gold operations in certain parts of the Amazon basin. It has been estimated that annually at least 25-50 tons of mercury are discharged to nature, and perhaps more. Fish from the gold mining areas contain anomalous values of mercury and there are preliminary indications of poisoning in the local population.
- Accumulation of solid waste, mine waste rock and tailings from concentration operations.
  The present annual production of solid waste from industrial mining is approximately 5 million tons, of which about one half is discharged to tailings dams or satisfactory construction and management. The remaining industrial mining operations will have to be equipped with such facilities at an estimated cost of US$ 6.5 million. The reclamation of old tailings, estimated to total about 104 million tons, implies a cost of US$72.5 million.

- Dust and gas emissions, particulate matter, dust containing heavy metals that can contaminate water and soils, gases from smelters which may contaminate the environment and affect human health.
- Occupational health and safety hazards, often extreme, specially in small scale and informal mining.

Unfortunately the trend in Bolivian mining during the recent years is not encouraging, as explained below.

- The change from tin mining to the exploitation of lead/zinc/silver has meant increased use of toxic process reagents such as xanthates and cyanide.
- The economic recession has caused a decrease in the ability and determination of management to control effluents and waste disposal.
- The surge of innumerable new small scale and artisan operations has given rise to serious problems, mainly of a social character but also affecting the environment. These activities are essentially outside any type of control, with little or no concern for health and safety or the environment.
- Closed operations are usually left without any type of clean-up or land reclamation, often resulting in acid drainage and other problems. Mines being closed often means less pollution from mine water and tailings, but it usually also ends the maintenance of existing tailing ponds. Another serious consequence is that closed mines attract disorderly operations by "cooperativistas".
- The proximity of some mining areas to urban areas (in some cases cities that have grown up around mining) presents a variety of serious public hazards.
- A positive development is the arrival in Bolivia of some foreign mining companies with a high degree of knowledge and standards in technical and environmental questions. However, there are also less responsible foreign investors entering the scene.

It is difficult to forecast the extension, character and location of future mining ventures and their implied environmental impacts prior to carrying out exploration activities. However, in the foreseeable future, the focus of conventional mining in Bolivia will continue to be in the Altiplano and to a large extent in areas of previous mining activities. The necessary environmental improvements in connection with such operations are reasonably well known and it can be assumed that, with the introduction of new policies and techniques, the problems can be managed in a satisfactory way.
New possible areas for mining in the eastern part of the country could present problems and possible conflicts with farming and other interests. Exploitation of mineral deposits in nature reserves should be legally prohibited; in this context, it will be important to define clearly the boundaries of nature reserves, which is still pending due to the recent change of Government.

Alluvial gold mining presents special problems, as noted before. The rest of current activity will probably remain the main exploitation areas for the nearer future. However, it is possible that new deposits will be found in other parts of the Amazon Basin, with a corresponding spread of contamination and destruction to new areas. This is the reason that an extensive inventory of all present alluvial operations, with projections for future development and its environmental consequences is now urgent.

5 Chile: Ventanas Environmental Project [3]

5.1 Background

The Ministry of Mining, aware of the serious repercussions that the mining sector had produced on the environment for many years, enacted the Decreto Supremo 185 in 1991. This decree was one of the first of its kind and emphasised the leading role taken by the mining sector in environmental protection in Chile.

One of the first of the plants to comply with the decree was the Ventanas smelter and refinery owned by the state mining company Empresa Nacional de Mineria (ENAMI).

5.2 The Environmental Setting

The Ventanas industrial complex is located within the V region of Chile about 150 km west of Santiago and is one of the main sources of atmospheric pollution in the coastal area. The complex comprises a copper smelter owned by ENAMI, and a coal-fired power station owned by the private company, Chilgener.

Previous studies of the Ventanas region have demonstrated that the main impacts of pollution from the complex appear to be confined to the municipality, or 'Comuna', of Puchuncaví located to the north and east of the smelter. Traditionally, the Comuna was an important area for livestock rearing and agricultural production however productivity has declined rapidly due to a combination of factors including industrial pollution.

The Comuna is separated from the Aconcagua valley system by the Sierra de la Costa mountain range which provides a natural barrier to the prevailing wind. Inland of this mountain range, the pollution impacts of the Ventanas complex are greatly reduced although acid rain damage to crops has been recorded at large distances from the plant.

In addition to the Ventanas plant, other industrial installations in the region include the Refinería Petróleo de Concón (RPC), the Chagres copper smelter and the Cemento Melón plant at La Calera. In comparison with the Ventanas plant, these industries contribute only marginally to pollution in the Puchuncaví valley. The main pollution impacts of these plants are on the irrigated agricultural lands of the Aconcagua valley system.

5.3 Environmental Management in Ventanas

The ENAMI smelter and refinery in Ventanas began operating in 1964 and has a current capacity of 313,396 tonnes of copper concentrate per year. The smelter was fitted with a sulphuric acid plant in 1990 to capture SO_2 and currently produces around 185,000 tonnes of acid per year. Gases from the reverberatory furnace and from the convertors are discharged via a single flue, 150m chimney.

The new law governing industrial air emissions was passed in Chile on 29 September 1991 (Decreto Supremo 185). This law stipulated that all industries, in areas where air quality standards were not being met, should install air quality monitoring networks in their zones of influence and provide data to be reviewed by the Ministry of Mining/Ministry of Health to ensure compliance. The law also stipulated that existing industries would be required to develop decontamination plans to reduce pollution to acceptable levels within a specified timeframe. The air quality standards contained within the Law are based on USEPA standards.

The objectives of the decontamination plan should be targeted at minimising emissions and mitigating impacts such that compliance with regulations and air quality standards are achieved. This may be accomplished by one or more of the following options:

- decreasing pollution (copper and power) to decrease emissions;
- investing in cleaner technologies to reduce emissions;
- modifying emission characteristics so as to improve dispersion;
- removing harmful pollutants from emissions prior to release (ie flue gas desulphurisation).

The new law formally enacted the ENAMI/Chilgener Protocol which had been agreed in 1990 in response to an initiative of the Ministry of Mining.

In 1992, several of the actions identified in the plan had already been carried out including the following.

- Commissioning of a plant for the production of sulphuric acid at the ENAMI copper smelter and refinery.
- Design and installation of an air quality and meteorological monitoring network to record continuously levels of SO_2 and particulates in suspension.
- Completion of a baseline survey on the quality of air, water, soils and ecosystems including an evaluation of the benefits of emissions controls on these receptors.

The Decontamination Plan represented the first definitive step towards the mid-term and long-term solution of environmental pollution in the Comuna of Puchuncaví. However, in response to the decline in the world copper prices, ENAMI has recently reviewed the directives laid out in the plan.

Although a revised plan is yet to be published, it is now likely that ENAMI will consider a reduction in total production as an option for emissions reduction in preference to the installation of a second sulphuric acid plant and the replacement of old polluting plant with more modern cleaner equipment. ENAMI believe that this policy will allow them to meet air quality standards without major investment in pollution control.

6 Concluding remarks

Mining in Latin America has been for centuries a source of interest for national and international investors. Such was this interest that the integrity and indeed the survival of a number of countries in the area have often depended on activities related to this sector. Although the mining production had declined lately due to lack of fresh investment, at present, mostly as a result of more stable conditions in the area and greater opening of the economy, this sector has regained its previous attraction.
Greater political stability has increased the confidence of foreign investors but, at the same time, democratisation has increased the awareness in the need to consider all interested parties in the mining activities: the miners, the land owners and the neighbouring population, often including indigenous people. The environment has also become an important source of concern and no major mining activity is now allowed without some consideration to the environmental implications which the development may have.

In the case studies presented there were three stages of development of the mining sector: Ecuador is taking initial steps to the large-scale development of the mining sector; Bolivia has a long mining history but technical capabilities are still limited; in Chile the mining sector has a high level of development and production of certain minerals is rated as one of the largest in the world.

The three countries have passed legislation to prevent and control environmental deterioration due to development projects including mining, and in Chile some measures have been taken to make a productive use of areas polluted by the mining industry. Besides their genuine interest in keeping pollution at bay, and their efforts to provide appropriate legislation to this end, the three countries have in common serious compliance difficulties.

The environmental situation in these countries and the trends in the mining activity suggest that needs for monitoring and control are much greater than the capabilities of the environmental agencies. At the same time, many mining installations are financially constrained in providing adequate pollution control measures.

As a result of various studies undertaken by the World Bank and national institutions, a number of options have been identified to ensure that the environmental dimension is systematically considered in the strengthening of the mining sector. These include the following.

- Provide a portion of the financial investment intended for the development of the mining sector to environmental control measures. The financial investment of new mining projects should consider an adequate amount for effectively preventing environmental deterioration.

- Introduce environmental considerations throughout the decision making process in order to ensure that the development of the mining sector is environmentally sound.

  Throughout the project cycle there are decisions that may have significant effects on the environment such as: the implementation of a new development, the expansion or reduction in production and the decommissioning of a mine or a refining plant. Therefore, the environment should be duly considered in all the stages of the project cycle.

- Strengthen the capabilities of the environmental agencies as "second tier institutions" with a good coordinating capacity but little direct involvement in monitoring and control.

  It is clear that environmental agencies cannot have a front-line involvement in environmental control any more. This is because this action reduces their effectiveness and requires excessive large amounts of resources. The alternative is for them to retain a coordinating role and use consultants or specialised sectoral institutions to provide guidance and effect monitoring and control activities on the agency's behalf.

- Encourage the continuing involvement of the mining authorities in helping to improve the sector's performance in environmental protection and control.

  The leading role that the mining institutions have in environmental protection has produced a considerable expertise in the field. This expertise gives mining institutions a good comparative advantage in providing guidance to the sector in preventative measures and clean technologies which should be capitalised upon.

The present trend towards the strengthening of the mining sector in various Latin American countries implies a renewed concentration of resources in this sector. Previous experience has shown that sustainable development can only be achieved if due consideration is also given to other sectors of the economy and, indeed, to the well being of the population and the protection of the environment.

7 References

1. Environmental Resources Limited, Overseas Development Administration (1993) **Ecuador Mining Development and Environmental Control Project: Appraisal Mission.** Report. London, January 1993.

2. SENMA, Ministerio de Minería y Metalurgia, Swedish Environmental Systems (1993) **Bolivia: Sectoral Environmental Assessment of the Mining**

and Industrial (Manufacturing) Sectors. Final Report. La Paz, June 1993.

3. Environmental Resources Management, Overseas Development Administration (1993) **Environmental Project - Ventanas Region, Chile.** Final Report. London, September 1993.

Reconciliation of mining with the concept of sustainable development: Minas Gerais, Brazil

J. A. Porter
L. S. Santana
K. D. Culbertson
Design Workshop, Inc., Denver, Colorado, U.S.A.

Abstract

This paper uses project examples, primarily associated with large iron ore mines in Brazil, to demonstrate opportunities for the mining industry to go beyond traditional compliance and impact mitigation to proactively operate within the ethics of environmental and economic sustainability.
Keywords: Brazil, Mining, Sustainability

Introduction

Sustainability is the ability to meet the needs of the present without compromising the ability of future generations to meet their needs. This simple, straightforward, common-sense notion was first put forth in 1983 by the United Nations World Commission on the Environment and Development, known as the Brundtland Commission. Today the concept is transforming entire global industries, making it common to hear of sustainable agriculture, sustainable tourism, sustainable manufacturing, sustainable forestry and any other industries that involve the use of renewable natural resources. The concept of sustainability becomes far more difficult to apply to development of non-renewable resources, particularly mining, which can result in ultimate degradation to the economy, environment, and social conditions when markets dwindle or resources are depleted. Developing strategies for mining within the concept of sustainability is a great challenge facing the industry today.

The question is how to combine the ethics of sustainability with the seeming contradictory development of non-renewable resources. The answer is not to leave resources in the ground according to World Bank economist Herman Daly and theologian John B. Cobb, Jr., who write, "How can a community avoid the absurdity of leaving its non-renewable resources forever in the ground doing no one any good, yet not allow their exploitation to deflect the community from the path of sustainable development?"[2]

This paper describes several recent planning and development efforts in Latin America and the U.S. that offer insights into how the principles of sustainability can be applied to mining. As a preface, several aspects of sustainability are worth noting.

Sustainability defines an essential relationship between economics and the environment. It is a term of art used by both environmentalists and economists to describe long-term objectives. It is becoming increasingly difficult for industry to operate in a manner that is environmentally acceptable to the public at large. The recognition of economics and environment in a single popular movement provides an opportunity for industry to demonstrate that development can be responsive to public values.

Sustainability is an ethic that has potential application to every aspect of our business and professional lives. Current literature cites examples of applying sustainability at all conceivable scales, including world population control, national energy policy, household recycling, policies that products remain the property of manufacturers once they are used and discarded, low energy light bulbs, and the use of green construction materials, as demonstrated in Walmart's new experimental store in Lawrence, Kansas.

The pervasiveness of sustainability makes it easy to use as a public relations term without content. It is important to acknowledge that sustainability must be an ethic, not merely a buzzword.

One of the few concepts for sustainable development of non-renewable resources has been advocated by another World Bank economist, Dr. Salah El Serafy, who suggests that renewable resources become sustainable when converted into a perpetual income stream. Dr. El Serafy suggests that a portion of income from non-renewable resource development be considered capital and reinvested each year "to create a perpetual stream of income that would provide the same level of true income, both during the life of the resource as well as after the resource has been exhausted."[1]The portion of the annual receipts that are to be considered capital depends on the life of the resource and the return on the reinvested capital. Although Dr. El Serafy's model was developed for application to gross national product (GNP) accounting, the principles can apply to a region or individual mine and could be of more value to communities impacted by the development if applied locally. His approach has even greater potential if capital can be reinvested in profitable enterprises directed at environmental restoration.

MBR

The Brazilian mining company, Minerações Brasileiras Reunidas (MBR), is initiating programs and projects that move toward sustainable mining. MBR is the second largest iron ore mining company in Brazil with annual sales of nearly (US) $400 million dollars. MBR is controlled by CAEMI Mineração E Metalurgia, one of Brazil's most significant companies in the areas of mining and metallurgy.

MBR and CAEMI are dedicated to efficient production and profitability. At the same time, the companies view mining as a temporary use of the land and recognize the importance of protecting the environment and the need to reinvest in strategies that will provide a sound economy for the time when ore reserves are depleted.

Like many of the world's major mining companies, MBR has long been responsive to environmental regulations. MBR's motives are the same as many other companies and represent trends in the mining industry.

Entitlements - The company's reserves of 1.5 billion tons of iron ore provide an opportunity to mine for generations. It is difficult to imagine another 100 years of mining without serious industry initiatives in the area of environment, community and economic diversity.

Corporate Standing - Ninety percent of MBR's production is exported internationally to Asia, Europe, elsewhere in the Americas, and the Middle East. Maintaining MBR's position in the international market requires that the company be recognized globally as an industry leader.

Asset Value - MBR owns approximately 60,000 acres of land in the region immediately adjacent to the City of Belo Horizonte (population three million). The long-term economic and environmental health of the region will have a significant impact on the value of these assets and the potential business and development opportunities associated with them.

MBR currently operates three large iron ore mines, (Pico, Aguas Claras and Mutucca in the small region south of Belo Horizonte), and the long-term potential exists for five additional mines. The company has been operating in the area since starting the Pico Mine in the early 1940's. The existing mines currently produce 23 million tons of iron ore per year with plans to increase annual production to 35 million tons a year by the end of this decade.

The company is currently expanding its planning process to prepare a comprehensive plan for all of its properties in the context of a regional master plan.

The purpose of the plan is to identify what is required to insure that mining remains a compatible use as the region develops. Steps include identifying areas of the environment to be protected and restored, providing for new industry and business to lead to economic diversification and sustainable economy, and delineating a regional development pattern that fits the environment, is efficient to service, and that maintains a sense of community

Profit is obviously the motive for protecting the future of mining and developing diverse business opportunities. The assumption is that doing business within the context of a sustainable regional plan will insure mining's future.

The master plan envisions that urban growth will take place in existing and new satellite villages as an alternative to the suburban sprawl that encircles much of Belo Horizonte and which is beginning to encroach into the region where mining is concentrated. Two new villages are proposed on MBR lands: one at the Aguas Claras mine site and the other at Lagoa Grande, a site with excellent transportation access and a lake amenity. The plan assumes that the new village centers will benefit from Belo Horizonte's economic and cultural infrastructure, but each center will be developed around new business activity and services that provide a high degree of self-containment and employment.

The vast areas of land between villages is to be dedicated to open space uses including mining, renewable resource development, preservation of sensitive environments and projects dedicated to restoring landscapes and bio-diversity destroyed in the past.

Mining is generously provided for in the plan but considered a temporary use. Mine sites are programmed for secondary development after mining is completed, providing land for future community and industrial uses, on already disturbed sites.

Information is the foundation of the planning process. A significant amount of information has been gathered over the years by government and MBR in the process of exploration, engineering, operating, and preparing environmental impact statements. As typically is the case, little of this information is available in a format suitable for planning or anything other than its original use. The first step in the planning process has been to consolidate existing information resources into a Geographic Information System (GIS) to provide easy information retrieval and the ability to combine and analyze information from a variety of sources for use in planning, development and operations.

The MBR master plan GIS database was compiled and analyzed using GRASS 4.0 software running on a Silicon Graphics workstation. A level I landcover classification was generated from a SPOT (Aug. 1987) and a 1/4 scene Landsat TM (July 1990) image. Each image was classified separately by using a supervised classification technique. The database also includes 20 meter contour information, scanned from 1:50,000 scale IBGE topographic maps, digitized geologic information mapped by the USGS and Brazil Ministry of Mining and Energy, hydrology digitized from 1:50,000 scale IBGE topographic maps, and roads and political boundaries digitized from 1:25,000 scale topographic maps. Design Workshop continues to maintain the system, which will ultimately be transferred to MBR.

The initial data base has been established. The next step will be to add information that is constantly being gathered through continued exploration, engineering, operations and permit applications. The strategy is to establish information resource areas for each project that correspond to watershed boundaries allowing the regional information base to be updated by watershed as individual projects are researched and developed.

Lack of well documented information is not unusual, particularly in developing countries. Regardless of the availability and quality of information, it is important to build an information base as early in the process as possible, to promote comprehensive thinking and provide a basis for gathering additional information.

The initial regional plan was generated by computer by assigning mining, community development and environmental protection values to each information category. The plan provides a context for the design of individual projects and for operation decisions. It will be updated as new information is gathered or new values established. The information base and GIS provide the mechanism to constantly test development and operation scenarios and identify potential conflicts between development, resource and community values.

Aguas Claras - Mining will be completed at the Aguas Claras mine in the next seven to ten years. MBR has spent considerable time investigating future uses for the mine site. Plans have been developed to build a village on the property in a manner that is consistent with sustainable economic and environmental values.

Aguas Claras is an ideal location for a village because the site is immediately adjacent to Belo Horizonte, which is experiencing growth pressures. Aguas Claras Village is planned as a model for accommodating the growth demands of the region in an efficient community pattern with minimum impact to the environment.

The village at Aguas Claras is intended to be a prototype for community development in the region. The heart of the development is a village center that includes shops, restaurants, a recreation club, a conference center, a church and an elementary school, creating a nucleus for growth. There will also be an employment center that, with the village center, will provide services and jobs within walking and bicycling distance of homes. This will maximize transportation efficiency and reduce the need for trips to Belo Horizonte. The scale of the village, and the massing and architecture are designed in the tradition of colonial hill towns of the region such as Ouro Preto and Congonhas do Campo. Thus, local heritage is combined with contemporary building technology to achieve the best of both.

The village is designed so that buildings will create interesting urban spaces. Streets and pedestrian plazas are oriented toward views of scenic features of the region. Sustainability is promoted through design by the use of local materials for building construction. Building orientation and the transit efficiency of the village will conserve energy. The open-space system provides for large produce gardens for the village, and the slopes of waste areas will be used to test agro-forestry plantings established as part of the mine revegetation. The facilities used to purify water runoff from the mine will also protect Aguas Claras Creek from pollutants caused by village construction or urban run-off.

The transition from a mine to a village will occur while mining is being phased out. The steps include constructing the MBR office building (which has just been completed as the first building in the office park); configuring the pit and waste areas during the last phases of mining to better accommodate village development; modifying of the reclamation plan to provide a beautiful, but functional garden setting for development; converting the mine infrastructure to urban uses; and slowly filling the pit with ground water and surface run-off to create a lake which will ultimately be the central feature of the village.

Implementation

One MBR executive's initial response to the first work program for the regional plan was a simple: "It's not mining", end of discussion. This straightforward statement symbolizes the change required for a highly focused and efficient industry to redirect energies to the less secure, loosely defined, target of sustainability.

These changes are underway throughout industry on a variety of fronts. Industry played an active role in the United Nations Conference on Development and the Environment in Rio de Janeiro. For the first time, industry was an official participant through the Business Council for Sustainable Development.

Sustainability requires standards of performance beyond typical minimum regulatory requirements; companies are imposing environmental policies and standards on themselves and their projects. This is particularly important in developing countries where regulatory standards are yet to be developed or are not enforced. MBR and the other companies within the CAEMI group operate under self-imposed environmental policies which have been developed and are enforced by management of each company.

Trillium, a timber company based in Bellingham, Washington, is planning a sustainable forestry operation of a six-hundred-thousand-acre site on the southern tip of Chile. To govern the operation, Trillium has adopted a set of stewardship principles that addresses every aspect of sustainability including environmental quality, Chilean values, Chilean employment, reinvestment and philanthropy, planning, scientific measurement and reporting, and project ethics and open process. Trillium has retained an independent steward to promote and monitor compliance with the stewardship principles.

There is a trend toward private sector leadership in promoting and sponsoring the type of context planning represented by the MBR Regional Plan. This corporate activism is particularly evident in areas of the western United States, where the local government lacks the information, management and financial resources required to manage the complex social, environmental and economic problems that often accompany large development projects.

One of the most dramatic examples of this type of planning is taking place in Flathead County, Montana, where citizens have organized the Cooperative Planning Coalition to prepare a master plan and to implement a strategy to protect the quality of the environment while sustaining and diversifying an economy which includes mining, forestry, agriculture, and tourism.

Another example exists at Keystone, Colorado, where Ralston Purina, the owner of Keystone, is providing the type of information found in the MBR plan to the Summit County Planning Department to provide the basis for updating the Snake River Master Plan which provides the context for the development of Keystone.

Public planning activism is applicable to Latin America. The Ministry of the Environment for the Brazilian state of Minais Gerais is preparing an Area de Proteção Ambiental (APA) which is an environmental plan for the region in which MBR operates. Initially the MBR Regional Plan was prepared to be an internal company planning document. MBR has since made the GIS information base available to the APA process through the Institute of Brazilian Mining, IBRAM. The information has been used by the industry to demonstrate a comprehensive planning process that can include mining with other economic, community, and environmental issues. As a result, the APA process is currently being reorganized to be more comprehensive.

The private sector's role in public planning must remain in perspective. Ultimately, sustainable planning must be public. Industry must be a proactive participant in a process it does not control. It is possible for industry to identify and fund experienced boards of directors, or experts, to act on the public's behalf, as Trillium has done, but ultimately, the public must be empowered to control the process.

Conclusions

There is a trend toward technically based companies and professions adopting broader perspective of business opportunities and public responsibilities. Companies are looking beyond compliance and establishing their own environmental and community performance standards.

Mining companies have enormous information and management resources that can be contributed to the public process at little cost to the companies and great benefit to the public and the company.

Dr. El Serafy's concept of a perpetual income stream from reinvested capital has the most potential as a truly sustainable strategy for non-renewable resource development.

Mining companies should consider voluntarily and privately applying this concept to the regions where they operate. Such an approach is a more efficient and beneficial application of the concept than would be a government imposed sustainability fee or tax, as some have proposed.

The movement toward sustainability is real and will have a profound impact on the way business is done in the world. The mining industry cannot afford to let others define the relationship between sustainability and non-renewable resource development. There can no longer be a distinction between development and conservation. Pulitzer Prize-winning biologist Edward O. Wilson, internationally considered the dean of bio-diversity studies, writes, "Except in pockets of ignorance and malice, there is no longer an ideological war between conservationists and developers. Both share the perception that health and prosperity decline in a deteriorating environment."[4]

The sustainability movement not only provides an opportunity for business and industry but requires the full energies of the private sector to succeed. Paul Hawken writes "No other institution in the modern world is powerful enough to foster the necessary changes. Perhaps during the many battles between environmentalists and business people we have been asking the wrong question all these years. As generally proposed the question is 'how do we save the environment?' As ridiculous as it may sound to both sides, the question may be 'How do we save business?'"[3]

[1] Ahmad, Y.J., El Serafy, S., Lutz, E., 1989, Environmental Accounting for Sustainable Development, The World Bank, Washington, D.C., pp. 13.

[2] Daly, H.E. and Cobb, Jr., J.B., 1989, For The Common Good, Beacon Press, Boston, Massachusetts, pp. 73.

[3] Hawken, P., 1993, The Ecology of Commerce, 1st ed., HarperCollins Publishers, Inc., New York, New York, pp. 17.

[4] Wilson, E.O., 1992, The Diversity of Life, The Belknap Press of Harvard University Press, Cambridge, Massachusetts, pp. 282.

Design Workshop, Inc., Snake River Basin Database and Resource Maps, 1994

Design Workshop, Inc., Flathead County Master Plan Draft, 1994

DW/Santana Associados S/C Ltda, Aquas Claras, Brazil, An example of a Sustainable Environment, 1992.

Smelting and Refining

Arsenic removal in anode refining by flux injection

G. A. Riveros
Dirección Convenio I&D, Centro de Investigación Minera y Metalúrgica, Santiago, Chile
R. I. Salas
Dirección Convenio I&D, Centro de Investigación Minera y Metalúrgica, Santiago, Chile
J. A. Zúñiga
Dirección Convenio I&D, Centro de Investigación Minera y Metalúrgica, Santiago, Chile
O. H. Jiménez
Sección de Refinación y Moldeo, Fundición de Concentrado, División Chuquicamata, CODELCO–Chile, Chuquicamata, Chile

Abstract

Refining operations present an opportunity for impurity removal from molten copper. Anode copper at Chuquicamata Smelter contains, at certain times, levels of As which is inacceptably high for the subsequent step of electrorefining. Several tests were carried out to implement the flux injection technique as a standard practice in the anode furnace. Test results for arsenic removal from molten copper and flux utilization as a function of several operational parameters are analyzed. Special reference is made to the effect of Na_2O and CaO content in slag on the removal of arsenic and antimony. It is shown that the rate of removal content is dependent on the flux rate of injection into the copper. On the basis of the above, the flux injection rate is now calculated using an appropiate model.
Keywords: Copper fire refining, distribution ratio, flux injection, operation of anode furnace, removal of arsenic, removal of antimony, slag, soda-lime flux.

1 Introduction

Copper concentrates contain a number of minor elements among which As, Sb and Bi must be limited in the final product to ensure its commercial viability. By example, Arsenic forms Cu_3As particles in annealed copper and can reduce the electrical conductivity of the metal by 23 pct for an arsenic content of only 0.1 wt pct. This situation highlights the importance of separation and capturing of impurities during the smelting and refining processes.

The main impurities to be avoided in electrolytic copper are tellurium, selenium, bismuth, antimony and arsenic. In general, arsenic, antimony and bismuth represent the greatest problem during copper electrorefining, because they are electrochemically dissolved and tend to accumulate in the electrolyte, reaching levels at which bismuth, antimony or arsenic compounds precipitate forming slimes. These slimes can cause severe contamination in cathode copper.

Division Chuquicamata of CODELCO-Chile produces high purity cathodes, quality that has been maintained for years. Table 1 shows the typical composition of the cathodes produced. The existence in the Chuquicamata mine of a zone rich in arsenic means that in certain periods an increase of this impurity is observed in the anode copper. In order to maintain the high purity CCC brand cathodes it was neccessary to implement a flux injection technique for impurity removal in the Chuquicamata Smelter as a standard practice.

This paper presents the more recent tests carried out at the anode furnace in order to improve the removal of arsenic from molten copper by flux injection. This industrial research project was jointly developed by the Chuquicamata Smelter and CIMM.

2 General information on the process

The Chuquicamata Smelter belongs to the mining complex called Chuquicamata Division, which is part of the National Copper Corporation of Chile (CODELCO-Chile). The main smelter furnace installations at Chuquicamata are one Outokumpu Flash Furnace, two Teniente Converter (the second Teniente Converter began operating in October 1993), one reverberatory furnace, six Peirce-Smith Converters, six anode furnaces and three casting wheels. The last reverberatory furnace is planned to be shut down in 1994 and the number of Peirce-Smith converters will be reduced by two. Thus, future smelting-converting equipment will consist of one Flash Furnace and two Teniente Converters plus four Peirce-Smith Converters.

The main product at this smelter is copper anodes. The production capacity is about 45,000 metric ton per month. A specifications of copper refining equipment are shown in Table 2.

Table 1. Cathode copper minor element analysis at Chuquicamata Refinery

Element	Content (ppm)
S	12
Te	5
Pb	2
As	2
Sb	2
Bi	1
Se	1
Te	1

Table 2. Specifications of copper refining equipment

Equipment	Number	Specification
Anode Furnace	4	240 t/cycle molten copper capacity Rotary type, 4.0 m-Diam x 9.1 m-Length
	2	380 t/cycle molten copper capacity Rotary type, 4.6 m-Diam x 9.1 m-Length
	2	tuyeres furnace, 38.1 mm ID each. Heavy fuel oil for heating Heavy fuel oil + steam/kerosene + steam for poling
Anode Casting Machine	3	Capacity : 60 t/h Wheel 12.8 m-Diam Moulds 24 Weight of anode for commercial cell: 382 kg Weight of anode for stripper cell: 382 kg

2.1 Flux addition in copper fire refining

The process to eliminate impurities from molten copper through basic flux additions has been applied in Chile for several years to produce fire refined copper (FRC)[1]. Moreover, during the period 1980-1990, CIMM in cooperation with the CODELCO-Chile and ENAMI smelters performed a number of experimental studies, both at the laboratory and the industrial scales, on the thermodynamic and kinetics of the transfer of arsenic and antimony to the slag through molten copper oxidation in the presence of basic oxides[2][3][4]. This was the basis for developing an industrial procedure for capturing arsenic and antimony during fire refining by flux injection[5][6][7]. A review of the advances on copper fire refining process in Chile is described in another publication in preparation[8].

2.2 Technological improvements on flux injection at Chuquicamata Smelter

A flux injection system was implemented in the area of copper fire refining to operate alternatively in the anode furnaces Nº 1 and 2. This pneumatic conveying system was designed to introduce a flow of fluxes together with the air through one tuyere of each anode furnace. In order to make the injection of fluxes into molten copper uniform, a suitable control system was prepared to prevent clogging of the pipe and tuyere. Other improvements were made to control the size of the particles, below 10 # Tyler and the coupling mechanism between the transportation pipe and the tuyere.

Schematic lay-outs of the anode casting system and the injection system are shown in Figure 1 and 2, respectively.

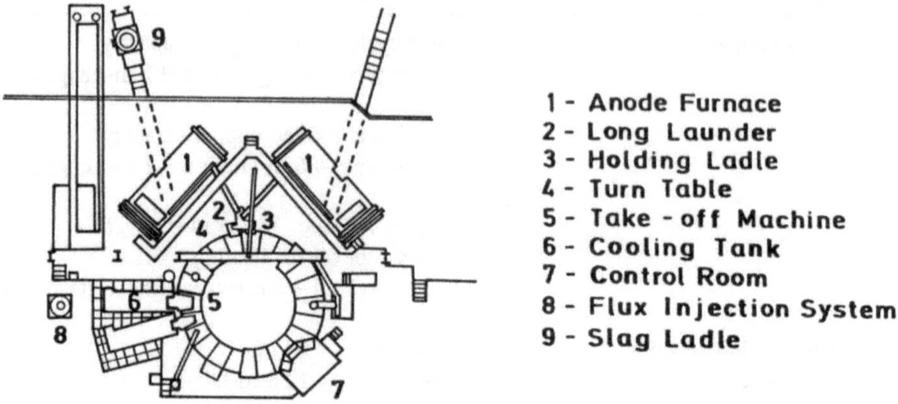

Fig. 1 Lay - out of Anode Casting System

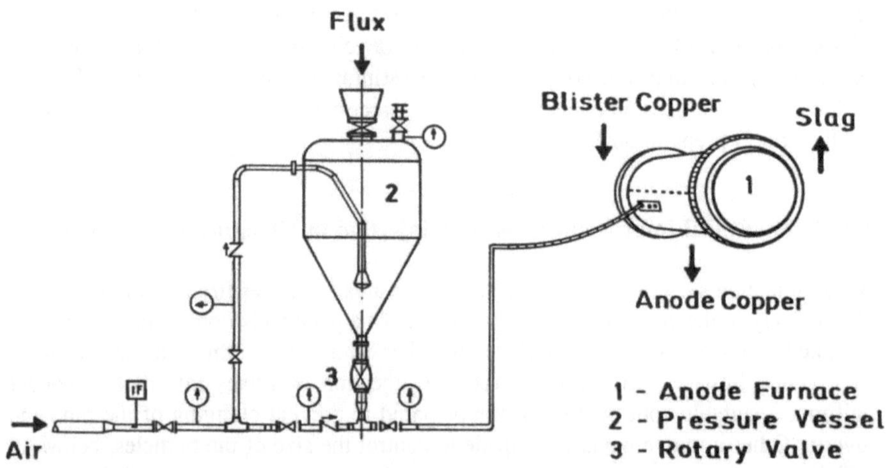

Fig. 2 Schematics of Tuyere Flux Injection Equipment

3 Operation of anode furnace

3.1 Charging of molten copper

Molten copper from the converters is charged into the 240-380 ton-anode furnaces by means of 9.2 m^3 ladles. One cycle operation of anode furnace corresponds to two cycles of converters.

3.2 Oxidation

Air (1000 Nm^3/h at 600 kPa) is blown for 0.3-1.5 hours into the molten copper through tuyeres and the sulfur is removed. The end of oxidation is controlled by visual inspection of copper samples.

3.3 1st slag skimming

Slag is skimmed through the central mouth by blowing air (at 300 kPa) into the molten copper through two lances which are located on both side of the central mouth. It requires about 0.5 hours to skim about 10 tons of slag which is returned to the smelter. It must be noted that about half of the slag is maintained in the furnace to protect the refractory from corrosion during the flux injection and to facilitate the second slag skimming.

3.4 Flux injection

Fluxes are injected (at a pressure of 400 kPa) into the molten copper through one of the tuyeres when the level of oxygen in copper reaches about 0.7%. It requires about 20 minutes to inject pneumatically 1600 kg of fluxes.

3.5 2nd slag skimming

The slag formed is skimmed through the central mouth following the same practice as before, taking special care that no slag is left in the furnace. The effectiveness of this final skimming prevents the reversion of the impurities to the metal. This operation requires about 15 to 30 minutes.

3.6 Poling

Poling is done by kerosene or heavy fuel oil mixed with steam which is blown into the molten copper through tuyeres. The flow of kerosene is about 14 l/min. The flow is controlled in order to maintain high reduction efficiency and proper off-gas temperature. It takes about 1.5 hours for poling. Overall reduction efficiency is about 60%. The Chuquicamata Smelter's oxygen content in anode for the Electrorefinery is lower than 0.2%.

4 Industrial flux injection experience

Industrial data for arsenic and antimony extraction from molten copper in rotary furnaces by the injection of soda ash and lime (or limestone) fluxes through tuyeres or lances is given in Table 3. It must be noted that the tuyere injection data are more recent and represent an optimized operational scheme, compared with the former lance injection refining. It can be seen in Table 3 that tuyere injection is characterized by larger tube diameter, lower injection pressure and higher inmersion depth. This results in somewhat lower air flowrates and much lower tuyere velocities. Finally, in the case of injection through tuyeres, injection time is longer, in spite of the use of about half the flux dosage than in the case of lances. However, since no reaction time was found necessary after injection when using tuyeres, total process time is from one fifth to one third of that utilized for lance injection, with very similar As and Sb removal efficiencies. Thus, the operation with tuyeres at optimized conditions resulted in large reductions in flux usage and process time, with comparable efficiencies in spite of increased initial concentrations of As and Sb, and represents the most favorable process results obtained to date.

By comparing the present data with former results obtained in reverberatory furnaces[1][9], it can be seen that by injecting a flux mixture formed by 50% Na_2CO_3 and 50% CaO at a dosage of around 6.5 kg/ton of copper, removals of 91-96% for arsenic and 61-62% for antimony are obtained. These results are similar to the former, with the advantage that flux injection required only 20 to 40 minutes of process time, as opposed to 300 to 350 minutes for flux addition over the bath in a reverberatory type furnace. The process may thus be considered to have been optimized for the given conditions. Other improvements are possible and will probably be achieved as continued experience is acquired.

4.1 Effect of the slag composition

Fluxing of impurities implies the transfer of the undesirable elements from metal to slag. An important criterion in the process is the distribution coefficient of the impurity between slag and metal, which is defined as follows[6]:

$$Dm = \frac{(\%M)_{slag}}{[\%M]_{copper}} \qquad (1)$$

where (%M) and [% M] are the concentrations of M in slag and metal, respectively. This distribution coefficient is primarily a function of the degree of oxidation of the system, measured by the oxygen content of the metal $(\%O)_{copper}$ or by the equilibrium oxygen pressure p_{O2}, and the activity coefficient of the impurity oxide MO in the slag, γ_{MO}, which is controlled by the chemical composition of the slag.

In a former publication[6] the effective slag/metal distribution coefficients for arsenic and antimony in the lance injection tests were correlated with three slag

Table 3. Comparison of Industrial Flux Injection Data through Lances and Tuyeres.

PARAMETERS			Lance Injection	Tuyere Injection – Test Number 4	16	19
Test Conditions:						
Weight of Copper	t		200	271	270	260
Number of tuyeres or lances			1	1	1	1
Tuyere Internal Diameter	cm		2.54	3.81	3.81	3.81
Depth of Inmersion	cm		50	70	70	70
Bath Temperature	°C		1180	1180	1200	1200
Injection Pressure	kg/cm^2		5.2	3.9	4.1	4.1
Tuyere Velocity	m/s		170	65	55	80
Air Flowrate	Nm^3/min		5	4.2	3.7	5.4
Flux Rate	kg/min		140	47.4	80	44
Solid/Gas Ratio	kg/kg		21.7	8.7	16.8	6.3
Na_2CO_3/CaO Ratio			37/63*	30/70	50/50	50/50
Flux/Copper Ratio	kg/t		13	6.7	6.5	6.9
Initial Concentrations	ppm,	As	900	1100	1300	1120
		Sb	70	95	70	145
		O	7300	7000	7050	7500
Injection Time	min		18	38	22	41
Reaction Time	min		98	-	-	-
Test Results:						
Final Concentrations	ppm,	As	20	45	120	40
		Sb	20	45	26	55
Extraction	%,	As	98	96	91	96
		Sb	57	53	61	62
Slag Weight	t		13	13	10	-
Slag Composition	%,	As	6.1	7.6	6.8	4.5
		Sb	0.12	0.09	0.09	0.21
		Cu_t	35.8	31.5	27.0	38.2
		Cu_{OX}	19.1	-	-	-
		Y_{Na2O}	0.22	0.19	0.19	0.19

* Flux used was Na_2CO_3-$CaCO_3$ mixture

composition parameters: X_{Na2O}, X_{CaO} and % Cu_{OX}, where X_{Na2O} = %Na_2O/(%Na_2O + %CaO + %SiO_2), X_{CaO} has a similar definition and % Cu_{OX} is the oxide copper content of the slag.

The tuyere injection data were found to correlate much better when modified composition variables for soda and lime in the slag were chosen, namely Y_{Na2O} = %Na_2O/(%Na_2O + %CaO + %SiO_2 + %FeO) and Y_{CaO} = %CaO/(%Na_2O + %CaO + %SiO_2 + %FeO), which take into account the total iron content of the slag expressed as FeO. In order to make valid comparisons, the older lance injection data were reanalyzed in terms of the Y variables. The linear regressions of log D versus Y_{Na2O}, Y_{CaO} and %Cu_{OX} for arsenic and antimony in both types of flux injection tests resulted in the following equations:

Lance Injection

$$\log D_{As} = 6.937\ Y_{Na_2O} + 1.177\ Y_{CaO} + 0.00804\ \%Cu_{OX} + 0.2804 \qquad (2)$$

$$r^2 = 0.93$$

$$\log D_{Sb} = 2.766\ Y_{Na_2O} + 0.986\ Y_{CaO} + 0.02646\ \%Cu_{OX} + 0.1769 \qquad (3)$$

$$r^2 = 0.89$$

Tuyere Injection

$$\log D_{As} = 7.238\ Y_{Na_2O} + 1.593\ Y_{CaO} + 0.00669\ \%Cu_{OX} + 0.6578 \qquad (4)$$

$$r^2 = 0.82$$

$$\log D_{Sb} = 3.110\ Y_{Na_2O} + 0.917\ Y_{CaO} + 0.00082\ \%Cu_{OX} + 0.6660 \qquad (5)$$

$$r^2 = 0.80$$

The experimental data for tuyere injection are shown in Figure 3, where the effective distribution coefficients are plotted versus Y_{Na2O}. Correlation lines are also shown in this Figure for both tuyere and lance injection. The lines correspond to the regression equations (2) to (5) given above, where average values of Y_{CaO} and $\%Cu_{OX}$ are replaced, and only Y_{Na2O} is retained as a variable. It should be noted that the fit of the data to the equations is better than is apparent from Figure 3, since as just mentioned, in this Figure the data poins are plotted as a function of just one variable, while the equations contain three independent variables. The r^2 values of 0.80 to 0.93, on the other hand, indicate a good to excellent fit.

From the correlation equations it may be seen that the effect of Na_2O on the arsenic distribution coefficient is the largest, with a seventh power dependence of D_{As} on Y_{Na2O}, while CaO shows only a 1.2 to 1.6 power relationship.

In the case of antimony, the corresponding power coefficients are about 3 for Na_2O and about 1 for CaO. In laboratory experiments[10][11] it had been found that, while Na_2O based slags were more effective to eliminate arsenic, slags with a higher CaO/Na_2O ratio were to be preferred to extract antimony. However, the industrial results for antimony, both in lance and tuyere injection tests, show that CaO has a very moderate effect on Sb elimination, even less than on arsenic, suggesting that for maximum impurity removal a flux with high soda content should be used. Notwithstanding this, there are several other considerations in choosing the most adequate flux formulation, such as: a) cost: soda ash is usually much more expensive than lime; b) slag fluidity, which is increased by soda and decreased by lime; and c) refractory life, which is decreased by the use of high soda slags that may be quite aggressive to the furnace refractory lining, especially at the slag line. The 50/50 Na_2CO_3/CaO flux mixture has been found quite adequate at the industrial scale, when all the above considerations are taken into account, with very good efficiencies, as seen in Table 3.

The effect of oxide copper content in the slag on the distribution coefficients are rather low in all cases. This may be taken as an indication that the oxidation was well controlled, that is, it was maintained within the range where it is close to optimum and, therefore, not very important in the final results.

4.2 Effect of flux rate and injection time

The behavior of soda-based slag is considerably more complex than silicate-based slags because of the instability of the carbonate ions. The presence of acidic oxides, such as SiO_2, Sb_2O_5 and As_2O_5 at high temperature decompose the soda by the formation of $2(Na_2O)\cdot SiO_2$, and $3(Na_2O)\cdot Sb_2O_5$, $3(Na_2O)\cdot As_2O_5$, respectively[10]. Furthermore, in the presence of lime the formation of similar compounds for As and Sb, $3(CaO)\cdot Sb_2O_5$ and $3(CaO)\cdot As_2O_5$, might be expected.

For stoichimetric calculations the removal reactions of As and Sb are:

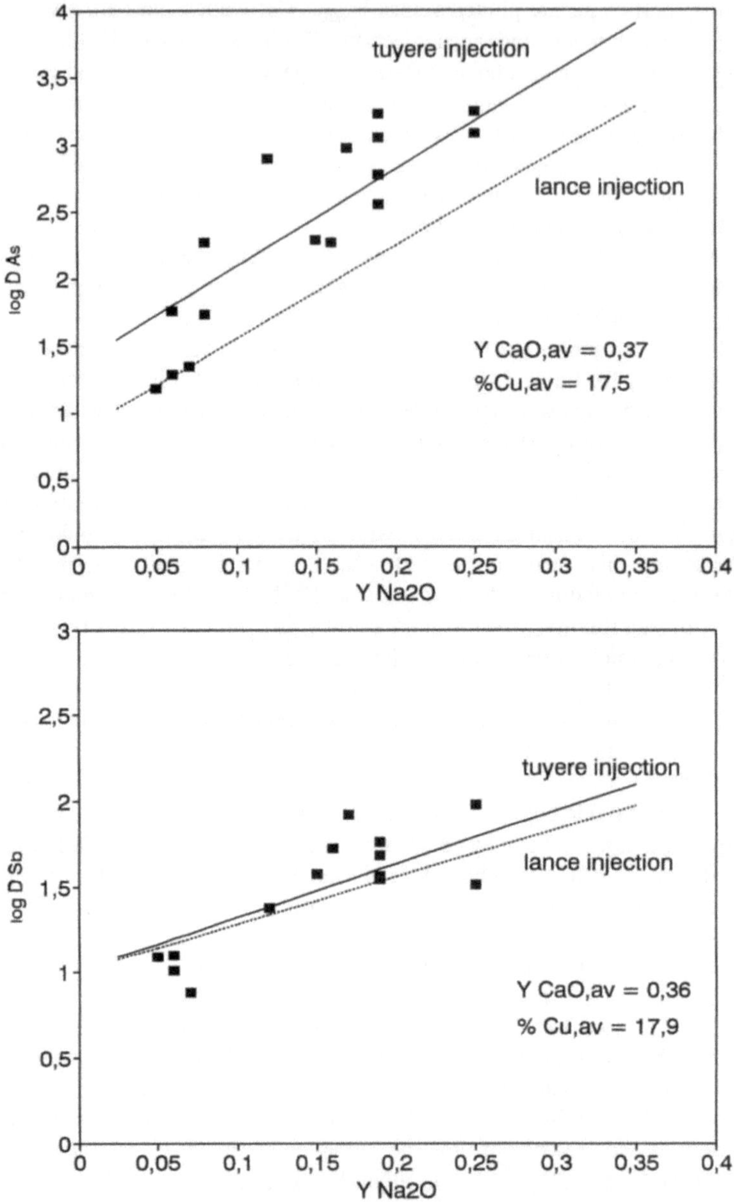

Figure 3. Arsenic and antimony distribution coefficients in lance and tuyere injection tests versus soda content of the slags.

$$3Na_2CO_3 + 2\underline{M}(As, Sb) + 5\underline{O} = 3(Na_2O) \cdot M_2O_5 + 3CO_2 \quad (6)$$

$$3CaO + 2\underline{M}(As\ Sb) + 5\underline{O} = 3(CaO) \cdot M_2O_5 \quad (7)$$

The amount of fluxes necessary for the process depends on the efficiency of utilization which is influenced by the injection conditions[6] and the amount of remaining slag in the furnace.

Figure 4 shows the behaviour of arsenic and antimony in molten copper as a function of time for two melts (Test Nº 4 and 16) where different flux rates were used. It is worth mentioning that these lines do not represent exactly the extraction kinetics of these impurities, because they are only based upon the values at the begining and at the end of the oxidation and injection periods. It is possible to see in this figure that the reaction rates increase with both arsenic and antimony content when the flux rate is increased. Unfortunately, the maximum capacity of the flux injection equipment was only 80 kg/min, so it was not possible to test over this value. Other tests carried out at lower flux rates showed the same pattern as can be seen from test Nº 19 (see Table 3) compared with test Nº 4.

From these tests, for the period May 1 to July 25, 1992, when a special program was conducted to follow As and Sb in anode furnace[12], it was concluded that the injected flux particles are melted instantly and react with the dissolved elements as they rise upwards to the top of the metal bath. This reaction pattern was considered more important than the permanent contact mode, through the slag-metal interface, because no reaction time was found necessary after injection.

Based upon this information, the slagging of impurities can be modelled assuming that there is no volatilization of the impurity and that during the refining process the variation of the metallic batch weight is negligible. The model and associated softwarc, called PYROREF, presented in user-friendly manner, provide the tool to calculate the flux neccessary before each injection. The flux dosage for As is calculated using the following parameters:

$$F_r = \frac{W_{Cu} \cdot (C^i_{As} - C^f_{As})}{E_{As} \cdot t_i} \quad (8)$$

where, F_r is the flux rate, kg/min; W_{Cu} is the weight of copper, t; $C_{As}{}^i$ and $C_{As}{}^f$ are the initial and final content of arsenic in the metal, respectively, ppm; E_{As} is the amount of arsenic extracted from the metal, kg/t of flux added; and t_i is the time of injection, minutes. E_{As} is an adjustable parameter which depends mainly on the injection system: flux dosage and injection pressure, the slag composition parameter,

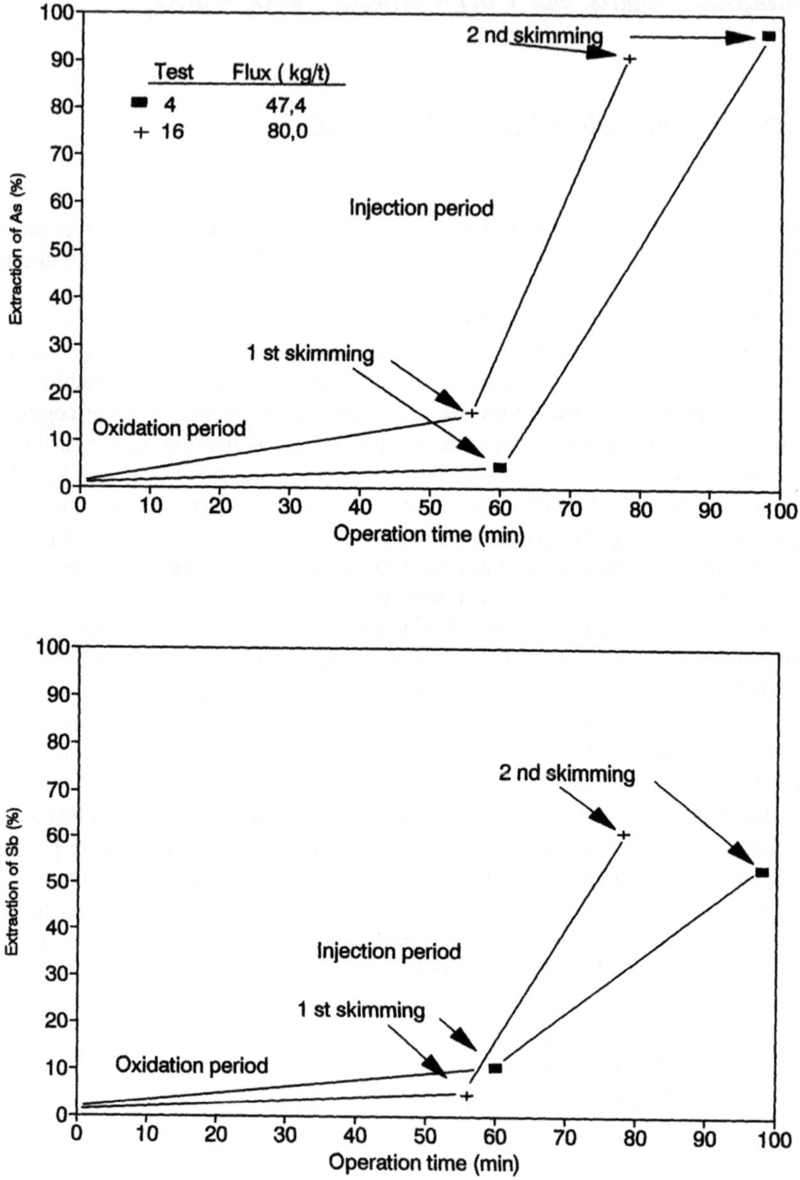

Figure 4. Arsenic and antimony extraction versus operation time.

(for $Y_{Na2O} > 0.15$), and oxygen content of molten copper, (for $\underline{O} > 5000$ ppm), and must be determined experimentally.

The above correlation was used to estimate the amount of flux required for reducing As to a target level for a specific operational condition. To do so, the initial As concentration in copper must be estimated before starting the injection.

At present, the anode furnace model of the PYROREF program is complete, and it has been adequately tuned to industrial results at Chuquicamata Smelter.

5 Conclusions

Based upon industrial scale tests the following conclusions can be drawn:

1) Plant tests have resulted in high flux utilization efficiency with short injection time. This is the result of a good control of the test conditions and good prediction of the required amount of fluxes injected.
2) Under industrial conditions, with oxygen dissolved in molten copper over 5000 ppm, and when more than stoichiometric amount of flux is used, the reaction products can be assumed to be $3(Na_2O)\cdot As_2O_5$, $3(CaO)\cdot As_2O_5$, $3(Na_2O)\cdot Sb_2O_5$ and $3(CaO)\cdot Sb_2O_5$.
3) The test performed showed that the addition of 6.5 kg of flux per tonne of copper (Na_2CO_3 50% + soda ash 50%) leads to 91-96% arsenic extraction and 61-62% antimony extraction in 20-40 minutes of processing time.
4) The effective slag/metal distribution coefficient for arsenic and antimony depends on Na_2O and CaO contents in slag.
5) The reaction rates of arsenic and antimony increase with increasing the rates injected under the metal bath.

6 Acknowledgements

The authors gratefully acknowledge permissions to publish this paper from the Chuquicamata Division of CODELCO-Chile and CIMM.

7 References

1. Basagoitía, A. (1974) As and Sb eliminations in copper fire refining. IIMCh, Santiago, Chile, pp. 219-231.
2. Devia, M. and Luraschi, A. (1984) Kinetics of arsenic and antimony elimination by slagging in copper refining. TMS Symposium on metallurgical slags and fluxes, Nevada, USA, pp. 643-648.
3. Takeda, Y. and Riveros, G. (1986) Equilibria among sodium carbonate, sodium ferrite slag and liquid copper. Metallurgical review of MMIJ, Tokyo, 3(3), pp. 60-71.
4. Riveros, G., Péndola, E. and Luraschi, A. (1988) Consideraciones básicas en el diseño de sistemas de transporte e inyección de sólidos para procesos metalúrgicos no ferrosos. CONAMET 88, Santiago, Chile, pp. 371-382.

5. Bassa, R., Del Campo, A. and Barría, C. (1988) Copper pyrorefining using flux injection through tuyeres in a rotary anode furnace. Copper 87, 4, Santiago, Chile, pp. 149-166.
6. Riveros, G., Gajardo, P. and Luraschi, A. (1988) Developments in fire refining of copper-flux injection into molten copper. Copper 87, 4, Santiago, Chile, pp. 239-254.
7. Riveros, G., Luraschi, A., et al., (1989) Procedimiento para la remoción de As, Sb y otras impurezas mediante inyección de fundentes en hornos de refino. Chilean Patent No. 37 876.
8. Riveros, G. and Luraschi, A. (1994) Advances in the copper fire refining process in Chile. To be presented at TMS Annual Meeting, 1994.
9. Luraschi, A., Riveros, G. and Urra, F. (1982) Puesta en operación del proceso de refinación a fuego con horno basculante. CIMM Report P-173 for El Teniente, Santiago, Chile.
10. Riveros, G. (1984) Investigación sobre la distribución de As y Sb en sistemas líquidos escoria-cobre. Procs. IV Congress of Mining Eng., Copiapó, Chile, pp. 413-422.
11. Recaval, H., Del Campo, A. and N. Santander, (1984) Estudio de la etapa de oxidación del refino a fuego mediante coeficientes de distribución escoria-metal. Procs. IV Congress of Mining Eng., Copiapó, Chile, pp. 443-454.
12. Riveros, G., Salas R. and J. Zúñiga, (1992) Suministro de tecnología de pirorefinación de cobre y eliminación de arsénico en horno de anodos. CIMM Report Nº 4, P-5008 for Chuquicamata, Santiago, Chile.

Scrap melting in a rotary furnace with use of oxy-fuel burners

Mario Castillo P.
CODELCO–Chile, División Chuquicamata, Chuquicamata, Chile

Abstract

This paper discusses technical and economic aspects of rotary furnace scrap melting using technical oxygen. Increasing fuel prices and inefficient energy use in the traditional smelting processes, are good reasons for trying to decrease the energy consumption and reduce operating costs. Other factors, such as melting rate, operating efficiency, productivity and flexibility of operation are factors that need improvement and thus favor modifications of the reverberatory furnace, vertical furnace and Peirce Smith converters (PSC). Rotary furnace specially equipped with oxy-fuel burners can lead to important advantages for the Chuquicamata smelter.

. Facilitate the heat transfer mechanism.
. Operational improvement.
. Productivity advantages.

Based on the aforementioned aspects, this scrap smelting option is currently more efficient and easier to implement than traditional equipment and methods. This report reviews and discusses general aspects of the new methods proposed, to be eventually used at Chuquicamata smelter.
Keywords: Casting capacity, experimental design, hand flame, oxy-fuel burners, rotary furnace, scrap, technical oxygen.

1 Foreword

The copper anode refining scrap generation is close to 20% of the tank-house's anode capacity. The resulting secondary copper scrap must be reprocessed due to its marketing difficulties and/or other difficulties such as handling.

This secondary copper scrap is melted and recovered as electro-refined grade-A cathode. The recycling process involves cost of handling, remelting, etc., and calls for some alternative solutions to minimize this cost, such as:

1. Use of the reverberatory melting furnace. This vessel is thermally inefficient and with high processing costs (roughly US$ 50/ton of secondary copper).
2. Use of vertical Asarco shaft furnace. This furnace is a continuous smelting unit at a relatively low cost (approximately US$ 40/ton of secondary copper).
3. Peirce Smith converters (PSC). This option is used worldwide because it is a cheap way and uses the excess heat developed during the copper blow. It is a very attractive way from an economical standpoint. It bears a cost not exceeding US$ 20/ton of scrap.
4. The proposed modified rotary furnace using oxy-fuel burners. The option in evaluation considers that three (PSC) cannot smelt all the scrap generated at the plant.

The application of this new technology is very interesting and can be applied at the Chuquicamata smelter.

2 Analysis of this proposed method

1. Use of the oxy-fuel burners provide proper control of the combustion flame, because it is very easy to switch from an oxidising flame to a reducing one, once the luminosity and convective effects are achieved.
2. Proper modifications of the loading-mouth to receive scrap to speed-up the charging steps.
3. Correct choice of insulating refractory to avoid heat losses in order to maximize the fuel efficiency.
4. Automatic control of the process parameters. Proper instrumentation to monitor the performance of the oxy-fuel burner cooling water system to avoid the explosion hazard (water leak).
5. Othe relevant aspects in connection with scrap hauling, handling, etc.

3 Discussion

Some relevant theoretical aspects

The equipment's energy consumption behavior is mainly defined by the implementation of the equation of the heat conductivity.

$$q_c = {}^{-}K \nabla \Delta T \qquad (1)$$

Where $q_c = \theta_1$ dependent area, insulation type operating temperature. Besides, it is noteworthy that radiation induced heat loss is defined by

$$q = C.A(E_1(T_1/100)^4 - L_1(T_1/100)^4) \qquad (2)$$

Where $q_r = \theta_2$ dependent on exhaust gases area, operating temperatures absorption and emissivity factors.

Proper control of function in the rotary furnace reveals significant differences among conventional methods and equipment.

4 Performance test

A test was carried out to determine the economic efficiency and flexibility of the proposed rotary furnace. Fifty tests were performed in a rotary furnace (4,0 meter wide, 7,6 meters long with a 3,6 m^2 loading month). These tests were planned under experimental design to figure out the value of three factors: Fuel consumption, scrap quantity and molten copper temperature. During the tests approximately 10.000 tons of secondary copper scrap were smelted. Since the start of this test, a total of 30.000 metric tons of scrap have been treated.

The following table summarizes the most relevant results obtained:

Table 1. Relevant results

Fuel comsumption	44 ± 2	(kg Enap-6/t Cu)
Oxygen comsumption	136 ± 11	(kg Oxygen/t Cu)
Smelting rate	19 ± 1	(ton Cu/hour)
Smelting cost	27 ± 3	(US$/ton copper)
Energy consumption	462 ± 21	(M Cal/ton copper)

The following diagram shows secondary copper smelting in a tilting furnace using oxy-fuel burners:

Figure 1. Rotary furnace cross-sections view

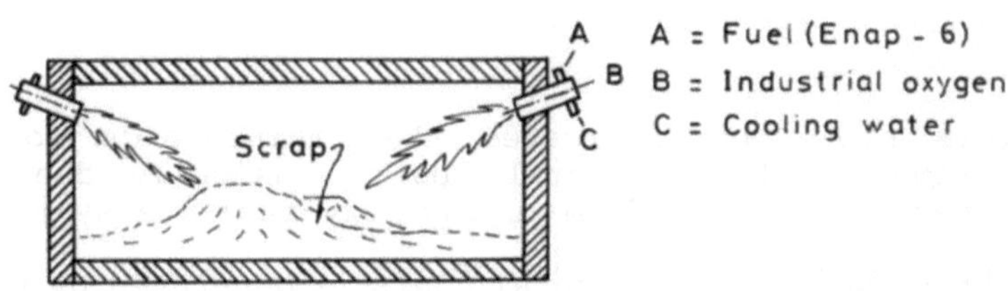

This process results in an energy breakdown shown in the following pie chart:

Pie chart 1. Energy breakdown

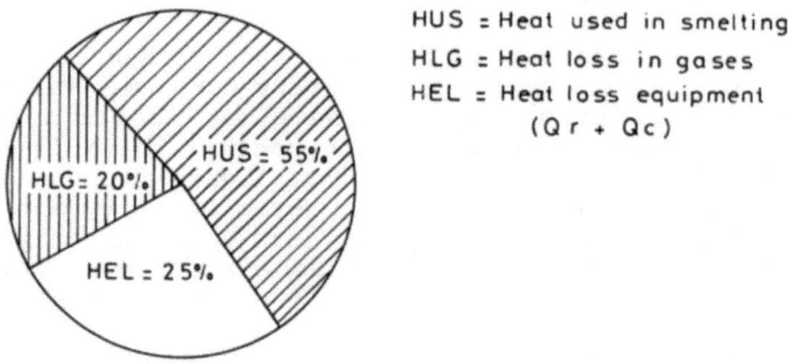

Pie chart 2 breaks down the costs involved in this process:

Pie chart 2. Cost breakdown

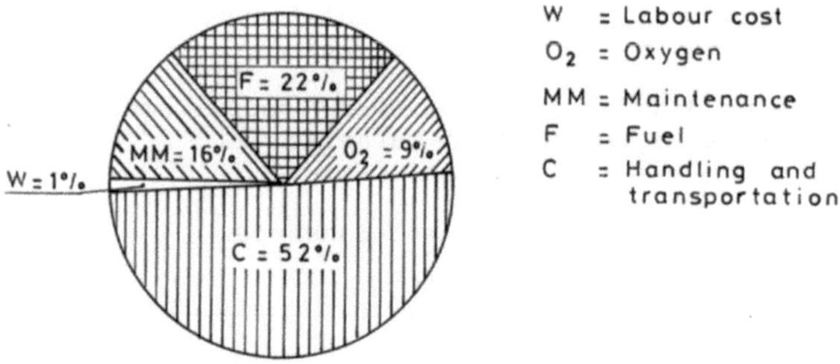

This pie chart indicates that maintenance costs are approximately 16% of total costs, basically due to the advantages derived from using oxy-fuel (easy-to-handle flame).

Fifty two percent of total costs is generated by variable casting expenses (handling and transportation). This is because the system is based on the concept of not properly using the refining and casting capacity (per utilization of equipment and workmanship).

Finally, 32% of the cost is generated by oxy-fuel and labour. Related expenses are in connection with the rotary furnace.

5 Conclusions

From the operating standpoint, the scrap smelting in a rotary furnace using oxy-fuel burners is affected by material handling problems that are common in the smelter.

The Chuquicamata smelter production will be increased to approximately 250 tons of copper per day by proper utilization of the refining and casting equipment.

6 Advantages of the proposed rotary smelting furnace

- Improvement of the energy efficiency: 55% of the energy input is used in smelting secondary copper scrap.
- The above mentioned energy efficiency leads to cost savings ranging between US$ 10 and US$ 20 per ton of copper, cheaper than the option of the Asarco shaft furnace and refine reverberatory furnace.
- Modifications to the rotary furnace are aimed at the best control of the vessel's heat losses, due to conduction and radiation.
- The operator enjoys great freedom regarding to production schedule. There is an increment in the smelting rate, sending more molten copper to the casting furnace.
- Increased smelter productivity by proper use of the casting capacity.
- The oxy-fuel burner provides a suitable flame for the furnace size and refractory life and less gas emissions to the environment.

7 Acknowledgement

The author wishes to thank to Chuquicamata Division of CODELCO-Chile for the permission to publish this paper.

8 References

1. Mc Adams Williams H., "Conducción de calor", **"Transmisión de calor"** 3ra. Edición, Imprenta Nacional S.A., Madrid España (1964).
2. Perry Robert, Green W., Moloney James O., "Combustión", **Manual del Ingeniero Químico**, 6ta. Edición (3ra. Edición en español), Editorial Mc Graw Hill (1992).

3. Vian Angel, Ocon Joaquín, "Radiación de calor", **Elementos de Ingeniería Química** (Operaciones Básicas), 3ra. Edición, Editorial Aguilar, Madrid, España, (1961).

Ausmelt's top submerged lance technology applied to copper smelting

K. R. Robilliard
W. E. Short
G. A. Guorgi
B. R. Baldock
Ausmelt, Ltd., Dandenong, Victoria, Australia

Abstract

Ausmelt has developed a process that allows the production of blister copper from copper concentrates, using either one or two furnaces depending on the scale of the operation.

Ausmelt's top submerged lance technology has been applied to numerous smelting processes including the recovery of most non-ferrous metals, and iron production. Based on its revolutionary top entry submerged lance, the application of Ausmelt technology to the production of blister copper introduces significant advantages in capital and operating costs, and in the environmental, safety and hygiene aspects of smelting.

The flexibility of the Ausmelt furnace system allows it to operate on a batchwise or continuous basis, depending on the scale of the operation. The capital cost savings and ability to operate in batch mode, allow operations using the Ausmelt process to become potentially economic projects at small scales. Continuous operation is generally applicable at larger scales.

The Ausmelt furnace system can also process complex copper concentrates to produce blister copper with low impurity levels, that will meet the requirements of a refinery.

This paper provides an illustration of Ausmelt technology applied to a hypothetical copper smelting operation, outlining the process route, the capital cost, operating requirements and other potential advantages available using this revolutionary technology.
Keywords: Ausmelt Technology, Lance, Furnace, Top Submerged Lance Smelting, Blister Copper Production, Complex Concentrate Processing.

1 Introduction

Ausmelt technology is a top submerged lance process and has a wide range of applications in the non-ferrous and ferrous metals industries. Over many years Ausmelt has performed successful laboratory and pilot plant scale studies on the recovery of copper and precious metals from concentrates of various compositions. These studies have shown that the production of blister copper is feasible using the

Ausmelt system. Moreover, use of Ausmelt technology has environmental benefits. The compact furnace system virtually eliminates fugitive off-gas emissions and this results in satisfactory plant hygiene. The slag produced contains such low levels of copper and precious metals that it is suitable for use as landfill or in cement manufacture without further processing or cleaning.

The technology also offers economic advantages. By submerging the lance tip in the liquid furnace contents, fuel combustion and process chemical reactions are localised within the bath. The turbulence in the bath results in rapid heat transfer, mass transfer and reaction rates which have economic benefits with respect to fuel requirements and plant throughput. The furnace's compact size and enclosed design mean that furnace heat losses are minimised. Oxygen enrichment of the process gases allows for low fuel consumption and in some cases autogenous converting is possible.

2 Ausmelt Copper Concentrate Smelting Technology

Submerged lance combustion technology is particularly suitable to the process of smelting copper concentrates because of the intense stirring and controllable oxidation of the sulphides that occurs in the bath. The products are blister copper, discard slag and a sulphur collection product such as gypsum from dual alkali scrubbing, elemental sulphur or sulphuric acid. Copper recoveries are high since correct fluxing and furnace operation control copper levels in discard slag to less than 0.6%. The slag requires no further treatment and is suitable for disposal as landfill or use as an aggregate in construction materials.

Ausmelt technology was first used to produce blister copper from concentrates in a single-furnace commercial scale operation in 1984. [1]

Central to Ausmelt technology is the top entry lance system. The concept is illustrated in figure 1. Ausmelt's specifically designed, latest generation lances provide efficient fuel combustion, as well as allowing for the reactions required in the various regions of the furnace. The stainless steel lance is not consumed and is protected from the furnace contents by a coating of frozen slag.

During operation, the tip of the lance is submerged below the static slag level of the furnace contents. Injecting the process gases deeply into the bath creates very turbulent conditions. Turbulence promotes very rapid reactions and the smelting capacity is high for the relatively small size of the furnace. Excellent contact between the bath and the process gases gives high fuel efficiency.

Combustion air, fuel, oxygen and afterburn air are injected through the lance. The technology can use any of a range of fuels including oil, LPG, natural gas or coal.

Enrichment of process gases to oxygen levels of up to 50% can be used with specially designed lances. Oxygen enrichment reduces fuel requirements and the volume of off-gas generated. The capital cost associated with gas handling equipment is therefore reduced. Oxygen enrichment also increases the concentration of sulphur dioxide in the flue gas, improving the operation of the acid plant.

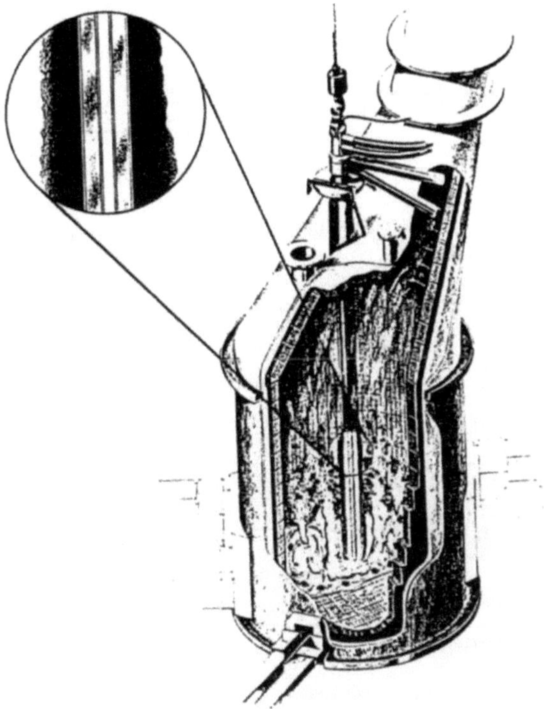

Fig.1 A Schematic Section through an Ausmelt Furnace

Fuel and fine materials can be conveyed down the lance and injected below the bath surface. Coarse, wet or lightly agglomerated wetted fine feed materials are dropped into the furnace through a feed port and are rapidly incorporated into the bath on the turbulent slag surface.

The furnaces are usually constructed as refractory-lined cylinders. Water cooling is used to ensure an acceptable period between vessel relines.

The simplicity of the Ausmelt system's configuration and operation has been discussed in previous publications.[2] [3] [4]

In the production of blister copper Ausmelt uses either a one or two furnace system, the choice is based on the required products and the scale of operation. For operations smelting greater than 100,000 tonnes per annum of concentrates, producing blister copper, a two-furnace operation is typical. Smaller smelters use a single furnace. If the required product is copper matte, a single furnace is used. This is the case for existing smelters upgrading their operations but retaining the conventional Peirce-Smith converter route.

2.1 Continuous Two Furnace Operation

The process flowsheet for smelting copper concentrates in the Ausmelt two furnace system is shown schematically in figure 2.

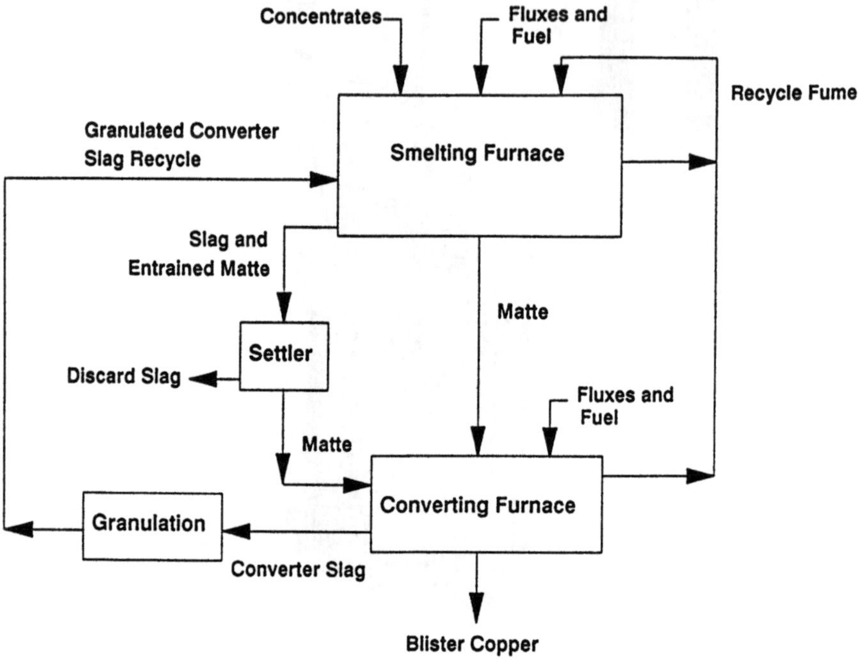

Fig.2 Copper Smelting Flowsheet

Concentrates, recycled materials, silica flux and limestone are mixed and agglomerated using a pelletiser or pugmill. The material is then fed to the furnace via a transfer conveyor. The level of moisture in feed is determined by the binding properties of the concentrates and fluxes. Drying or roasting of the feed is not required.

The smelting furnace operates between 1250 and 1300°C, producing copper matte, containing approximately 60% copper, and discard slag containing less than 0.6% copper. The copper losses in slag are attributed to entrained matte and oxidised copper. The high degree of turbulence in the smelting furnace sustains low levels of copper oxidation according to the matte-slag equilibrium conditions. Effective control of slag chemistry keeps losses of copper dissolved in slag to a low level.

Matte and slag exit the smelting furnace in separate systems. The slag continuously overflows into a launder and flows to a settling furnace to allow separation of any entrained matte. The temperature of the settler is maintained by oil or gas burners in the free-board region above the molten bath. Molten slag enters through an opening located in the roof and off-gas ducting connects into the gas handling equipment.

Matte tapped from the settler is transferred to the converting furnace. Slag is tapped from the settler and granulated. Further cleaning or treatment of the slag is not required.

Matte produced in the smelt stage is either granulated and fed to the converting furnace or transferred molten by launders. Transfer of solid matte has advantages in simplifying matte handling, while molten matte transfer provides advantages in reducing fuel requirements and ease of converter operation. The converting furnace can be designed to operate with either feed arrangement depending on the application.

Operating the Ausmelt furnace system as described above avoids spillage and therefore the quantity of reverts produced is negligible.

The Ausmelt converting furnace produces blister copper by selectively oxidising the iron and sulphur components of the matte to the slag and gas phases respectively. Limiting slag oxidation, and therefore magnetite formation, is achieved with flux and coal additions. The converting temperature is controlled at 1300 to 1350°C. Furnace design will usually allow for autogenous operation of the converting stage and adding recycled scrap anode metal is possible.

Blister copper can either be tapped continuously or batchwise from the furnace without interrupting the converting operation. The metal product is suitable for further processing in an anode furnace and electrolytic refinery.

In view of the high oxidation rates present in the converter, and the fact that the slag phase is in contact with the metal, high levels of copper exist in the converter slag. This slag is therefore granulated, de-watered and returned as feed to the smelting furnace for further copper recovery.

Process gases are extracted from the top of the furnace through a flue off-take. The continuous streams of gases from the smelting and converting furnaces are rich in sulphur dioxide and require cleaning before discharge. At these larger scales the gas handling system typically consists of a waste heat boiler and electrostatic precipitator for each furnace. The waste heat boilers cool the gases to approximately 360°C and produce steam that can be used for heating purposes or to run a steam turbine for electricity generation.

The electrostatic precipitators collect carry-over feed and fume generated from the volatile components of the feed, either for recycling or further treatment. Ausmelt designs the feed handling equipment associated with the smelter and, when correctly operated, the amount of carry-over feed can be kept as low as 1% of the feed rate. This small amount does not cause an energy burden in the furnace. Fume and dust do however act as a sink for lead, zinc and other volatile materials in the feed. Levels will build up to a point where a bleed of this material will be required to maintain low levels of contamination of the blister copper.

After the electrostatic precipitator, the gases are drawn through an induced draft fan, combined and ducted into a desulphurising unit before being discharged to atmosphere via a stack. Both streams are expected to contain sulphur dioxide levels between 8 and 15%. Various alternatives exist for reducing the level of sulphur dioxide in the process gas streams to comply with local environmental discharge regulations.

At large scales the production of sulphuric acid is often economically attractive, depending on the proximity of a suitable market. Alternatively, should acid production be uneconomical, a scrubbing unit for gypsum manufacture is possible. If the plant is located at a mine site then the gypsum produced can be discarded with mill tailings or sold.

2.2 Single Furnace Operation

At smaller scales a single furnace operation is possible. In this case the process is as described for the two-furnace operation, except, it is operated in batch mode. During the smelt stage discard slag is tapped intermittently. Concentrate smelting is continued until the furnace matte capacity is reached. Feeding of concentrate is then stopped and matte batch-converted to blister copper. The slag produced in the converting stage remains in the furnace after tapping the blister copper and copper recovery takes place in the subsequent smelting stage.

Alternatively, matte can be granulated and stockpiled. The furnace is then operated on a smelting campaign followed by a campaign of matte converting. Converting stage slag is granulated and recycled for copper recovery in the smelt campaigns. The method of operation used depends on the specific requirements of particular projects.

The smallest scale of operation likely to be economic is production of between 5,000 and 10,000 tonnes per year of blister copper; however, this will be highly dependent on specific project factors such as transport costs, concentrate grade, precious metal content and consumable costs. For a small scale operation, a waste heat boiler is not likely to be justified. The gas handling system at smaller scales typically consists of an evaporative gas cooler, baghouse and sulphur dioxide scrubber.

2.3 Complex Concentrates

The Ausmelt furnace system can also be used to process complex copper concentrates containing elements considered undesirable in conventional copper circuits. Bismuth, arsenic, lead and zinc can be recovered to a fume that can be sold or further processed for metal recovery.[5] The blister copper produced from the Ausmelt furnace system contains suitably low levels of these impurities to meet the requirements of a refinery.

These results are achieved in the Ausmelt furnace system because of the vigorous gas flushing conditions, the ready control of process conditions and low dust carry-over.

Complex concentrates are typically available at a lower price than clean, high grade concentrates. Consequently, long term supply contracts under favourable conditions are possible. Plant design needs to include the process steps necessary to achieve these multi-metal separations in order to provide the required capacity and product quality assurances.

3 Environmental Benefits In An Ausmelt Copper Smelter

The Ausmelt system is designed to produce minimal impact on the local environment. The furnace is a well-sealed unit; the escape of gases and fume is controlled by ensuring that a constant negative pressure is maintained. Positive pressure to the furnace only occurs during abnormal operations. Each furnace is relatively small in size and is tightly sealed. These aspects of furnace design allow for minimisation of fugitive gas emissions, maintaining a clean and safe working environment in the furnace area.

All Ausmelt plants are designed to satisfy local safety standards and working practices. This includes the wearing of safety clothing where necessary (head and eye protection, respirators, safety footwear and non-flammable clothing). Operators tapping the furnace also wear leather or woollen jackets, aprons, gloves and gaiters.

3.1 Particulate Collection

Draughted hygiene hoods at the slag weirs and transfer launders minimise gaseous emissions. A dedicated hygiene baghouse can be used to collect the fume and dust from the hygiene hoods, or they can be ducted into the gas handling equipment associated with the furnaces.

3.2 Dust Handling

As previously discussed, fume and dust generated in the Ausmelt furnace system are collected, either in a baghouse or electrostatic precipitator, depending on the scale of the operation, off-gas volume and cooling method used. If the dust/fume generated in both stages of smelting and converting are recycled, the dusts are dropped onto a sealed conveyor and taken to the feed handling and preparation plant where the dust is agglomerated along with the concentrates and fluxes to be fed to the smelting furnace. If the dusts are not recycled, they are transferred by sealed conveyor to their place of treatment or stockpile.

4 Typical Copper Smelting Example

The following illustration describes an Ausmelt plant that will produce 60,000 tonnes per annum of blister copper containing more than 97% copper, and discard slag containing less than 0.6% copper. For this production rate a two-furnace system is required. Process flows to and from the furnace are provided in figure 3.

The system will process the copper concentrates as follows;

- smelt copper concentrates, fluxes, revert materials, recycled converter slag and reductant coal to produce a medium grade copper matte and a slag containing less than 0.6% copper which will be discarded, and
- convert the matte to blister copper, producing a slag that is recycled to the smelt stage.

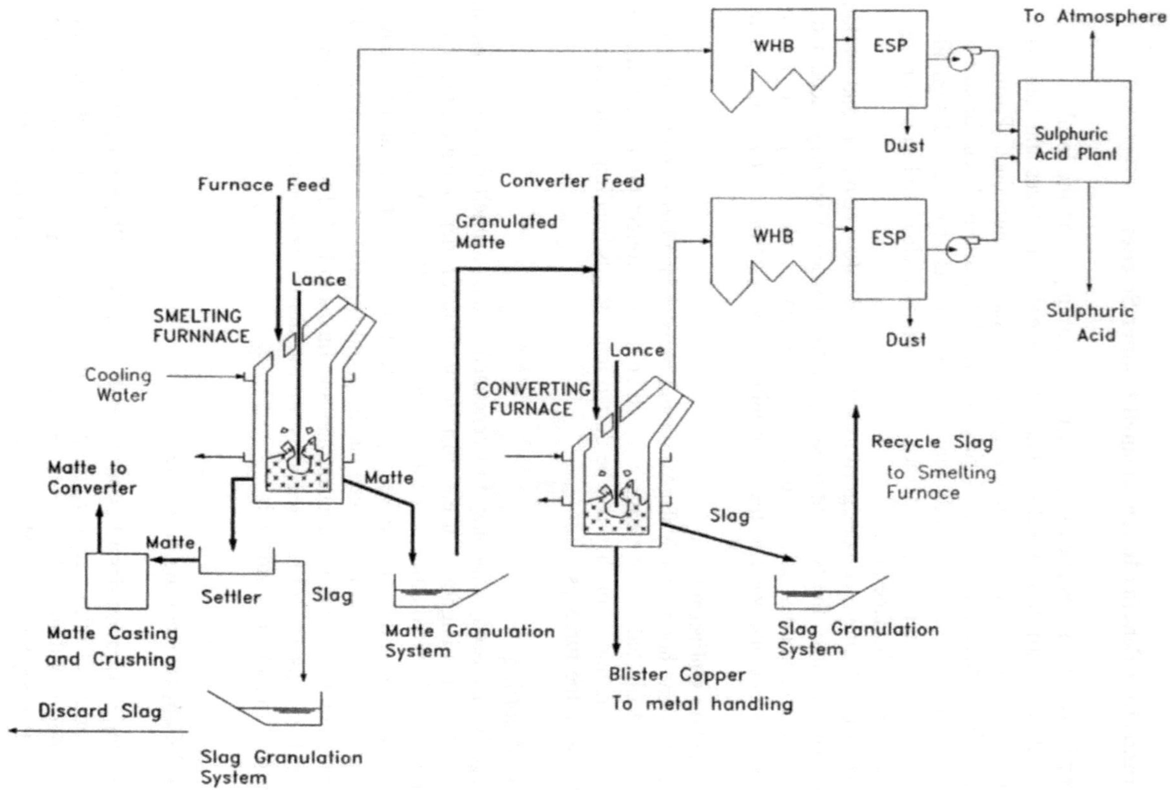

Fig.3: Process Flows To and From the Ausmelt Furnace

Following is preliminary data for the required furnace system. The data provided in this paper is an indication only.

The composition of the concentrate in this illustration is provided in table 1. A typical coal containing 5% ash has been assumed as fuel.

Table 1. Concentrate Assay

Element	Copper Concentrate (%)
Cu	20
Fe	30
S	30
SiO_2	5
CaO	2
Al_2O_3	1
MgO	1

At this scale Ausmelt recommends the use of oxygen enrichment of the process gases to minimise the size and cost of equipment, including the smelt furnace, off-gas ducting, waste heat boiler and electrostatic precipitator. In this example the smelter is designed to use a 40% oxygen level in injected process gases. The furnace off-gas compositions are estimated in table 2.

Table 2. Furnace Off-Gases

	Smelt Furnace	Convert furnace
Volume (Nm^3/h)	45,000	25,000
Temperature (°C)	1250	1250
Analysis (Vol %):		
N_2	65	80
H_2O	10	5
CO_2	10	5
SO_2	15	10

It is assumed that at this scale the production of sulphuric acid is economically justified. The combined gas stream to the sulphuric acid plant will be approximately 70,000 Nm^3/h at 300°C and will contain 12 to 13% sulphur dioxide.

Each furnace has a mild steel shell and is lined with high quality chrome-magnesite refractory bricks. The smelting furnace has an external diameter of 5 meters and a total height of 12 meters. The converting furnace is a similar size. Both furnace shells are water cooled. Hot water is collected in a sump and cooled before returning to the furnace.

There are three openings in the top of each furnace; a feed port, lance entry port and a sample/inspection/standby burner port. Water cooled copper blocks are fitted at the tapholes in the side of each furnace, near the base of the shell, and above the maximum matte level for slag tapping.

The smelter will have two gas handling systems, one for each furnace. Each system will include a waste heat boiler, electrostatic precipitator and fan. The gas streams will be combined and treated in a single sulphuric acid plant before discard.

The estimated capital cost of this smelter is approximately US$60 to 70 million. This estimate includes all the equipment required for smelter operation as described above.

In order to operate the process optimally the smelter will require;

- Fluxes — supply of 40,000 tonnes per annum of silica and 30,000 tonnes per annum limestone fluxes to control slag chemistry.
- Fuel Supply — 20,000 tonnes per annum of pulverised coal as fuel to the smelter.
- Air Supply — compressed air for injection into both furnaces to provide agitation of the contents. Compressed air is also required in both furnaces for afterburn of volatiles above the bath.
- Reductant Supply — 20,000 tonnes per annum of lump reductant coal.
- Water Supply — 100,000 tonnes per annum of water is required at the plant boundary and will be distributed to the furnace shell cooling systems, slag and matte granulation systems and to the gas cooling systems.
- Electricity Supply — the total installed electrical load of the plant is approximately 10 MW, excluding the power requirements of the waste heat boilers and sulphuric acid plant.
- Labour — labour requirements for the smelter depend on the integration or separation of labour from other facilities on the site. In addition to normal supervisory, metallurgical, engineering and administrative overview, the labour required in Australia for each shift would be;

6	Furnace Operators
2	Feed Operators
1	Front End Loader Driver
1	Fitter
1	Analyst
2	Security

A boilermaker/welder and an electrician is required on day shift and to provide 24 hour coverage when needed.

5 Conclusions

Top submerged lance technology has been proven to be ideal to produce blister copper from concentrates of various compositions. The environmental impact of the operation is minimised as the slag produced requires no further processing before discard and fugitive gas emissions are reduced because of the compact, sealed furnace design.

Recoveries of metal to matte or blister copper are high since feed carry over is minimised and discardable slags are achieved. Treatment of concentrates with high levels of heavy metals is also possible using the Ausmelt process because of the ability to remove these elements to fume.

6 References

[1] Anderson, L.E., Floyd J.M., Lightfoot B.W. and Muller R. (1985) Smelting of Olympic Dam copper concentrates using Sirosmelt Technology. TMS-AIME Fall Extractive Metallurgy Extractive Meeting on **Complex Sulphides; Processing of Ores, Concentrates and By-Products,** San Diego, pp 69-76.

[2] Floyd, J.M. (1992) Sirosmelt - The Emerging Role of New Bath Smelting Technology in Non-Ferrous Metals Production. The Minerals, Metals & Materials Society. **Savard/Lee International Symposium of Bath Smelting,** Quebec, pp 103-123.

[3] Floyd, J.M. (1992) Sirosmelt/Ausmelt Technology - Australian Innovation at the Cutting Edge. **Mervyn Willis Symposium and Smelting and Refining Course,** Melbourne, Australia, pp 13.1-13.51.

[4] Floyd, J.M. and Short W.E. (1993) Ausmelt Development of Top Submerged Lance Technology. Presented to: IMM London **High Intensity Pyrometallurgy,** London, United Kingdom.

[5] Swayn, G.P., Robilliard K.R. and Floyd J.M. (1993) Processing Complex Copper Smelter Dusts Using Ausmelt Technology. Publication of **The Journal of The Minerals, Metals and Materials Society**, pp 35-38.

Investigation of silver extraction from a silver-sulphur flotation concentrate from the electrolytic zinc plant of Companhia Paraibuna de Metais, Juiz de Fora, Brazil

A. L. Almeida
Companhia Paraibuna de Metais, Juiz de Fora, Minas Gerais, Brazil
W. J. Bruckard
CSIRO Division of Mineral Products, Port Melbourne, Victoria, Australia
G. J. Sparrow
CSIRO Division of Mineral Products, Port Melbourne, Victoria, Australia
J. T. Woodcock
CSIRO Division of Mineral Products, Port Melbourne, Victoria, Australia

Abstract

A silver-sulphur flotation concentrate produced from the hot acid leach residue in the electrolytic zinc plant of Companhia Paraibuna de Metais was investigated for hydrometallurgical silver extraction. This concentrate assayed 8320 g/t Ag, 20.6% Pb, and 43.7% S. It consisted mainly of lead sulphate and elemental sulphur with numerous minor components. The form of the silver was not precisely established and several forms appeared to be present.

Cyanidation of the concentrate for 48 h extracted only 42% of the silver with high cyanide and lime consumptions. A thiourea leach extracted only 24% of the silver, partly because silver re-precipitation occurred. It was evident that some form of pretreatment was needed to make the silver amenable to leaching.

Pressure oxidation with sulphuric acid at 200°C dissolved 56% of the silver in 4 h provided an initial acid strength of 90 g/L was used. Pressure oxidation with 190 g/L nitric acid at 125°C for 1 h dissolved 58% of the silver. It was found that the temperature needed to be above the melting point of sulphur (118°C) to achieve these extractions.

Acid brine leaching at 125°C, using 300 g/L NaCl, extracted over 98% of the silver in 1 h. Operation below the melting point of sulphur gave much lower extractions.

By conditioning with a solution containing 5-30 g/L NaCl and 36 g/L H_2SO_4, less than 1% of the silver was dissolved, as expected, but cyanidation of the resultant solid product extracted about 93% of the silver present. However, cyanide and lime consumptions were high, even when the pellets of elemental sulphur formed during conditioning were first removed.

Results of the work indicate that acid brine leaching or brine conditioning followed by cyanidation are the preferred options for extraction of silver from the flotation concentrate but optimisation studies are required.

Keywords: Acid pressure leach, brine conditioning, brine leaching, Companhia Paraibuna de Metais, cyanidation, electrolytic zinc plant concentrate, silver extraction, silver-sulphur flotation concentration, thiourea leaching.

1 Introduction

Companhia Paraibuna de Metais (CPM) operates an integrated electrolytic zinc plant at Juiz de Fora, Minas Gerais, Brazil, to produce cathode zinc as zinc ingots and a variety of other products. Feed to the plant is comprised of zinc flotation concentrates sourced from Peru, Canada, Mexico, and Brazil.

These concentrates contain silver and it is estimated by CPM that 60 per cent of the silver in the plant feed reports in a silver-sulphur flotation concentrate produced from a hot acid leach residue as indicated on Fig. 1. This concentrate assays about 8000 g/t Ag and recovery of the silver in a saleable form is attractive to CPM.

This paper reports and discusses the results of a preliminary study of some hydrometallurgical routes for silver extraction from the concentrate. These routes were

1. direct cyanidation,
2. thiourea leaching,
3. oxygen pressure leaching with sulphuric acid,
4. hot oxidative leaching with nitric acid,
5. acid brine leaching using high brine levels to dissolve the silver and lead, and
6. conditioning with low brine levels to convert the silver to silver chloride and extraction of the silver chloride by cyanidation.

2 Experimental

Experimental work involved characterisation of the concentrate received, and treatment by the six hydrometallurgical routes noted above, together with associated sampling and assaying.

2.1 Sample preparation

The silver-sulphur flotation concentrate used in this work was air-freighted from Brazil to Melbourne and received no heat treatment during quarantine clearance in Australia. The 20 kg sample was air-dried, brushed through a 10 mesh screen (aperture 1.651 mm), and riffled into 50 g, 100 g, 400 g, and 500 g lots. One 100 g lot was pulverised for analysis.

2.2 Cyanidation

Cyanidation tests were conducted on 500 g samples of the original concentrate and on the residue from brine conditioning with 20 g/L NaCl.

The dry solids were pulped with distilled water to a liquid:solid (L:S) ratio of about 2:1 by weight and sufficient powdered lime added in small increments to give a stable pH in the range 9.5-10.0. Stock cyanide solution (25 g/L NaCN) was added to give 0.05-0.15% NaCN w/v in solution and the L:S ratio was adjusted to 3:1 by weight. Leaching was then conducted in a rolling bottle with free access of air for 24 h or 48 h. Periodic checks were made of lime and cyanide strengths and any appropriate additions made.

After agitation, the pulp was weighed to check the final L:S ratio on which calculations were based, and then filtered to recover the pregnant solution. This solution

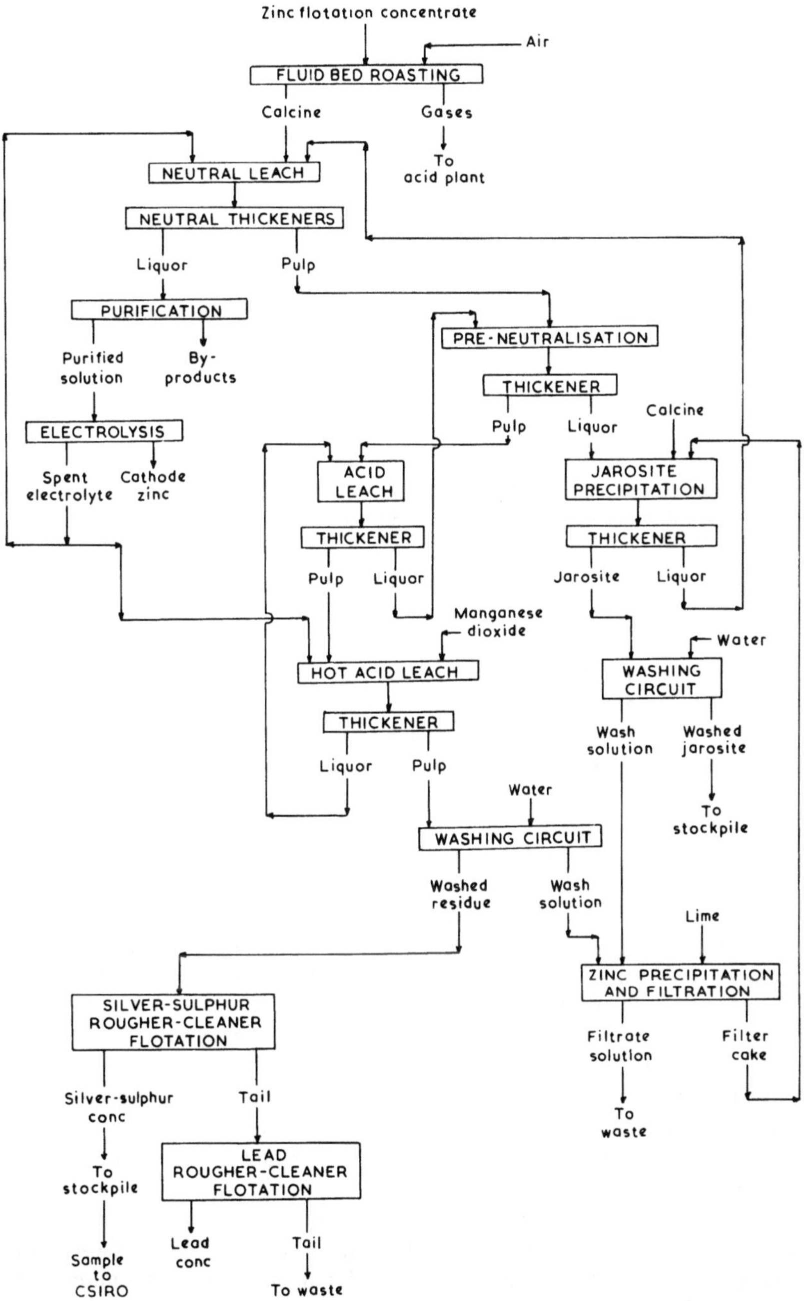

Fig. 1. Simplified schematic flowsheet of the electrolytic zinc plant of Companhia Paraibuna de Metais, Juiz de Fora, Brazil, showing the production route of the silver-sulphur flotation concentrate used in this work.

was assayed for cyanide, free CaO, silver, zinc, copper, and iron. The filter cake was washed by repulping with tap water and refiltering and then dried, weighed, and assayed for silver, zinc, copper, iron, and lead.

2.3 Thiourea leaching

Thiourea leaching tests were conducted with 400 g samples of the concentrate at an L:S ratio of 3:1 by weight for 5 h under controlled pH and redox potential and partial control of thiourea concentration. A pH-stat, consisting of a pH electrode (a combined glass electrode and a calomel reference electrode) and a Radiometer PHM82 auto-titrator system, was used to control pH by addition of 2.5 *M* sulphuric acid. Redox potential was controlled with an Eh-stat, consisting of a platinum/calomel electrode pair and a Radiometer PHM82 auto-titrator, by addition of 10 volume hydrogen peroxide (100 volume stock solution diluted 10:1). All redox potentials given in this paper are reported as the potential of a bright platinum electrode relative to a saturated calomel electrode (SCE). Thiourea levels were maintained at 0.4% by taking samples at 1 h and 2 h, titrating for thiourea, and then adding sufficient thiourea, as a slurry in water, to return its concentration to the desired level.

Solution samples were taken at 1, 2, and 5 h for silver determination. After leaching, the slurry was filtered to recover pregnant solution for assay. The filter cake was washed by repulping with water and refiltering. The washed cake was dried at 70°C, weighed, sampled, and assayed.

2.4 High temperature acid leaching

Most high temperature acid leaching tests were conducted with sulphuric or nitric acid in a 1 L titanium autoclave with an overpressure of oxygen or nitrogen. One test was conducted under reflux conditions at 108°C in a Quickfit glass flask.

For the autoclave tests the start-up procedure involved adding a 50 g or 100 g sample of concentrate to the autoclave followed by 500 mL of the lixiviant, which was either sulphuric or nitric acid of the desired strength or an acid brine solution (5-30 g/L NaCl or 300 g/L NaCl with 0 or 36 g/L H_2SO_4). A surfactant solution (1% w/v) of calcium lignosulphonate (10 mL) and Quebracho ATO (5 mL) was added to all tests to give 0.2 g/L and 0.1 g/L respectively in the pulp to inhibit molten sulphur from occluding unreacted sulphide particles (Doyle et al. [9]). Agitation was commenced and air in the autoclave purged with nitrogen at 700 kPa pressure to limit oxidation prior to the desired operating temperature being reached. When the operating temperature was reached the nitrogen overpressure was released through a condenser to prevent loss of water and the autoclave flushed several times with oxygen. The desired oxygen overpressure of 700 kPa was then applied and the leach commenced. The oxygen overpressure was taken as being the total pressure of the system less the steam pressure at the operating temperature.

The temperature was maintained within ±2C° of the desired value during the leach by an electric heating coil. For the more exothermic reactions, especially those with nitric acid, the outside of the autoclave was cooled with an air blast; temperatures were then maintained within ±5C°.

At the end of the leach the autoclave and contents were cooled with an air blast and the pulp withdrawn by suction. The pulp was filtered using a buchner funnel and

Whatman papers and the filtrate refiltered through a 0.5 μm Millipore membrane filter to obtain a clear solution for analysis.

The filter cake was washed three times with 200 mL lots of distilled water and then air-dried. Dried residues were brushed through a 14 mesh screen (aperture 1.168 mm), sampled, and analysed.

2.5 Acid brine leaching and conditioning

Acid brine leaching, using 300 g/L NaCl plus 0 or 36 g/L H_2SO_4, was conducted in the autoclave, as noted above, at 125° and 200°C for periods ranging from 10 min to 4 h. A separate test was conducted at 108°C for 4 h under reflux. The 10 min leach was achieved by simply heating the charge to temperature at 125°C and then cooling over a 45 min period. The products from these tests were treated as noted previously.

In some tests, in which the brine solution contained only 5-30 g/L NaCl and 36 g/L H_2SO_4, the aim was to condition the solids at 125°C with acid sodium chloride solution rather than to leach the silver and lead. These tests are therefore termed conditioning tests. In these conditioning tests, pellets of elemental sulphur were formed. These were removed by wet screening the residue on a 250 μm aperture screen, and the pellets and residue prepared separately for analysis. For one set of conditions, ten replicate conditioning tests were conducted and the screen undersize fractions combined for cyanidation.

2.6 Metal analyses

Leach liquors from all tests were diluted as necessary with an appropriate solvent and analysed for Ag, Zn, Cu, and Fe by atomic absorption spectroscopy (AAS). For cyanide leaches water was used for dilution. For acid leaches (sulphuric, nitric, thiourea), 0.1 *M* H_2SO_4 was used. For acid brine leaches, 3 *M* HCl was used to avoid precipitation of lead or silver, but this was not always successful with respect to lead. No lead assays are therefore reported on solutions.

For analysis of feed and residues, it was found preferable to take a weighed aliquot of a sample and first heat it in a muffle furnace at 500°C for 1 h to burn off the elemental sulphur present. The resultant product was then fused with sodium peroxide, and the melt taken up in concentrated hydrochloric acid and made up to volume for AAS. Because of the presence of elemental sulphur, especially in the concentrate received and some residues, some samples were too reactive for direct fusion with sodium peroxide as a prelude for analysis.

Gold was determined on the concentrate by digestion of a sample with aqua regia, extraction of the gold into methyl isobutyl ketone, and an AAS finish.

2.7 Miscellaneous analyses

Total sulphur in feed was determined using an induction furnace combustion technique with an infra-red spectrophotometric determination of the sulphur dioxide evolved. Sulphate sulphur was determined gravimetrically after extraction with sodium carbonate solution, acidification, and precipitation as barium sulphate. Elemental sulphur was determined gravimetrically after extraction with carbon disulphide and evaporation of the solvent. Sulphide sulphur was calculated by subtracting the elemental sulphur content plus the sulphate sulphur content from the total sulphur content.

Free acidity (as g/L H_2SO_4) in leach liquors was determined by taking an aliquot of the solution, reducing ferric ions in solution with thiourea, and then titrating with standardised sodium carbonate solution using methyl orange as an indicator.

Free cyanide was determined by titration with standardised silver nitrate solution using potassium iodide as an indicator. Free CaO in cyanide solutions was determined by taking the product from free cyanide determination and titrating with standardised sulphuric acid using phenolphthalein as an indicator. For some solutions the pH was determined.

Thiourea was determined by titration of an aliquot with 0.1 *M* potassium iodate solution, using an iodine indicator, after acidification of the test solution with phosphoric acid and dilution with water.

2.8 Sizing

A particle size analysis of the sample was determined with a Leeds and Northrup Microtrac Model 7995-12 particle size analyser. The sub-sample used was dispersed in ethanol before the measurements were made as the material could not be dispersed satisfactorily in water. The instrument produces a particle size analysis based on particle volumes in the sample.

2.9 X-ray diffraction

X-ray powder diffraction (XRD) patterns were recorded on a Philips PW 1050 goniometer with a PW 1710 diffraction controller using Cu Kα radiation. Compounds present in the sample were identified by comparison of the peak positions with data in the files of the Joint Committee on Powder Diffraction Standards (JCPDS).

2.10 Electron microscopy

A sub-sample of the concentrate was mounted in epoxy resin, cured, polished, coated with carbon, and examined under a Cambridge S100 scanning electron microscope (SEM). X-ray spectra of specific particles were recorded with the energy dispersive (ED) spectrometer on the SEM to identify the particular elements present.

3 Results and discussion

Results presented in this paper include characterisation by chemical analysis, sizing, XRD, and a limited SEM investigation of the silver-sulphur concentrate supplied; silver extraction from the concentrate by direct leaching with cyanide or thiourea; simultaneous pressure oxidation and leaching of silver with sulphuric or nitric acid; hot acid brine leaching of silver and lead; and hot dilute brine conditioning of the concentrate followed by cyanidation of brine-conditioned residues for silver extraction. Some of the chemistry and literature of silver leaching with cyanide, thiourea, and brine is summarised by Wyslouzil and Salter [30] and is not repeated here.

3.1 Characterisation of concentrate

Two important features of the concentrate are its general phase composition and the form of occurrence of silver.

3.1.1 General composition

As noted earlier, the material used in this work was a silver-sulphur flotation concentrate produced from a hot acid leach residue in a typical electrolytic zinc plant circuit (Fig. 1). The original feed to the plant was a zinc flotation concentrate derived from various mines; it consisted of sphalerite, galena, various silver minerals, various gangue minerals, and a number of minor constituents.

This feed is first roasted in a fluid bed roaster and the calcine is leached with return spent electrolyte (a sulphuric acid solution containing zinc sulphate) in the neutral leach circuit to dissolve part of the zinc. The residue from the neutral leach is partly washed and then sent to the acid leach section where it is leached with strong sulphuric acid to dissolve most of the easily-soluble zinc.

Acid leach residue is then treated by hot sulphuric acid leaching to dissolve as much zinc as possible and the residue is thoroughly washed. The washed residue is sent to the flotation section where a silver-sulphur concentrate is produced by differential rougher-cleaner flotation. This concentrate is currently stockpiled. A representative sample of the concentrate was used in this investigation.

A chemical analysis of the concentrate is given in Table 1. Notable features are that the concentrate, which assayed 8320 g/t Ag and 1.7 g/t Au, contained nearly 44% sulphur (33% elemental sulphur, 8% sulphur as sulphate, and 3% sulphur as sulphide), about 21% lead, and a few per cent of other base metals and constituents.

Table 1. Analytical, phase, and sizing data on silver flotation concentrate.

Chemical analysis		Calculated phase distribution†		Sizing analysis	
Constituent	Value	Phase	Wt %	Size range (μm)	Vol %
Ag (g/t)	8320	Elemental sulphur	32.8	-102+75	1.8
Au (g/t)	1.7	Anglesite (lead sulphate) ($PbSO_4$)	30.2	-75+53	8.7
Zn (%)	3.4			-53+38	4.2
Cu (%)	0.13	Gypsum ($CaSO_4.2H_2O$)	10.8	-38+27	12.8
Fe (%)	1.75			-27+19	11.9
Pb (%)	20.6	Sphalerite (ZnS)	5.1	-19+13	13.5
CaO (%)	3.52	Iron compounds	3.5	-13+9.4	10.3
SiO_2 (%)	3.8	Quartz	3.8	-9.4+6.6	10.1
As (ppm)	<3	Silver	0.83	-6.6+4.7	8.0
Total sulphur (% S)	43.7	Total	87.0	-4.7+3.3	8.7
				-3.3+2.4	6.2
Sulphate sulphur (% S)	7.9			-2.4	3.8
Elemental sulphur (% S)	32.8			Total	100.0
Sulphide sulphur (% S)*	3.0				

* By difference.
† See text.

An XRD pattern of the concentrate is shown in Fig. 2. This indicates that the major phases present were elemental sulphur (S^0), anglesite (lead sulphate, $PbSO_4$), and gypsum (hydrated calcium sulphate, $CaSO_4.2H_2O$) with small amounts of quartz (SiO_2) and sphalerite (ZnS). Only a few, very minor, peaks were unassigned in the pattern.

By making various assumptions (eg. that all the elemental sulphur reported by assay did in fact occur as elemental sulphur, all the lead was present as anglesite, all the calcium was present as gypsum, all the zinc was present as iron- and manganese-free sphalerite, the silica was all present as quartz, and the iron was present as phases averaging 50 per cent iron) the phase proportions shown in Table 1 were derived. These total 87%. This is consistent with the analysis and the XRD pattern and with the general rule that the limit of detection for XRD is about 5% by weight (ie. constituents representing more than 5% of the total would usually have assignable peaks in the XRD pattern and conversely that the absence of unassigned peaks indicates that individual unidentified constituents represent less than 5% of the total). One unresolved issue is that the sulphate analysis indicates that sulphur equivalent to about 8% sulphate (as SO_4) is not specifically accounted for.

The concentrate was fine, as shown in Table 1, with all the material being finer than 106 µm and 85 per cent minus 38 µm. However, many of the particles in the sample had formed aggregates which were not readily dispersed in water (hence conventional wet screening was not successful) but may have been dispersed during the acid or alkaline leaching tests.

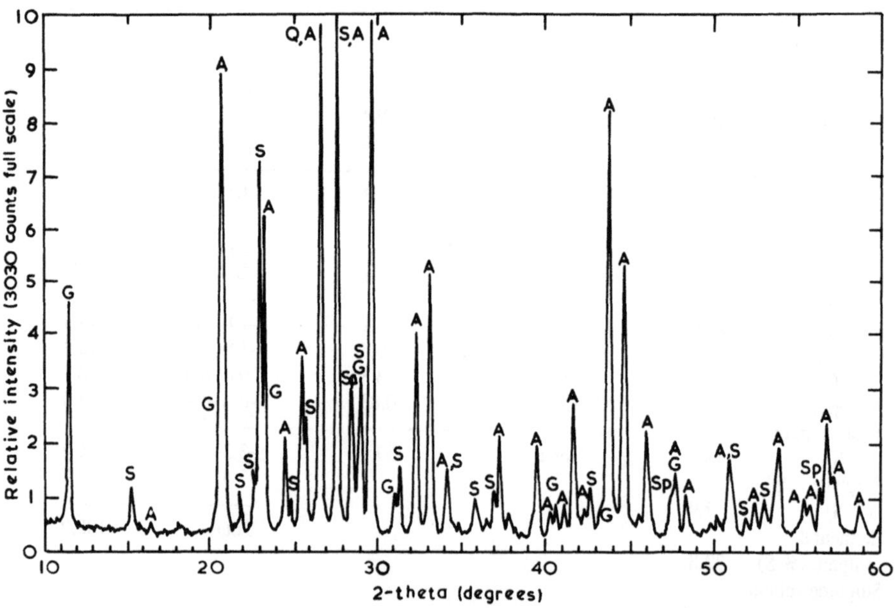

Fig. 2. XRD pattern of silver-sulphur flotation concentrate. A, anglesite; G, gypsum; Q, quartz; S, sulphur; Sp, sphalerite.

3.1.2 Occurrence of silver

The form of occurrence of the silver in the concentrate received is not clear. It is thought that silver in the original zinc flotation concentrate occurred as acanthite, tetrahedrite, as solid solution in galena and possibly in sphalerite, and probably in other silver minerals. However, in view of the treatment that the original concentrate has been given, as shown in Fig. 1, many possibilities exist for the form of the silver in the silver-sulphur flotation concentrate as discussed by Dutrizac [12] and Ildefonse et al. [21]. Silver behaviour during jarosite formation has also been discussed by Dutrizac [11] and Dutrizac and Jambor [14, 15].

The brief SEM examination of the sample located the silver-containing particle shown in Fig. 3. The ED spectrum of this particle (Fig. 4) indicated that it contained copper, zinc, lead, iron, sulphur, and silicon as well as silver. It may have been tetrahedrite initially and been partly fused during roasting and then protected by a film of precipitated silicic acid or a silicate, or possibly elemental sulphur, during subsequent leaching. Correia, Carvalho, and Monhemius [8] have shown that even under autoclave leaching with chloride solutions, tetrahedrite is a relatively refractory mineral.

Chen and Dutrizac [7] found that silver occurred mostly as Ag_2S or $(Ag,Cu)_2S$ associated with sphalerite or as fine particles disseminated in elemental sulphur/ silica gel patches, as well as other forms, in a silver flotation concentrate produced from zinc neutral leach residues. Dutrizac and Chen [13] examined the superhot acid leach residue and silver flotation concentrate from Vieille-Montagne's zinc plant in Belgium. In the superhot leach residue, silver-impregnated sulphur grains, isolated patches of Ag_2S on sphalerite, and silver-bearing jarosite were observed. The major silver carrier in the flotation concentrate was silver chloride, commonly associated with sulphur, with some

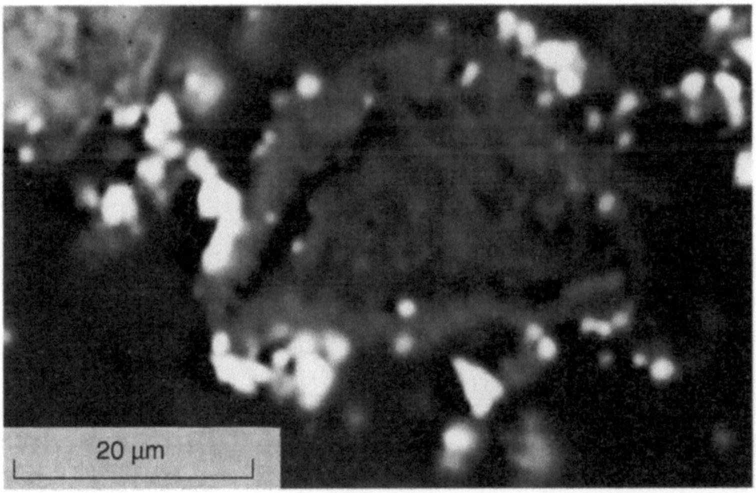

Fig. 3. Electronmicrograph of a silver-containing particle (the relatively large dark particle in the centre) in the silver flotation concentrate. The white particles are lead sulphate.

silver-bearing jarosite. The presence of chloride ions in the Vieille-Montagne electrolyte was suggested to be the reason for the formation of silver chloride.

It is therefore possible that in the CPM silver flotation concentrate, apart from the complex compound shown in Fig. 3, silver occurs as silver sulphides associated with the sulphur grains, as an oxidised compound in lead sulphate grains, as silver-bearing jarosite, silver-silica gel aggregates, and possibly as silver chloride. Some of this silver was expected to be directly soluble in a suitable lixiviant but the remainder would need appropriate treatment to make it soluble.

3.2 Direct cyanidation

Direct cyanidation of the concentrate for 24 h at an initial cyanide concentration of 0.15% NaCN and pH 10 resulted in dissolution of 41.9% of the silver. Extending the cyanidation time to 48 h resulted in only another 0.6% of the silver dissolving. This showed that although some 40% of the silver was readily soluble, the remaining 60% was refractory to cyanidation.

Reagent consumptions were high. Cyanide consumption was about 17.5 kg/t NaCN for 24 h cyanidation and this was caused by the elemental sulphur and the zinc and copper present. A high concentration of thiocyanate (CNS^-) was found to be present in solution after cyanidation and about 20% of the zinc and 37% of the copper had been dissolved. No ferrocyanide was detected in solution. In the second 24 h cyanidation there was little change in the amount of zinc and copper dissolved but the cyanide consumption increased to 22.5 kg/t NaCN indicating that reaction of cyanide with the elemental sulphur was continuing.

Lime consumption was about 50 kg/t CaO for 24 h and 48 h cyanidation. This was probably caused mainly by reaction of the lead sulphate present with OH^- ions according to reaction 1.

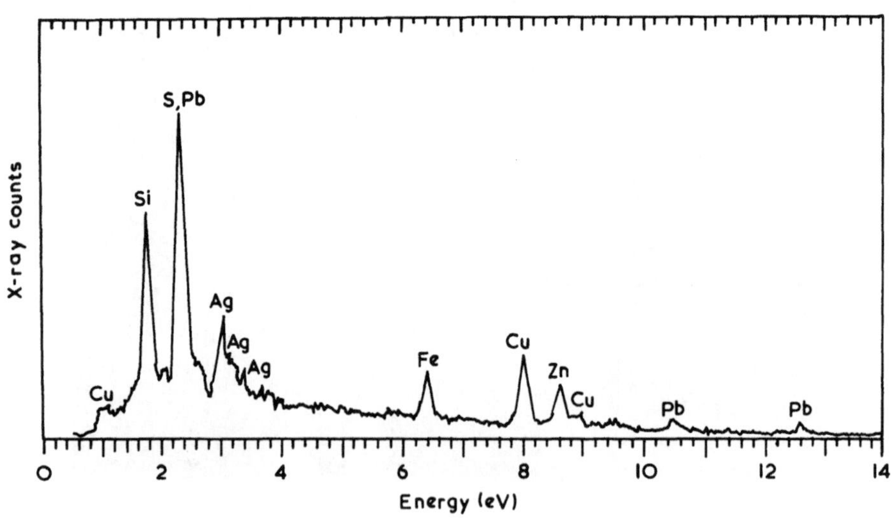

Fig. 4. Energy dispersive x-ray spectrum of the silver-containing particle in Fig. 3.

$$PbSO_4 + 2OH^- = Pb(OH)_2 + SO_4^{2-} \quad (1)$$

The lead hydroxide formed either dissolved in solution as plumbite ions (9-11% of the lead was found in solution after cyanidation) or precipitated in an amorphous form (the XRD pattern of the cyanidation residue showed no trace of the anglesite present in the feed and no other lead salt was identified).

Cyanidation under the conditions used probably extracted any silver present as silver chloride and silver present in anglesite but little of that present as a sulphide or associated with sulphur or occurring in the unreacted sphalerite present. Ek [16] obtained 50-90 % extraction of silver by cyanidation of zinc jarosite residues depending on the product treated. Pre-treatment of the CPM flotation concentrate by hot (90°C) conditioning with lime, as done by Berezowsky et al. [3] on a zinc pressure leach residue, could give improved silver extraction by cyanidation.

3.3 Thiourea leaching

Thiourea has been suggested as an alternative lixiviant to cyanide for silver and has been used by CSIRO on many Australian materials [eg, 4, 5, 6]. It has been used by Rosato et al. [26] for treatment of a zinc plant residue at the plant of Canadian Electrolytic Zinc, and some experimental work has been reported by Ek [16]. Thiourea was therefore used in this work in three tests as summarised in Tables 2 and 3.

The first test (351) was at pH 1.7 and the natural redox potential of the system. This potential was about 300-325 mV in the first hour, but dropped to 250 mV on addition of extra thiourea at 1 h and remained in the range 220-250 mV for the remainder of the test. No further thiourea was added. This test was conducted under conditions which had been found previously to be satisfactory for dissolution of silver (ie, a pH of 1.5-2.5, a redox potential in the range 180-250 mV, and a thiourea concentration of 0.4%).

Results in Table 2 show that only 3% of the silver had dissolved in the first hour and 24% after 2 h, but reprecipitation then occurred and only 12% was in solution after 3 h. This result was much poorer than that with cyanide.

Two further tests (352 and 353) were conducted (Table 3) in which the redox potential was controlled at a higher level using either ferric sulphate alone or ferric sulphate and hydrogen peroxide. Silver extractions were very poor after 2 h being under 2.5% for both tests. Continuation of the tests for a further 1 h or 3 h, respectively, resulted in reprecipitation of silver.

XRD patterns of the residues from all three thiourea leaches were similar to that of the silver concentrate. The intensities of the peaks due to lead sulphate and sulphur were almost the same while those due to gypsum were slightly more intense. Sphalerite was still present in the residues although the relative intensities of its peaks indicated that some of it had reacted.

It is believed that the main cause of the poor silver extraction was the generation of hydrogen sulphide by oxidative acid attack on the sphalerite or elemental sulphur present which led to precipitation as silver sulphide of any silver dissolved. Iron present or added could have formed an insoluble silver jarosite. This effect could possibly be lessened by operating at a lower redox potential (eg, < 200 mV) by addition of ferrous sulphate or possibly sodium sulphite. However, this may not provide sufficiently oxidising conditions to dissolve the silver. Furthermore, some of the silver could have

Table 2. Results of thiourea leaching of the flotation concentrate at natural redox potential.

Time (h)*	pH	Eh (mV)	Solution assay Ag (ppm)	Ag extraction (%)	Cumulative thiourea used (kg/t)
0	1.69	469	0	0.0	0.0
1	1.68	303	80	2.9	7.2
2	1.69	246	650	23.6	7.2
5	1.68	222	330	11.8	7.2

Weight of residue was 88.0% of feed weight.
Calculated silver feed assay was 8760 g/t Ag.
* Thiourea was added at 0 h and 1.0 h.

Table 3. Results of thiourea leaching of flotation concentrate under different conditions.

Leach No.	Oxidant	pH	Eh (mV)*	Ag extr. after 2 h (%)	Thiourea used (kg/t)
351	Nil	1.7	250	23.6	7.2
352	Fe^{3+}	1.2	350	1.9	17.3
353	$Fe^{3+} + H_2O_2$	1.8	300	2.3	12.6

* Value at 2 h.

been present in fine associations with various compounds or present in minerals which did not react with the solution and therefore was not accessible to thiourea leaching. These compounds would need to be changed in form to liberate the silver.

Other conditions could be worth investigating as Rosato et al. [26] obtained 85-95% silver extraction by thiourea leaching of neutral zinc leach residue using 4 g/L thiourea and pH 2-3 at 75°C for 15 min. However, Ek [16] obtained poorer results on similar material with similar conditions but with 7 h leaching time.

3.4 Oxygen pressure leaching with sulphuric acid

The silver concentrate was subjected to oxygen pressure leaching in sulphuric acid solution with the aim of dissolving silver and/or making the silver available for subsequent dissolution in cyanide or thiourea. Results for leaches at initial acid strengths of 0, 36, and 90 g/L sulphuric acid at a liquid:solid ratio of 10:1 are given in Table 4 (tests 9, 7, and 8 respectively) and are plotted in Fig. 5. These show that at 90 g/L sulphuric acid about 56% of the silver was dissolved in 4 h but only 22-23% at the lower acid strengths. A similar effect of acid strength was reported by Onozaki, Nagata, and Ishimori [24].

Silver sulphate is a sparingly soluble salt (about 1 g $AgSO_4$ per 100 mL water at 25°C). However, the highest concentration of silver encountered in the leaching tests reported in Table 4 (test 1) was 836 mg/L Ag (or 0.17 g $AgSO_4$ per 100 mL) which was well below the solubility limit even allowing for the common ion effect and the presence of other salts. Moreover, changing the L:S ratio in tests 1 and 8 made little significant difference to the silver extraction. It appears that there may be a critical acid strength which must be exceeded to attack a silver-bearing compound in this concentrate, although changes in other conditions may be more beneficial. On the other hand the conditions were already reasonably severe.

Very little acid was generated, apart from test 9, which suggests that little elemental sulphur was oxidised in these tests. If silver was occluded with this sulphur, this may be the reason for the low silver extraction. However, in test 2, in which ferric sulphate was added to increase the oxidising state of the pulp, a very low silver extraction was also recorded, probably because of the formation of a silver jarosite. Formation of such products is likely to be a major problem in attempting to leach silver from the concentrate in acid sulphate solutions containing iron.

Extractions of zinc and copper from the concentrate (Table 4) were high under all conditions used, and the iron extractions ranged from 12% to 77% depending on the initial acid addition. An average of only about 9% of the lead was extracted because of the low solubility of lead sulphate in the solutions used.

The XRD patterns of all the leach residues were similar. They indicated that the elemental sulphur content of the products had decreased but that there was still substantial sulphur present. The concentration of lead sulphate had increased because of the loss of weight in the solid residue from the leach. There was no sphalerite present because it had all reacted. There was evidence for the formation of a jarosite or hematite in some tests. The XRD patterns also indicated that the gypsum ($CaSO_4.H_2O$) had been dehydrated to anhydrite ($CaSO_4$) during the leaches. Hence the loss of weight during the leaches to give a residue containing about 70% of the feed weight was caused by dehydration of gypsum, oxidation of some elemental sulphur to a soluble form, and the dissolution of zinc, copper, and iron compounds.

In view of the results in Table 4, it was felt that the sulphuric acid leaching route did not offer sufficient promise to pursue further. Only 56% of the silver was extracted and

Table 4. Results of oxygen pressure leaching of silver concentrate with sulphuric acid at 200°C for 4 h.

Test No.	Initial L:S ratio	Free acidity		Residue weight (% of feed)	Metal extraction				
		Initial H_2SO_4 (g/L)	Final H_2SO_4 (g/L)		Ag (%)	Zn (%)	Cu (%)	Fe (%)	Pb† (%)
9	10:1	0	18.4	73.8	21.9	82.4	92.3	12.4	4.7
7	10:1	36	38.9	72.8	22.9	83.9	96.3	45.0	8.5
8	10:1	90	91.3	68.6	56.2	95.2	96.8	71.3	9.1
1	5:1	90	85.5	70.6	52.0	94.6	97.7	77.0	10.9
2*	5:1	36	43.7	75.9	6.7	92.0	95.8	§	9.4

* Addition of 50 g/L $Fe_2(SO_4)_3$ at the start of the leach.
† Based on residue and head assays.
§ Could not be calculated because of ferric sulphate addition.

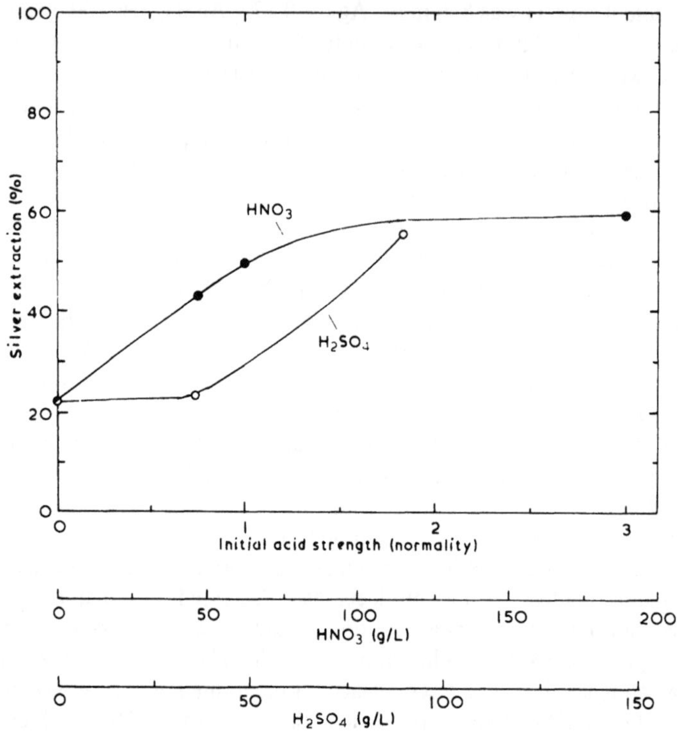

Fig. 5. Effect of acid strength during pressure leaching on extraction of silver from the silver flotation concentrate. Sulphuric acid tests conducted at 200°C; nitric acid tests conducted at 125°C.

although there was a possibility that after silver recovery from the leach solution the resultant silver-free solution could be integrated into the main plant for zinc and copper recovery, the problem of extracting the remaining silver from the leach residue still remained. Several possibilities were apparent, but the overall process did not appear to be attractive.

3.5 Nitric acid oxidation and leaching

Two nitric acid processes, the Arseno process [2] and the Nitrox process [18, 28] have attracted a great deal of attention for the oxidation of refractory sulphide gold ores, although these processes are usually followed by cyanidation. However, Kunda [23] has applied direct nitric-sulphuric acid leaching to the extraction of silver and other metals from jig and flotation concentrates obtained from a sulphide silver ore, and achieved over 90% silver extraction in a two stage process.

In view of the oxidising properties of nitric acid, and the fact that silver nitrate is more soluble than silver sulphate, the silver flotation concentrate used in this work was treated with nitric acid (47-189 g/L) at temperatures between 100°C and 150°C. During oxidation with nitric acid, the nitric acid is reduced to nitric oxide (NO) which is subsequently re-oxidised to nitric acid via nitrogen dioxide (NO_2) as an intermediate. Brown fumes of nitrogen dioxide were observed in gases vented from the autoclave after the nitric acid leaches.

Results of nitric acid oxidation and leaching are shown in Table 5 and on Fig. 5. Table 5 shows that with 63 g/L nitric acid, less than 15% of the silver dissolved in 1 h at 100°C (test 3), but at 125°C (test 4), almost 50% of the silver dissolved. The higher temperature is above the melting point of sulphur (118°C), indicating that when the sulphur had melted the occluded silver compounds were released and dissolved faster. At 125°C, increasing the acid strength increased the silver extraction (tests 5, 4, and 6) and at 189 g/L nitric acid initially, 58% of the silver was extracted in 1 h (test 6). However, as indicated on Fig. 5, not much higher extractions can be expected. These silver extractions were similar to those achieved with hot sulphuric acid leaching.

The XRD patterns of the residues from the nitric acid leaches indicated that at 100°C only some of the sphalerite had reacted while at 125°C almost all the sphalerite had reacted during the leach. The patterns also indicated that the elemental sulphur content of the residue decreased with increased temperature and acid strength, while the lead sulphate content increased because of the decrease in residue weight. Gypsum was present in the residues from the leaches at 125°C with 47 g/L and 189 g/L nitric acid (tests 5 and 6) but not in the others. Also there was no anhydrite present in any of the residues. The weight loss recorded for the leaches was due to the oxidation and dissolution of the sphalerite and some of the elemental sulphur, and dissolution of some of the gypsum.

Increasing the reaction temperature to 150°C in test 16 increased the extent of the reactions with various components in the concentrate as shown by the lower residue weight and higher extraction values for zinc, copper, and iron given in Table 5. The XRD pattern of the residue from the leach at 150°C indicated that only anglesite and anhydrite were present. There was no elemental sulphur, sphalerite, or gypsum. Although all the elemental sulphur was oxidised, and presumably released any occluded silver compounds, only 58.6% of the silver was extracted, which was no better than that at 125°C or that with sulphuric acid when only part of the sulphur was oxidised. It may be noted that there was some difficulty in controlling the temperature in the test at 150°C because of the rapid exothermic oxidation of elemental sulphur which occurred.

Table 5. Results of nitric acid leaching of silver concentrate for 1 h at a liquid:solid ratio of 5:1.

Test No.	Temp (°C)	Initial free acidity HNO_3 (g/L)	Residue weight (% of feed)	Metal extraction Ag (%)	Zn (%)	Cu (%)	Fe (%)	Pb* (%)
3	100	63	71.6	14.8	24.5	75.9	36.3	12.1
5	125	47	79.6	42.6	32.9	89.4	40.3	6.5
4	125	63	71.1	49.8	41.9	90.7	44.2	10.2
6	125	189	73.0	58.2	76.0	87.8	42.8	5.0
16	150	189	44.3	58.6	94.5	92.8	86.3	4.8

* Based on residue and head assay values.

Although there were no sulphate ions in solution at the start of the nitric acid leaches the oxidation of elemental sulphur and dehydration of gypsum gave substantial concentrations of sulphate at the end of each leach and this may have affected the silver extraction. Since the best silver extraction obtained with nitric acid was no better than that with sulphuric acid, and the same problems of having silver in two products for further treatment remained, there seemed to be no special merit in persisting with nitric acid treatment of the concentrate.

3.6 Acid brine leaching

Acid brine solution is known to be a good lixiviant for oxide silver minerals such as silver chloride and silver sulphate [25, 27] but not for silver metal or silver sulphides. It also dissolves oxide lead minerals but not sulphide lead minerals. In such solutions, silver is dissolved as argento chloro complexes ($AgCl_2^-$, $AgCl_3^{2-}$, and $AgCl_4^{3-}$) and lead as similar chloro complexes. Results of acid brine leaching obtained in this work are summarised in Table 6. The variables studied briefly were sodium chloride concentration, initial free acidity, temperature, time, and oxygen concentration.

Sodium chloride concentration is an important variable in acid brine leaching, because high chloride levels are needed to get adequate silver and lead extractions, as illustrated by the results for tests 14 and 22 in Table 6. At 20 g/L NaCl, only 0.5% of the silver and 10% of the lead were extracted whereas at 300 g/L NaCl, over 98% of the silver and 97% of the lead were extracted. At low acidity (tests 10 and 11), the silver extraction values for 200 g/L NaCl were substantially below those for 300 g/L NaCl. Ralston [25] has shown that silver is significantly less soluble in 200 g/L NaCl than in 300 g/L NaCl. A saturated brine solution contains about 360 g/L NaCl.

Table 6. Results of acid brine leaching of silver concentrate at a L:S ratio of 10:1.

Test No.	Temp (°C)	Time (h)	Initial free acidity H_2SO_4 (g/L)	NaCl conc (g/L)	Residue weight (% of feed)	Metal extraction Ag (%)	Zn (%)	Cu (%)	Fe (%)	Pb† (%)
10	200	4	0	300	46.6	96.8	87.0	93.5	27.3	88.4
11	200	4	0	200	56.8	28.0	84.1	94.7	23.8	86.2
12	125	4	0	300	58.0	77.6	64.8	9.2	15.8	71.3
14	125	4	36	20	73.8	0.5	76.9	87.3	47.8	10.4
17	125	2	36	300	45.2	94.6	83.1	93.1	80.6	92.2
18	125	1	36	300	47.4	98.2	70.1	94.0	76.0	88.8
19	125	0.5	36	300	46.8	98.4	56.2	93.4	71.4	97.0
20	125	1	36	300	47.2	98.3	74.6	93.9	78.3	97.3
21	125	0.5	36	300	49.2	98.3	57.2	92.2	77.1	95.7
22	125	4	36	300	46.8	98.3	86.3	92.2	86.6	97.2
23§	125	0.1*	36	300	49.2	97.9	21.7	74.1	77.1	96.0
24	108	4	36	300	47.4	42.4	19.0	33.8	86.3	96.0

* Sample simply heated to temperature and immediately cooled.
† Based on residue and head values.
§ Conducted under nitrogen; other tests conducted under oxygen.

Free acidity is important for silver and copper extraction as shown by tests 12 and 22. Zero added acid results in relatively low extractions whereas with 36 g/L free H_2SO_4, high extractions of silver and copper (ie, well over 90%) were obtained together with over 85% extraction of zinc and iron.

Temperature was important with respect to silver, zinc, and copper extraction, but less so with respect to lead. High extractions for all metals were obtained in test 22 at 125°C when using 300 g/L NaCl, 36 g/L H_2SO_4, and 4 h leaching time. However, at 108°C (test 24), which was below the melting point of sulphur (118°C), poor extractions of silver, zinc, and copper were obtained. A similar effect of temperature was obtained with nitric acid.

Leaching time was not an important variable when using 300 g/L NaCl, 36 g/L H_2SO_4, and 125°C as shown by tests 17, 18, 19, and 22 (Table 6). Decreasing the leaching time from 4 h in test 22 to 0.5 h in test 19 had no effect on the extraction of silver, copper, and lead. In fact, times as short as a few minutes seem to be adequate for silver and lead as shown by test 23 in which the leach charge was simply heated up to 125°C and then cooled down.

Most brine leach tests were conducted under an over-pressure of 1050 kPa of oxygen to aid the dissolution of any elemental silver or silver-bearing sulphides present. However, test 23 was conducted under a nitrogen atmosphere for a short leaching period. The results indicate that silver and lead extractions were unaffected but zinc and copper extractions were markedly lower when leaching was conducted under nitrogen instead of oxygen.

A study of the XRD patterns of the residues from the leach tests with acid brine confirmed the above comments. At high sodium chloride concentrations and acidities there was little sign of lead sulphate. However, there was more lead sulphate in the residues from the leaches at 200°C compared with those at 125°C both with and without 36 g/L acid initially. At 125°C the lead sulphate appeared to dissolve rapidly. Even in the leach that was just brought to temperature and allowed to cool (test 23), very little lead sulphate remained in the residue. A decrease in intensity of the lead sulphate peaks was associated with an increase in the elemental sulphur peaks, because the proportion of sulphur in the residue was higher than in the feed.

The XRD patterns indicated that sphalerite had reacted in the tests with oxygen present at 200°C and 125°C after 4 h, while at 125°C there was still some sphalerite present after 2 h. Little sphalerite had reacted when the charge was just heated to 125°C under nitrogen and then cooled down. Hematite (tests 10 and 11) and a jarosite plus hematite (test 12) were identified in some residues. No jarosite or hematite could be identified in the patterns of residues when 36 g/L acid was added initially to the brine leaches. Anhydrite was present and gypsum was absent in all brine leaching residues at 125°C, even in the leach which was just raised to temperature and cooled.

In several tests, particularly those at low sodium chloride concentrations, pellets composed mainly of elemental sulphur were visible in the residues. This is considered in more detail in the section below on acid brine conditioning.

3.7 Dilute brine conditioning and cyanidation

In the above work, which indicated that most of the silver present in the silver concentrate could be extracted by acid brine leaching at temperatures above the melting point of sulphur, it was apparent that the silver had been converted to silver chloride before or during dissolution. It was of interest, therefore, to investigate if this conversion could be achieved by conditioning at a relatively low concentration of sodium chloride and the resultant silver chloride extracted by cyanidation. In addition, since it had been noted that pellets of sulphur up to several millimetres in diameter had been formed during leaching at low brine concentrations, it was of interest to ascertain if this sulphur could be removed by screening before cyanidation and thereby minimise cyanide consumption. Eropkin et al. [17] subsequently reported that conditioning of a zinc plant residue with a solution containing 20-25% NaCl before flotation converted the silver to silver chloride.

3.7.1 Brine conditioning

A summary of the experimental conditions for dilute brine conditioning is given in Table 7 together with the residue weight for each test. Table 8 gives a summary of the silver assays and silver distributions in the various test products as well as the extractions of the main base metals into the dilute brine solutions used for conditioning.

During dilute acid brine conditioning up to 30% of the feed dissolved in the brine as shown by the residue weights given in Table 7. Minimum dissolution (14%) occurred in test 25 in which no sodium chloride or acid were added and the test was conducted at a low L:S ratio under nitrogen for only 1 h. The greatest dissolution (about 28%) occurred in tests 27 and 28 at an L:S ratio of 10:1 with 5 g/L NaCl and 36 g/L H_2SO_4 under either nitrogen or oxygen. Insufficient tests were conducted to differentiate clearly between all the factors involved.

Similarly, the proportion of the product reporting in the sulphur pellet fraction and the corresponding non-sulphur (non-S) fraction varied. This was probably affected by the manual screening technique used to make the separation since visual estimation was

Table 7. Contact conditions and residue weights for dilute acid brine conditioning of concentrate at 125°C.

Test No.	NaCl conc (g/L)	Leach time (h)	Initial H_2SO_4 (g/L)	L:S ratio	Atmos-sphere	Residue weights (as % of feed)		
						Total pellets	Sulphur fraction	Non-S fraction
14	20	4	30	10:1	Oxygen	73.8	*	*
15	5	4	36	10:1	Oxygen	76.0	*	*
25	0	1	0	5:1	Nitrogen	86.2	*	*
26	5	1	0	10:1	Nitrogen	74.1	*	*
27	5	1	36	10:1	Nitrogen	70.9	25.1	45.8
28	5	1	36	10:1	Oxygen	71.6	28.4	43.2
29	10	1	36	5:1	Oxygen	73.0	23.2	49.8
30	10	1	36	5:1	Oxygen	75.8	24.9	50.9
32	30	1	36	5:1	Oxygen	78.8	15.4	63.4
34	20	1	36	5:1	Oxygen	78.0	30.3	47.7

* Pellets not separated.

Table 8. Results of dilute brine conditioning of silver concentrate at 125°C.

Test No.	Assays			Silver distribution			Extraction of base metals			
	Non-S fraction Ag (g/t)	Sulphur pellets Ag (g/t)	Brine soln Ag (g/t)	Non-S fraction (%)	Sulphur pellets (%)	Brine soln (%)	Zn (%)	Cu (%)	Fe (%)	Pb (%)†
14	10600*	n.d.	4.2	n.d.	n.d.	0.5	76.9	87.3	47.8	10.4
15	10200*	n.d.	1.2	n.d.	n.d.	0.2	70.6	77.3	46.8	7.8
25	9200*	n.d.	0.3	n.d.	n.d.	0.0	11.3	13.8	6.9	4.6
26	10600*	n.d.	0.4	n.d.	n.d.	0.0	14.9	7.6	21.2	7.2
27	16100	1440	0.3	95.3	4.7	0.0	23.9	54.1	61.2	13.3
28	16300	2100	0.2	92.2	7.8	0.0	40.9	91.8	58.3	12.6
29	14300	3160	0.4	90.7	9.3	0.0	40.0	84.5	43.2	10.4
30	13900	3690	6.6	88.2	11.4	0.4	47.9	77.4	43.4	9.9
32	13100	490	11.6	98.4	0.9	0.7	59.6	89.3	41.9	9.1
34	13200	5460	1.3	79.1	20.8	0.1	53.7	85.6	35.5†	8.0

* Assay of total residue because sulphur pellets were not separated.
† Based on residue and head assay values.
n.d.: Not determined.

applied to determine the end point. However, on average, about 25% of the feed reported in the sulphur pellet fraction and 45-50% in the non-sulphur fraction.

Generally speaking, 90-98% of the silver reported in the non-sulphur fractions (Table 8) (the result from test 34 is anomalous in that the proportion of silver contained in the non-sulphur fraction was relatively low), and the product assayed 13,000-16,000 g/t Ag. Less than 10% of the silver reported in the sulphur pellets which assayed 2000-3000 g/t Ag. Less than 1% of the silver reported in the dilute brine solution.

Low copper extractions (7-14%) were obtained in the dilute brine during conditioning in the absence of oxygen or acid (tests 25 and 26). However, high copper extractions (77-92%) were achieved when oxygen and acid were present. This copper is potentially recoverable. Zinc extractions were also low in the absence of oxygen, but were reasonable at 40-77% in the other tests. Lead extractions were low (up to 13%) in all conditioning tests because of the low sodium chloride concentration used. Since the lead extractions shown were estimated from the feed and tailings assays and weights, the values given may not be accurate. The lead assays of the tailings were of the order of 38% Pb.

3.7.2 Cyanidation of non-sulphur fraction

Metallurgical results of cyanidation of the non-sulphur fraction from brine conditioning in test 34 (Tables 7 and 8) are summarised in Table 9. It was not realised at the time that the cyanidation was conducted that the conditioning results were anomalous in relation to the other conditioning tests in that the proportion of the silver contained in the non-sulphur fraction was relatively low.

Table 9. Results of cyanidation of non-sulphur fraction obtained in test 34 after brine conditioning of silver flotation concentrate.

Product	Wt. %	Silver assay (g/t)	Silver distribution	
			Based on cyan. feed (%)	Based on orig. conc (%)
Preg. cyan. soln	381	3200	92.8	73.4
Cyanide residue	112	840	7.2	5.7
Calculated feed	100	13100	100.0	79.1
Head by assay	100	13200	100.0	-

However, Table 9 shows that a good silver extraction was obtained by cyanidation of the product. About 93% of the silver in the cyanide feed was extracted, and this was equivalent to about 73% of the silver in the original concentrate. If the same extraction can be obtained by cyanidation of the other non-sulphur fractions, which contain on average about 95% of the silver in the original concentrate, an overall extraction of about 88% of the silver would be obtained.

In this test the cyanide consumption was unexpectedly high at 30 kg/t, although it could probably be decreased by better removal of sulphur before cyanidation. Lime consumption was also high (80 kg/t) because of reaction with acid salts and lead compounds as noted earlier. Partly because of this, the residue weighed more than the feed. An XRD pattern of the residue showed that no lead sulphate was present, but calcium sulphate occurred together with unidentified products. Lead salts were not noted because they were poorly crystalline or amorphous.

4 General discussion of hydrometallurgical treatment methods

Preliminary investigation of six possible hydrometallurgical treatment routes for the extraction of silver from a silver-sulphur flotation concentrate from the CPM plant showed that two (acid brine leaching and dilute acid brine conditioning plus cyanidation of the conditioned product) gave good silver extractions (over 90%) and justify further investigation. These two are discussed in more detail below in relation to tentative flow-sheets. However, the other four (direct cyanidation, direct thiourea leaching, pressure leaching in sulphuric acid, and pressure leaching in nitric acid) gave poor results in that they extracted only 40-60% of the silver. Although it is possible that adjustment of the leach conditions in these latter four methods might give better results, the methods did not appear to offer as much promise. It appears that perhaps half of the silver is readily extracted by each method, but it is not known whether or not the same silver compounds are being extracted.

Direct cyanidation resulted in the extraction of only 42% of the silver, probably because most of the silver was locked in sulphides and other compounds. Moreover, a very high cyanide consumption was obtained because of the high proportion of elemental sulphur present, and there was a high lime consumption because of the lead sulphate present. Thiourea leaching extracted up to 24% of the silver, but substantial re-precipitation of the silver occurred even after short leach times. In addition, thiourea consumptions were high.

Oxygen pressure leaching in 90 g/L sulphuric acid solution extracted 56% of the silver in 4 h. Although this silver could be readily recovered after a liquid-solid separation by addition of chloride to give insoluble silver chloride and recycling of the barren solution to the leach or integration into the main zinc plant circuit, the problem of recovering silver from the leach residue remains. Since this acid leach residue still contained elemental sulphur, cyanidation or thiourea leaching would probably be inapplicable, and in any case the silver compounds present may have been coated with silica or converted to silver jarosite.

Leaching with nitric acid above the melting point of sulphur gave a similar silver extraction to that with sulphuric acid. Silver recovery from the leach solution by chloride precipitation could be relatively straightforward, but the barren solution could not be integrated with the main zinc plant. Recovery of silver from the nitric acid leach residue would present similar problems to those with the sulphuric acid leach residue, although most of the sulphur was removed by the nitric acid. However, another leach (ie, double leaching overall) does not appear to be attractive although, in principle, silver chloride could be precipitated in the leach residue and then extracted by cyanidation or thiourea leaching.

4.1 Hot acid brine leaching

Hot acid brine leaching at 125°C was found to readily extract over 98% of the silver and 97% of the lead as well as much of the zinc, copper, and iron present. A simplified schematic flowsheet for recovery of silver from the concentrate, based on this finding, is shown in Fig. 6. Figueiredo, Silva, and Silva [19] have discussed a similar flowsheet. However, much work is needed to optimise the conditions at each stage. Other methods for the recovery of silver and lead from solution could be investigated and the extensive literature on ferric chloride leaching of complex base metal ores and concentrates should be taken into account.

It is envisaged that the hot acid brine leaching stage would be conducted at 125°C (ie, above the melting point of sulphur) using 300 g/L NaCl and 30 g/L H_2SO_4 at an L:S ratio of 5:1 under a slight overpressure of oxygen (to ensure oxidising conditions) for a relatively short time (say 1 h). Some iron would be leached from the feed and ferrous iron would be present from recycled solution; this would need to be oxidised to the ferric state during leaching. Jarosites should not be a problem provided the sulphuric acid level is kept above 0.3 *M* (30 g/L H_2SO_4) [10]. It could be preferable to use calcium chloride in place of sodium chloride (Bahr and Priesemann [1]). The addition of cupric chloride could help to control the redox potential of the leach and help to dissolve the small amount of gold present [29]. Wilson [29] also noted that oxidising brine solutions do not attack elemental sulphur.

The leach pulp would then go to a liquid-solid separation and washing stage (probably by counter-current decantation) where make-up acid brine would be added as the wash solution. The products would be a washed residue for discard or possible

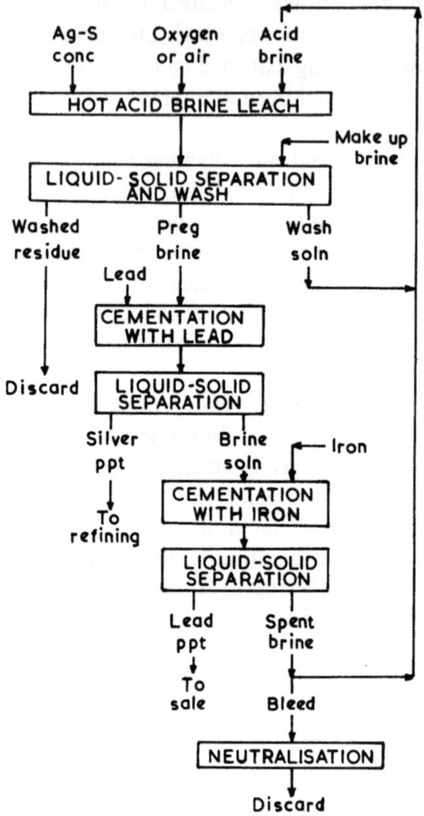

Fig 6. Simplified schematic flowsheet proposed for the recovery of silver and lead from the silver-sulphur flotation concentrate by hot acid brine leaching and cementation.

treatment for sulphur recovery (Doyle et al. [9]), a wash solution high in brine and acid but low in dissolved metals which would be recycled to the leach, and a silver-lead pregnant solution.

The pregnant solution would go to the first stage precipitation where lead would be added to precipitate the silver (Kim, Kang, and Sohn [22]) and probably the copper. After a liquid-solid separation stage to recover the silver precipitate for shipment to a refinery, the brine, which would still contain lead and other metals, would go to the second stage precipitation where iron would be added to precipitate the lead [20, 25] and any remaining silver and copper. After another liquid-solid separation, the precipitate would be recovered and sent for sale. Part of the barren brine, containing acid, zinc, and iron, would be recycled to the leaching stage and part would be bled off, sent to a neutralisation stage, and discarded.

Some 95% of the silver and of the lead could be obtained in separate saleable products and reagent consumptions should be reasonable.

4.2 Dilute brine conditioning plus cyanidation

It was also found that by conditioning the silver-sulphur flotation concentrate with hot dilute acid brine, screening out the sulphur pellets formed, and then cyaniding the non-sulphur product, over 93% of the silver could be extracted. A simplified schematic flowsheet for the recovery of silver, based on this finding, is shown in Fig. 7. As for acid brine leaching (Fig. 6), much optimisation work is still required on the various stages.

It is envisaged that the silver-sulphur concentrate would be conditioned at 125°C and an L:S ratio of 5:1 with a dilute acid brine (10 g/L NaCl and 30 g/L H_2SO_4) for 1 h

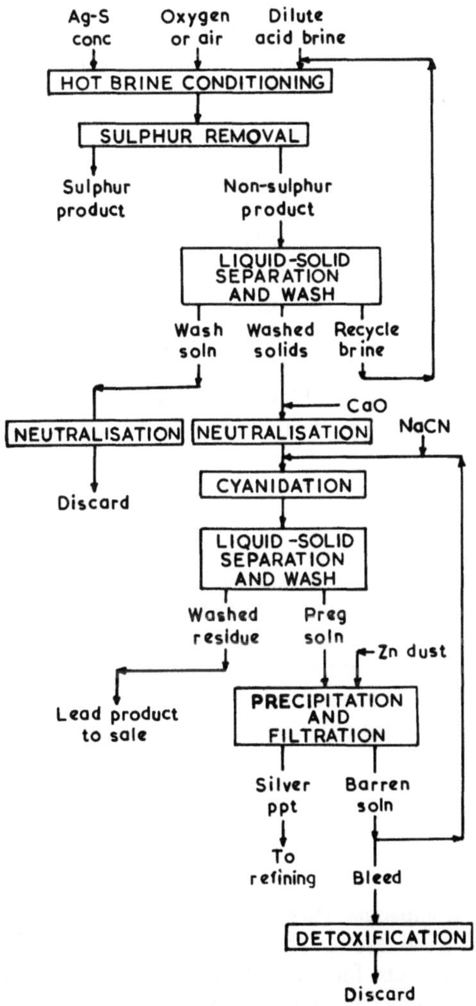

Fig 7. Simplified schematic flowsheet for the recovery of silver and lead from the silver-sulphur flotation concentrate by dilute acid brine conditioning and cyanidation.

under a slight overpressure of oxygen or air. This would convert the silver to silver chloride and generate molten sulphur globules. The discharge from conditioning would go to a sulphur removal stage, which possibly might require only cooling and screening (Doyle et al. [9]). The sulphur product, which could assay several thousand grams per tonne silver, but which would contain only a few per cent of the silver, might warrant further treatment for silver recovery. The non-sulphur product (screen undersize) would go to a liquid-solid separation and washing stage, primarily to remove free acid and dissolved base metal salts. Wash solutions would be make-up acid brine and then water. The main brine solution would be recycled to the conditioning stage and the water wash would possibly go to a cementation stage to recover copper and then to neutralisation and discard.

Conditioned washed solids would go to neutralisation where lime would be added to raise the pH to 9.5. This would involve a substantial addition of lime as all the lead sulphate or lead chloride present would have to be converted to lead hydroxide (Bilston et al., [4]). Neutralised solids would then go to cyanidation and be agitated for sufficient time to dissolve the silver.

Cyanidation discharge would be washed in a liquid-solid separation circuit and the washed residue would be a lead product for sale. The silver pregnant solution would go to precipitation where zinc dust would be preferable to the application of carbon-in-solution or carbon-in-pulp because of the high silver grade of the solution. Silver precipitate would be refined to produce silver bars. Barren cyanide solution would be partly recycled to the cyanidation stage and partly bled to a detoxification circuit and discarded.

Well over 90% of the silver should be recovered in bullion form and nearly all of the lead should report in a saleable product.

5 Conclusion

This work has shown that of six hydrometallurgical treatment routes investigated for silver extraction from the silver-sulphur flotation concentrate produced in the CPM plant only two routes (hot acid brine leaching and hot dilute brine conditioning followed by cyanidation) appear to offer much promise for commercial application. The other four routes (direct cyanidation, thiourea leaching, hot sulphuric acid pressure leaching, and hot nitric acid pressure leaching) resulted in poor silver extractions and had other disadvantages.

6 Acknowledgements

The authors thank Companhia Paraibuna de Metais of Juiz de Fora, Brazil, for permission to publish these results. They also thank Mr N.F. Dyson, formerly of CSIRO Division of Mineral Products, for help with the experimental work and assays. One of us (ALA) thanks Companhia Paraibuna de Metais and CSIRO for the opportunity to work at the CSIRO Division of Mineral Products, Melbourne, for 12 months as a visiting scientist.

7 References

1. Bahr, A., and Priesemann, Th., 1988. Recovery of silver from refractory ores, in **Proc. XVI International Mineral Processing Congress**, Stockholm (Ed. K.S.E. Forssberg), pp. 1121-1134 (Elsevier: Amsterdam).
2. Beattie, M.J.V., Raudsepp, R., Foo, K.A., and Balderson, G.F., 1989. Application of Arseno (Redox) process technology to refractory ores and concentrates, in **Precious Metals 1989**, Montreal, pp. 327-333 (International Precious Metals Institute: Allentown).
3. Berezowsky, R.M.G.S., Stiksma, J., and Kerfoot, D.G.E., 1990. Silver and gold recovery from zinc pressure leach residue, in **Lead-Zinc '90** (Ed. T.S. Mackey and R.D. Prengaman), pp. 135-150 (The Minerals, Metals & Materials Society: Warrendale).
4. Bilston, D.W., Bruckard, W.J., McCallum, D.A., Sparrow, G.J., and Woodcock, J.T., 1990. Comparison of methods of gold and silver extraction from Hellyer pyrite and lead-zinc flotation middlings, in **Sulphide Deposits – Their Origin and Processing** (Ed. P.M.J. Gray), pp. 207-221 (Institution of Mining and Metallurgy: London).
5. Bilston, D.W., LaBrooy, S.R., and Woodcock, J.T., 1984. Gold and silver leaching from an oxidised gold ore with thiourea under controlled conditions, in **Extractive Metallurgy Symposium**, pp. 51-60 (Australasian Institute of Mining and Metallurgy: Melbourne).
6. Bruckard, W.J., Sparrow, G.J., Woodcock, J.T., and Tun Zhu, 1991. Effect of iron removal and reduction on gold and silver extraction by thiourea leaching of a pressure oxidation product, in **World Gold '91**, pp. 99-108 (Australasian Institute of Mining and Metallurgy: Melbourne).
7. Chen, T.T., and Dutrizac, J.E., 1988. Mineralogical characterization of silver flotation concentrates made from zinc neutral leach residues, **Metall. Trans. B**, 19B: 803-817.
8. Correia, M.J., Carvalho, J.R., and Monhemius, A.J., 1993. Study of the autoclave leaching of a tetrahedrite concentrate, **Minerals Engineering**, 6: 1117-1125.
9. Doyle, B.N., Masters, I.M., Webster, I.C., and Veltman, H., 1978. Acid pressure leaching of zinc concentrates with elemental sulphur as a by-product, in **Proc. Eleventh Commonwealth Mining and Metallurgical Congress, Hong Kong, 1978** (Ed. M.J. Jones), pp. 645-653 (Institution of Mining and Metallurgy: London).
10. Dutrizac, J.E., 1981. Jarosite formation in chloride media, **Proc. Australas. Inst. Min. Metall.**, No. 278: 23-32.
11. Dutrizac, J.E., 1983. Jarosite-type compounds and their application in the metallurgical industry, in **Hydrometallurgy – Research, Development and Plant Practice** (Ed. K. Osseo-Asare and J.D. Miller), pp. 531-551 (The Metallurgical Society of AIME: Warrendale).
12. Dutrizac, J.E., 1985. Silver behaviour in hydrometallurgical zinc circuits, in **Zinc '85**, Tokyo (Ed. K. Tozawa), pp. 707-727 (The Mining and Metallurgical Institute of Japan: Tokyo).

13. Dutrizac, J.E., and Chen, T.T., 1988. Mineralogical characterization of silver flotation concentrate produced by Vieille-Montagne, Balen, Belgium, **Trans. Instn Min. Metall. (Sect. C: Mineral Process. Extr. Metall.)**, 97: C180-C190.
14. Dutrizac, J.E., and Jambor, J.L., 1984. Formation and characterization of argentojarosite and plumbojarosite and their relevance to metallurgical processing, in **Applied Mineralogy** (Ed. W.C. Park, D.M. Hausen, and R.D. Hagni), pp. 507-530 (The Metallurgical Society of AIME: Warrendale).
15. Dutrizac, J.E., and Jambor, J.L., 1987. Behaviour of silver during jarosite precipitation, **Trans. Instn Min. Metall. (Sect. C: Mineral Process. Extr. Metall.)**, 96: C206-C218.
16. Ek, C.S., 1990. Recovery of silver from residues of the zinc jarosite process, in **Lead-Zinc '90** (Ed. T.S. Mackey and R.D. Prengaman), pp. 227-238 (The Minerals, Metals & Materials Society: Warrendale).
17. Eropkin, J.I., Khodov, N.V., Kuznetou, O.K., Malinovskaja, N.D., Maximov, I.I., Otrozdenova, L.A., and Sataev, I.S., 1993. Description of effective technology for silver recovery from zinc cakes obtained as by-product in metallurgical zinc plants, in **XVIII International Mineral Processing Congress**, Vol. 4, pp. 1061-1066 (Australasian Institute of Mining and Metallurgy: Melbourne).
18. Fair, K.J., and Basa, F.J., 1989. Treatment of Agnico Eagle's silver-bearing flotation concentrate by the Nitrox process, in **Processing of Complex Ores** (Ed. G.S. Dobby and S.R. Rao), pp. 411-420 (Pergamon Press: Toronto).
19. Figueiredo, J.M., Silva, A., and Silva, J.M.L., 1985. Application of hydrometallurgy to bulk concentrates, in **Complex Sulfides (Processing of Ores, Concentrates and By-products)** (Ed. A.D. Zunkel, R.S. Boorman, A.E. Morris, and R.J. Wesely), pp. 581-592 (The Metallurgical Society: Warrendale).
20. Hamdorf, C.J., 1961. Precipitation of lead and silver from brine solutions by metallic iron, **Proc. Australas. Inst. Min. Metall.**, No. 199: 19-50.
21. Ildefonse, J.P., Perrotel, V., Darcy, M., and Gabis, V., 1984. Process mineralogy and silver behaviour in zinc hydrometallurgy, in **Applied Mineralogy** (Ed. W.C. Park, D.M. Hausen, and R.D. Hagni), pp. 547-560 (The Metallurgical Society of AIME: Warrendale).
22. Kim, K.H., Kang, T., and Sohn, H-J., 1989. Electrochemical study of silver cementation on lead in chloride solution, in **Productivity and Technology in the Metallurgical Industries** (Ed. M. Koch and J.C. Taylor), pp. 639-649 (The Minerals, Metals & Materials Society: Warrendale).
23. Kunda, N., 1984. Hydrometallurgical processing of silver concentrate, in **Precious Metals: Mining, Extraction, and Processing** (Ed. V. Kudryk, D.A. Corrigan, and W.W. Liang), pp. 397-423 (The Metallurgical Society of AIME: Warrendale).
24. Onozaki, A., Nagata, H., and Ishimori, N., 1990. Recovery and subsequent treatment of gold and silver in zinc processing at the Iijima refinery, in **Lead-Zinc '90** (Ed. T.S. Mackey and R.D. Prengaman), pp. 211-226 (The Minerals, Metals & Materials Society: Warrendale).
25. Ralston, O.C., 1924. Hydrometallurgy of lead, **Trans. Amer. Inst. Min. Metall. Engrs.**, 70: 447-470.

26. Rosato, L.I., Stanley, R.W., Bérubé, M., Blais, M., Leroux, G., and Shink, D., 1990. Precious metal recovery from zinc plant residue by thiourea leaching, in **Lead-Zinc '90** (Ed. T.S. Mackey and R.D. Prengaman), pp. 109-120 (The Minerals, Metals & Materials Society: Warrendale).
27. Sandberg, R.G., and Huiatt, J.L., 1986. Ferric chloride, thiourea and brine leach recovery of Ag, Au and Pb from complex sulphides, **J. Metals**, 38(6): 18-22.
28. Van Weert, G., Fair, K.J., and Schneider, J.C., 1986. Prochem's Nitrox process, **CIM Bull.**, 79 (No. 895): 84-85.
29. Wilson, A.T., 1985. An economical method for the recovery of gold from the sulphur containing residue of a hydrometallurgical process, in **Complex Sulfides (Processing of Ores, Concentrates and By-products)** (Ed. A.D. Zunkel, R.S. Boorman, A.E. Morris, and R.J. Wesely), pp. 143-148 (The Metallurgical Society: Warrendale).
30. Wyslouzil, D.M., and Salter, R.S., 1990. Silver leaching fundamentals, in **Lead-Zinc '90** (Ed. T.S. Mackey and R.D. Prengaman), pp. 87-107 (The Minerals, Metals & Materials Society: Warrendale).

Subject Index

MIX
Papier aus verantwortungsvollen Quellen
Paper from responsible sources
FSC® C105338

If you have any concerns about our products,
you can contact us on
ProductSafety@springernature.com

In case Publisher is established outside the EU,
the EU authorized representative is:
Springer Nature Customer Service Center GmbH
Europaplatz 3, 69115 Heidelberg, Germany

Printed by Libri Plureos GmbH
in Hamburg, Germany